Islandology

Islandology

GEOGRAPHY, RHETORIC, POLITICS

MARC SHELL

STANFORD UNIVERSITY PRESS   STANFORD, CALIFORNIA

Stanford University Press
Stanford, California

Printed in the United States of America on acid-free, archival-quality paper

Library of Congress Cataloging-in-Publication Data

Shell, Marc, author.
   Islandology : geography, rhetoric, politics / Marc Shell.
     pages cm
   Includes bibliographical references and index.
   ISBN 978-0-8047-8629-4 (cloth : alk. paper)—ISBN 978-0-8047-8926-4
   1. Islands.   2. Islands in literature.   3. Cultural geography.   I. Title.
   GB471.S57 2014
   910.914′2—dc23

                                                    2013040186

Designed by Bruce Lundquist

Typeset by Newgen in 10/15 Adobe Garamond Pro

For Jacob Adam Shell

It is well to have some water in your neighborhood, to give buoyancy to and float the earth. One value even of the smallest well is, that when you look into it you see that earth is not continent but insular.

—Thoreau, *Walden*

Contents

List of Illustrations xi

Preamble 1

PART ONE: DEFINING ISLANDS AND ISOLATING
DEFINITIONS

1. Defining Islands and Isolating Definitions 13
    Definition 13   Epistemology 16   Dialectic 18
    Nesology 20

2. Horizontal and Vertical 26
    Horizontals 26   Verticals 30

3. Animate Swimmers and Inanimate Floaters 35
    Big Fish 35   Ship 38   Floating Island 40

4. Material Substance and State of Matter 42
    Ice Floe and Lava Flow 42   Northwest Passage 45
    Feeling for Snow 48   Bergman's Island 50

PART TWO: PLACEMARKS AND CULTURES

5. Cities in Straits 57
    Straits 57   Island-Cities 59

6. Naming and Sovereignty 81
    What's in a Name 81   The Trouble with Islands 85

7. Utopias and Laboratory Hypotheses 93
    One Place as Another 93   Hypotheses 94
    Three Island Utopias 97

8. Politics, Philosophy, Epic Drama 104
    War of Civilizations 104   Poets' Contest 106   Islands Razed
    and Raised 107   If Athens Were an Island 109   Aristotle's
    Drowning 114

PART THREE: HAMLET'S GLOBE

9. The Distracted Globe 121
    The Globe 121   North Sea Empire 124   *Holmgang* 132
    Plots and Patches 135   Thin Ice 141   Sublime Coast 146

10. Island Words 151
    Ham 151   Elsinore 153   Orisons 154

11. Dire Straits 160
    Passages, Geological and Zoological 160   Hven in Øresund 163
    Hamlet's Egg 166   The Goodwin Sands 169   Breaking Out 170

12. Liberty 172
    Exile, Theatrical and Corporeal 172   *Þing* and King 173   Magna
    Carta Island 178

PART FOUR: SEA AND LAND

13. Hamlet *Is* Germany 183
    Rule, Germania! 183   Planted on a Free Island 188
    Germany Is Hamlet! 192

14. The Region of Illusion 194
    The Ring in the Reef 194   *S.O.S. Eisberg* 202   Airships 207

15. Building for a Future 215
    Beyond the Pale 215   Floating the British Empire 218

    Postamble 231
    Speaking on the Shore 231   Terra Infirma and Critical
    Topography 235   Geography as Institution 237
    Where to Sit 245

    Acknowledgments 249
    Notes 251
    Name Index 333
    Place Index 345

Illustrations

Illustration 1   Animal. Four Legs. Mineral 14

Illustration 2   Islands Defined as Venn Diagrams 15

Illustration 3   Map of the World as Archipelago 22

Illustration 4   *Terra Firma* 27

Illustration 5   Planet Earth as Island and Milky Way as Archipelago 29

Illustration 6   Just Suppose 30

Illustration 7   Erasmus Darwin's Insignium 32

Illustration 8   Nukuoro Atoll, Micronesia 33

Illustration 9   Behemoth and Leviathan 37

Illustration 10   *Saint Brendan on the Back of a Whale* 39

Illustration 11   Floating Village near Nasiriyah, Iraq 40

Illustration 12   *The Sea of Ice* 43

Illustration 13   Sealers on Pan Ice 44

Illustration 14   Island of Water on Iceberg in Foreground with Islands of Land
                   in Background 45

Illustration 15   Tenochtitlán 58

Illustration 16   Rialto Bridge 62

Illustration 17   Map of Venice 64

Illustration 18   Map of Venice 65

Illustration 19   Map of Hormuz 67

Illustration 20   *Marco Polo with Elephants and Camels Arriving at Hormuz on the
                   Gulf of Persia from India* 68

Illustration 21   Japanese Postage Stamp, 1931 75

Illustration 22   The World 80

Illustration 23   *Europa Regina* 88

Illustration 24   Hans Island Viewed from the South 91

Illustration 25   Coral Island and Encircling Coral Reef Creating a Lagoon 95

Illustration 26   Lisca Bianca 97

Illustration 27   *Utopiae Insulae Figura* 101

Illustration 28   *Gulliver Discovers Laputa, the Flying Island* 102

Illustration 29   Bridge of the Euripus 105

Illustration 30   Iphigenia in Aulis Commemorative Stamp 106

Illustration 31   *Insulae Maris Aegaei* 108

Illustration 32   Xerxes at the Hellespont 109

Illustration 33   Swan Theatre 122

Illustration 34   Farnese Atlas 123

Illustration 35   Archimedes' Lever 124

Illustration 36   Kalmar Union 130

Illustration 37   Duel at Ganryu Island 133

Illustration 38   Map of the North Sea Region 137

Illustration 39   *Group of the Principal Forms of Barrows* 139

Illustration 40   Tinghøjen 140

Illustration 41   Polacks on the Ice 142

Illustration 42   Fortinbras's Soldiers on the Ice 143

Illustration 43   *Battle of the Ice* 144

Illustration 44   *The Battle of the Lake* 145

Illustration 45   Crossing the Red Sea 146

Illustration 46   Hamlet and the Ghost at the Shore 1 149

Illustration 47   Hamlet and the Ghost at the Shore 2 149

Illustration 48   Hamlet and the Ghost at the Shore 3 150

Illustration 49   Sir Walter Scott Visiting Smailholm Tower 152

Illustration 50   *The Scandinavian Peninsula Looks Just Like a Whale* 154

Illustration 51   Anthropocentric Map of Denmark Showing the "Sound" 155

Illustration 52   Ophelia 1 156

Illustration 53   Ophelia 2 157

Illustration 54   Map of Denmark and Sweden 158

Illustration 55   *Foetus in the Womb* 162

Illustration 56   Øresund 164

Illustration 57   *Archimedes' Lever* 166

Illustration 58   Excavation of Sutton Hoo Burial Ship, 1939 171

Illustration 59   Althingi, Iceland 174

Illustration 60   Ting Holm Island 177

Illustration 61   Magna Carta as Kings Embracing in the Waves 178

Illustration 62   Britain as Ship: The Heneage Jewel or Armada Jewel 185

Illustration 63   Opening Scene of *The Rhine Gold* 198

Illustration 64   *Toteninsel* (Isle of the Dead) 198

Illustration 65   *Rocky Reef on the Sea Shore* 199

Illustration 66   Rhine Maidens with Alberich Postage Stamp 201

Illustration 67   Where Lower and Upper Worlds Meet 203

Illustration 68   Airplane Encircling the Globe 204

Illustration 69   Airplane and Iceberg 205

Illustration 70   *Graf Zeppelin* Postage Stamp 210

Illustration 71   *The Bridges between the Worlds* 211

Illustration 72   *The Brick Moon* 212

Illustration 73   Strabo Holding the Globe in His Hands 213

Illustration 74   Noah's Ark Landing on the Insular Mount Ararat 228

Illustration 75   *Demosthenes Practicing Oratory* 232

Illustration 76   World Tectonic Plates as Islands 234

Illustration 77   Continental Drift or Plate Tectonics 237

Illustration 78   Map of the World Island According to Strabo 239

Illustration 79   World Island and Natural Seats of Power 241

Illustration 80   Collegium Albertinum on Kneiphof Island 244

Illustration 81   Map of Königsberg 245

Illustration 82   Moving from Land to Sea 246

Color plates follow page 120.

Islandology

Preamble

"Islands have always fascinated the human mind," perhaps because fascination "is the instinctive response of man, the land animal, welcoming a brief intrusion of earth in the vast overwhelming expanse of sea." So wrote Rachel Carson in her best-selling book *The Sea around Us* (1951).[1] Islandology argues that there is more to it than that.

In Chapter 1, we begin this argument by defining islands and isolating certain definitions, including the definition of *definition*. After pinpointing the meaning of what logician John Venn calls an "island of meaning," we explore ways of speaking about actual islands and consider how human imagination of islandness has variably informed cultures. Islandness, we discover, resides in a shifting tension between the definition of *island* as "land as opposed to water" and the countervailing definition as "land as identical with water." This tension is linked with notions of "social space," both positive and negative.

The "critical topography" or "philosophical topography" of place (*locus*) involves more than just the real estate slogan "Location, location, location."[2] The leading modern geographer, Immanuel Kant, in *Physical Geography*, defines *geography* in terms of nature and politics, distinguishing among the physical objects of study: geography (the entire world), topography (single places), and chorography (regions), as well as orography (mountains) and hydrography (water areas).[3] Spatiality, as we will see, influenced Kant's thinking in general,[4] including his epistemology, his topography of mental faculties,[5] and his notion of worldly unity and ownership, as discussed in the *Metaphysics of Morals*.[6]

Said the Sicilian islander and mathematician Archimedes, in the third century BC, "Give me a place to stand on and I can move the earth."[7] Understanding islandness requires that place to stand (*pou stō*). Pappus of Alexandria, who reported this statement of Archimedes, was a specialist in projective geometry with a focus on points at infinity on horizons. The limiting beach, which everywhere surrounds dry land on Earth, likewise defines the sea's coasts.

Suppose oneself, then, at a beach. The coastline marks the cutoff where land ends and water begins. If one believes that one can walk or sail around the land perimeter and end up where one began, then one is probably on an island. (In this sense, an island is an *insula*: "solid earth [*terra firma*] surrounded on the horizontal plane by liquid water [*aqua liquida*].") If one believes one cannot go all around, or circumambulate, that land, then one probably does not call it "island." One does not always know, of course, whether one is on an island or on something else, maybe a peninsula or mainland. That uncertainty was especially common before the exploration of the world was complete. On the Europeans' first sighting of Guanahani (modern Bahamas) or Maracaibo (modern Venezuela), who really knew for sure whether or not the "terra firma" where they might land would be circumnavigable? Floating? Animate? The world, as we will see, remains much unexplored. Just so, we will see how naming a place like Guanahani—or, indeed, any land or water place on Earth—remains much vexed.[8]

ΔΩΣ ΜΟΙ ΠΑ ΣΤΩ ΚΑΙ ΤΑΝ ΓΑΝ ΚΙΝΑΣΩ

Give me a place [*pou*] to stand and I will move [*kinaso*] the Earth [*gē*]. For this dictum of Archimedes, there are many translations. (1) Some translators rely on the Greek-language version passed down to us by John Tzetzes. Francis R. Walton thus translates Archimedes' *kinēsis* as "move with a lever," and Bram Stoker renders *pou* as "fulcrum."[1] Political and social theorists like to emulate this interpretation. Thomas Paine, in *The Rights of Man* (1791), thus calls the American Revolution an Archimedes moment, and the media theorist Marshall McLuhan, in *Understanding Media: The Extensions of Man* (1964), calls electronic technology the Archimedean fulcrum of the modern world.[2] Such preemptory simile between the natural earth and the human world seems too soon to drive out modern geography's attempt to understand the logical links between earth and world. (2) Among those who seek to head off the dangers of a mechanical interpretation is René Descartes in his *Meditations*. He translates *gē* as *terra integra*, which the Duc de Luynes (who translated Descartes's Latin into French) represents geologically as *globe terrestre*.[3] (3) My translation of Archimedes' *pou* as "place," in the passage quoted, evokes Archimedes' topography (from *topos*, meaning "place"). At the same time, it recalls such other translations of Archimedes' *pou* as "somewhere" and "where."[4] (4) There is also the more abstract geometrical meaning of *pou* put forward by Descartes in his Latin-language *Meditations* and in the French-language translation by the Duc de Luynes. There *pou* is rendered substantively as *punctum* or *point*. Descartes modifies this *point* with two words lacking in the Greek as passed down to us by Pappus of Alexandria (which is my source text), but suggested in the later versions offered by Plutarch. (4a) The first word is *immobile* (immovable). Plutarch's interpretation of the dictum already had it that "if there were another

From the viewpoint presented in the last paragraph, an island is "land on which, when one walks along its coastline in one direction, one eventually gets back to where one started." This perambulatory viewpoint distinguishes sharply the "edge" or "coast" between land and sea, but usually ignores how the difference between earth and water already implies their identity and, in fact, how the word *island* already also means "sea-land" (*is-land*), or the place, no matter how small or large, where water and earth are one and the same.

Islandness, in this sense of identity confronting difference, informs primordial issues of philosophy: how, conceptually, we connect and disconnect parts and wholes, for example, and how we connect and disconnect one thing and another. Whether islandness, and hence geography, is fundamental to philosophy and its history or is merely contingent or exemplary is a question we pursue in *Islandology*. If there were not islands already, as we will see, it would it be necessary for human beings—the logical and political creatures that we are (or strive to be)—to

earth, by going into it [Archimedes] could remove this [earth]" (as rendered by John Dryden). The hypothesis involves a mariner's notion of infinite space. In islandic terms, that would entail a partly otherworldly bridge linking a movable object with an immovable one; in planetary terms, that would mean a spaceship space-traveling between a kinetic object and a nonkinetic one. (4b) Descartes likewise modifies the Archimedean *point* as "firm." This modification suggests the heavenly firmament even as it requires a "firm spot,"[5] which is how many later translators render "pou." *Firm* serves to suggest an imagined *terra firma*: a "continent," "mainland," or "dry land" in the midst of an infirm universe or all but boundless ocean. The Sicilian island mathematician Archimedes, famously concerned with issues of buoyancy, flotation, and horizon in the service of Syracusan tyranny, put forth a vision of a universal archipelago on which, thanks to cartographical coordination and perspectival calculus, human understanding might really come to stand.

1. Diodorus Siculus, *Library of History*, trans. Francis R. Walton, Loeb Classical Library 409 (Cambridge, MA: Harvard University Press, 1957), vol. 11, bk. 26. Walton is working with his own translation of John Tzetzes, *Chiliades* 2:129–130; see also Mary Jaeger, *Archimedes and the Roman Imagination* (Ann Arbor: University of Michigan Press, 2008), 104. Bram Stoker, *Dracula* (New York: Grosset & Dunlap, 1897), chap. 25.

2. Thomas Paine: "What Archimedes said of the mechanical powers, may be applied to Reason and Liberty: 'Had we,' said he, 'a place to stand upon, we might raise the world.' The revolution of America presented in politics what was only theory in mechanics." Paine, *The Rights of Man* (London: H. D. Symonds, 1792). Marshall McLuhan: "Archimedes once said, 'Give me a place to stand and I will move the world.' Today he would have pointed to our electric media and said, 'I will stand on your eyes, your ears, your nerves, and your brain, and the world will move in any tempo or pattern I choose.' We have leased these 'places to stand' to private corporations." McLuhan, *Understanding Media: The Extensions of Man* (New York: McGraw-Hill, 1964).

3. René Descartes, *Meditationes de prima philosophia* (Paris: Michel Soly, 1641), Meditation II, para. 1. For the French translation: Duc de Luynes, trans., supervised by Descartes, *Méditations métaphysiques* (Paris: J. Camusat & P. Le Petit, 1647).

4. For "somewhere": Ivor Thomas, *Greek Mathematical Works: Aristarchus to Pappus*, Loeb Classical Library 362 (Cambridge, MA: Harvard University Press, 1941), 2:35. For "where": John Bartlett, *Familiar Quotations*, ed. Emily Morison Beck, 14th ed. (Boston: Little, Brown, 1968), 105. This version of the Archimedes dictum comes from Pappus of Alexandria, *Synagoge* [Collection], bk. 7.

5. See, for example, *Oxford Dictionary of Quotations*, 2nd ed. (London: Oxford University Press, 1953).

invent them. This book thus names "islandology" the discourse that marks off human beings not only as children of the main, understood as both "land" and "sea," but also as creatures of the natural shore who inhabit, at once, both positive and negative space.

In the previous discussion, we considered a patch of land when we are standing on it, so that it seems possible to begin immediately its attempted circumambulation. Consider now a patch of land, seen at a distance from across the waters, as if we were on another patch of land, or imagine a ship (or a "floating island"), or picture a peninsula that, without our knowing, is connected horizontally with the land whereon we stand. For all we know, we cannot get there without going underwater (like seals, submarines, or passengers in underwater tunnels) or without traveling on the surface of the water (like water striders, surface ships, or pedestrians on pontoon bridges) or without flying above that surface (like birds, airships, or passengers on airships). We dream of swimming now instead of walking.

Swimming is understood here as *natation*, an English-language term that is cognate with the ancient Greek *nēsos* (usually translated as "island").[9] The term emphasizes the sense in which main and mainland are one and how all stations, including Earth and the place where the little boy sits in *A Child's Geography of the World* (Illustration 6), are equally insular and mainland. To understand islandology after the first scientific Age of Exploration means not only looking out to sea from the viewpoint of land but also looking out to land from the viewpoint of sea. It means wondering whether there is any safely stable harbor, *pou stō*, wherefrom even to look out.

The study of islands, as isolates known and unknown, is not new. There have been dozens of approaches to the topic. Some focus on particular colonial and postcolonial settings—as does Rebecca Weaver-Hightower in *Empire Islands* (2007).[10] Others speculate on how thinking about islands encourages scientific hypotheses and literary fictions—as does Jill Franks in *Islands and the Modernists* (2006).[11] A few provide psychological examinations of persons who suffer from island mania—as does Jill Franks in "Men Who Loved Islands" (2008).[12]

Professional geographers study the smallness of islands in relation to the largeness of mainlands,[13] examine the effect of bridging islands with mainlands, scrutinize the sociology of modern tourism,[14] investigate specific environmental issues,[15] and study the characteristics of insular cartography.[16] Richard Grove, in *Green Imperialism* (1991), shows how global politics exacerbates islandic environmental issues.[17] The anthropological historian Marshall Sahlins, in *Islands of History* (1985), stresses the intellectual advantages of an island-centered historiography of mobility.[18] Fernand Braudel argues in *The Mediterranean* (1949) that "the events of history often lead to the islands."[19] And John R. Gillis, in *Islands of the Mind:*

How the Human Imagination Created the Atlantic World (2004),[20] discusses how conceiving islands in terms of long distance helps explain the historical process of continental discovery. Islandology, in its study of how we speak about islands, recognizes these approaches—and many more to be cited in the chapters that follow—and, at the same time, builds on them.

Part 1 of *Islandology* includes, in Chapter 1, a study of definitions and isolations, with special attention given to the horizontal plane. Chapter 2 moves the focus to the vertical plane, and an examination of animate and floating islands follows in Chapter 3. The difference made by differing material substances and differing states of matter is the subject of Chapter 4.

Part 2 focuses on kinds of geographic places and concomitant human constructions. Chapter 5 concerns island-cities, among them Venice and Hormuz. Chapter 6 focuses on the politics of island toponymy and sovereignty. Chapter 7 considers how islands, real and imaginary, provide scientific, literary, and political hypotheses for thinking about the world (Thomas More, Jonathan Swift, Daniel Defoe, and Charles Darwin). Finally, Chapter 8 investigates the ways that ancient Greek geography informs foundational epic poetry (Hesiod and Homer), tragedy (Euripides, Sophocles, and Aeschylus), and dialectical thinking (Plato and Aristotle).

Parts 3 and 4 provide a broadly based double "case study" for many of the subjects introduced in Parts 1 and 2, even as they redevelop them. Part 3 provides a historical, textual, biographical, and geographical analysis of *Hamlet*, Shakespeare's Scandinavian play. *Hamlet* is arguably the best-known work in world literature, but its islandological structure and meaning heretofore have been unrecognized. Its contribution to global thinking—and to the study of islandness—has not yet been digested by geographers and other theorists. These chapters on *Hamlet* answer for the first time the question, Why does Shakespeare move the distinctly continental setting of the old Hamlet story, mainland Jutland, to a definitively different setting, island Zealand (sea land)? They show what this particular difference between mainland and island makes, not only to our understanding of this one work of literature, however brilliant and influential in its own right, but also to the more general geographic comprehension of the polity and nature of human beings.

The specific viewpoints of these chapters include a study of the *holmgang*—an originally Scandinavian English-language word referring to a struggle for island possession or a struggle that takes place on an island—and an examination of how islandness informs human conceptualization of the body and organization of the family. When it comes to dramatic stagecraft itself, we will see that the peninsular aspect of Shakespeare's stage is crucial. When it comes to politics as statecraft, we discover Shakespeare's needfully esoteric meditation on the origins of British parliamentary democracy in the *ting*—another originally Scandinavian English word that indicates a popular meeting held for legislative purpose or political election. The essential location of the *ting*, as we will see, is the *tingholm*.

Part 4, likewise a series of representative case studies for understanding islandology, discusses two interrelated ways in which mainland German national thinkers, mostly in the nineteenth century, sought to discover or create a unified German nation. First, they sought the "German future" reflected in the history of island Britain, a search that included defining the German character in terms of Shakespeare's Danish island play. Second, they discovered a true "German past," not so much on the islands of ancient Greece, where eighteenth-century thinkers had focused their attention, as on the islands of the Baltic Sea, especially Rügen and other islands long inhabited by various Scandinavian groups.

Among philosophical islandic thinkers we consider in Part 4 are Johann Gottfried von Herder and Friedrich Nietzsche, as well as the Nazi ideologue Carl Schmitt. Among painters is Caspar David Friedrich, and among composers is Richard Wagner, whose opening scene of *The Ring of the Nibelung* at an island reef, taken together with Charles Darwin's island-centered geological and biological theories of evolution, marks a turning point in modern islandology.

Islandology engages problems of political import: the modern tendency to confuse circumferential natural borders with political ones and the ancient inclination to except circumferential seas from imperial sovereignty. Both problems focus on issues of pressing environmental concern. A reexamination of the Darwinian theory of coral island reefs and volcanic islands in relation to insular plate tectonics conceives anew the pressures of "global warming," for example. Likewise, contextualized interpretations of movies, among them the Danish *Smilla's Sense of Snow* and the German-American *S.O.S. Eisberg*, rethink the melting of the polar ice caps in terms of both different states of matter and different material substances.

Nineteenth-century thinkers, both American and German, often relied on tendentious and needless theories of climatic and geographic determinism; this reliance, no matter how productive in its way, brought with it needless and unhappy political consequences. Most likely, the extensive closings of departments of geography worldwide—especially in the United States—during the latter part of the twentieth century had some of the "value-neutral purposes"—beneficent at least in the short term—that backers of the then-competitive disciplines (international politics, comparative literature, earth and planetary sciences, linguistics, and environmental studies) often articulated. Yet none of these disciplines has recovered the global and philosophical vision of geography, now so much required, that sees all lands and seas on Earth as participants in a single archipelago.

THE WORD *islandology* provides this volume with its title. It refers both to the *rhetoric* of speaking about islands and to the *science* of islands. (The suffix *-logy* indicates no less a way of speaking, as for *brachyology* [a condensed expression][21] and *tautology* [a proposition which is unconditionally true . . . by virtue of its logical form], as a field of study *theology* [the study of God].)[22] Where the subject matter is the definition of *definition*, as it is in Chapter 1, the rhetoric and the science verge on the same. How the logical definition of *definition* merges with the geographic definition of *island* is part of the science of rhetoric. *Islandology*, in this context, is one of those neologisms that, no matter how awkward, has its place in the language. In Thomas Hardy's *Tess of the d'Urbervilles* (1891), Angel Clare comments on Tess's imaginings in this way:

What are called advanced ideas are really in great part but . . . a more accurate expression, by words in *logy* . . . of sensations which men and women have vaguely grasped for centuries.[23]

The introduction of the word *islandology* combined institutional aspects with a geopolitical impetus. In 1945, Raine Edward Bennett founded the American Institute of Islandology (Washington, DC), partly in response to his island experiences during the two world wars. The institute's first purpose was determining whether Australia was an island or a continent. While Bennett said that Australia was "the world's largest island," an Australian newspaper reporter probably had it right when he said, "We [Australians] [wi]ll want [the nomenclature] both ways . . . as the smallest of the large [continents] and the largest of the small [islands]."[24] This droll impasse caused the institute's founders to stumble out of the starting gate, which explains why the institute's second goal was never accomplished: assembling and publishing a fifteen-volume encyclopedia of islands with a worldwide focus.

Half a century later, other scholars published an *Encyclopedia of Islands*,[25] which presented no general "islandology" of a philosophical and historical nature. The editors of this modern encyclopedia use *island* loosely to mean "any discrete habitat isolated from other habitats by inhospitable surroundings." For them, it seems to mean *biosphere*. Yet the word *island* has, as we will see, cross-cultural political, geographic, and cultural baggage, in a different sense from that of the presumably value-neutral word *biosphere*, whose inventor, the geologist Eduard Suess, defined in his study of the Alps (1875) as "the place on Earth's surface where life dwells."[26] (Vladimir Vernadsky, in his 1926 *Biosphere*, teased out of Suess's notion the idea that the *geosphere* is where there is only inanimate matter.)[27] Such ways of defining *island* have no determinate reference either to the interaction of land with water (geology) or to the different ways of understanding that interaction among cultures and logical systems. In that sense, these scholars avowedly apply an island "metaphor" to a palpably noninsular setting, whether biospherical or otherwise characterized.[28] (In much the same way, for example, the geographer David Harvey uses *island* to mean "a group of people living in wealthier fashion than their neighbours" and passes over the traditional meanings of the word, as if the traditional discipline and its geopolitical aspect were of no consequence.)[29]

Edmund Burke, in *A Philosophical Inquiry* (1759), says, "When we define, we seem in danger of circumscribing nature within the bounds of our own notions."[30] The logical definition of *island* is linked with the logical circumscription of definition in a way that cannot avoid the linguistics and natural history of islands.

Part One

M. Canaria.

Insula Fortunata.

Defining Islands and Isolating Definitions

Defining Islands and Isolating Definitions

DEFINITION

Horizon . . . f[rom] *horos*, [meaning] boundary, limit.

—*OED*, s.v. "horizon"

Generally speaking, a scholarly work begins (or should begin) with asking whether its subject *can* be defined—and, if so, whether it *should* be defined overtly—and, if so, whether the definition should be at the beginning, middle, and/or end of the scholarly work or should suffuse it. Most studies avoid defining *definition* itself, as if they feared becoming bogged down in terminology even before the journey starts.

Thus *Islandology* cannot afford to avoid defining *definition*. It is already inescapably concerned, from the beginning, with the "coast"—a term understood in this context as the "cut" where one kind of thing is supposed to begin and another kind is supposed to end (*finir*). This cut, or limit—Hamlet calls it a *bourn*, meaning "horizon"[1]—is crucial to any definition of *island*. The commonsense understanding of *island* as "insulet"—meaning "land circumferentially bordered or insulated by water and entirely defined horizontally by its shoreline"[2]—already raises questions about definition.

The logician John Venn, originator of the Venn diagram ("a group of circles that may or may not intersect according as the logical sets they represent have or have not elements in common"),[3] has something to say about such questions of definitional islandness.[4] In his *Principles of Empirical or Inductive Logic* (1889), Venn considers the debate between John Stuart Mill and William Whewell about whether the "discovery" of the elliptical orbit of the planet Mars circumnavigating the sun was a case of induction or deduction.[5] He brings up the parallel case where "a navigator sail[s] round an island and then pronounce[s] it to be an island."[6] If circumnavigation alone makes a "land" an "island," says Venn, then the eighteen members of Ferdinand Magellan's crew who made it back to Portugal after their three-year voyage should have concluded that the planet Earth was an island. Richard Eden, in the preface to his translation of Sebastian Münster's *Treatise of the Newe India* (1553), writes, "The [w]hole globe of the world . . . hath been sayled aboute."[7] One thinks of Jules Verne's *Around the World in Eighty Days*, published in 1873, with its nicely named character Passepartout.

The meanings of most words, not only *island*, often seem to dissipate a bit around the edges. In *Methods of Logic* (1952), the philosopher Willard V. Quine claims that the "whiteness of a region in a Venn diagram means nothing but lack of information."[8] A concept, writes the logician Friedrich Ludwig Gottlob Frege in *Foundations of Arithmetic* (1884),

must have a sharp boundary. If we represent concepts in extension by areas on a plane, this is admittedly a picture that can be used only with caution, but here it can do us good service. To a concept without sharp boundary there would correspond an area that had not a sharp boundary-line all around, but in places just vaguely faded into the background. This would not really be an area at all; and likewise a concept that is not sharply defined is wrongly termed a concept.[9]

Of this viewpoint of isolation, Ludwig Wittgenstein makes a critique in his post-humously published *Philosophical Investigations* (1953).[10]

This way of speaking about groups obtains as well for nations as for words. Thus Johann Gottfried von Herder envisions a nation as a closed autonomous island, each corresponding to a people's territorial area and linguistic extent[11]—or so such recent works as *Spaces of Culture: City, Nation, World* (1999) seek to envision his thought. In fact, Herder knows well that many cultures exist on small islands. He often contrasts the cultural uniformity of large islands with the cultural diversity of small ones. Whether the word *island* represents a particularly telling *locus* of thought—where parts become wholes and wholes become parts—is an ancient question. Plato raises the problem of the potential link between islandic geography and thinking when he discusses the waters of the Euripus Strait, the narrows that eventually took Aristotle's life.[12]

Consider, then, how a Venn diagram presents the cut between a finite group of parts and their wholes (see Illustration 1). Venn illustrates set theory in terms of geographic regionalism. An archipelagic logic shows islands of the British Empire (including the "British Isles") seeming to match up to islands of meaning[13] (see Illustration 2)—an appearance that informs both his essay "On the Employment of Geometrical Diagrams for the Sensible Representation of Logical Propositions" (1880)[14] and his book *Symbolic Logic* (1881).[15] Here the geographer's task of separating island from mainland and one island from another precedes the philosopher's dialectical ambition to define parts and wholes. It is no wonder that twenty-first-century schoolteachers still use a Venn diagram to compare the Sargasso

ILLUSTRATION 1
Animal. Four
Legs. Mineral.
Standard (Leonhard)
Euler diagram.
Source: Collection
Selechonek.

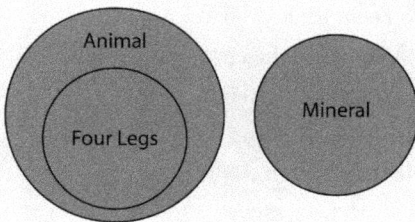

Sea—"a sea within a sea,"[16] "the sea with no shores,"[17] "a sea without a coastline"[18]—with the fictional Sargasso Sea of Jules Verne's science fiction novel *Twenty Thousand Leagues under the Sea* (1870).[19]

If islands constitute an instigating and typical case for philosophical definition, then philosophy and geography are probably interrelated in such a way as to raise questions. Might it turn out to be more than a "mere metaphor" that John Robert Ross, in *Constraints on Variables in Syntax* (1967),[20] with his Chomskyan generative, focuses on the "marooned," or insulated, location of *wh*-words in languages? Is it an "accident" that Benoit Mandelbrot poses his question, "How Long Is the Coast of Great Britain?" (1967),[21] in terms of the fiction of measurable circumambulation of islands? After all, Mandelbrot's *Fractal Geometry of Nature* (1982) links mathematical forms with natural objects in the way of geography, as in his argument that "clouds are not spheres, mountains are not cones, coastlines are not circles, and bark is not smooth, nor does lightning travel in a straight line."[22] Does Venn's having hailed from the Humber—the estuary on the North Sea where the tidal rivers Ouse and Trent daily reveal and conceal mudflat islands[23]—suggest how biography can become of formative significance? Or are there other concerns at work in delimiting the intersection of island thinking with thinking in general?

Venn's work was attractive to a Victorian empire then ruling over a great part of the Earth: the British Empire, as he knew it, was the largest in world history. Writes Carl Schmitt, a Nazi opportunist who set himself up as that empire's enemy: "It was only by turning into an island, in a new sense previously unknown, that England could succeed in conquering the oceans and win the first round of the planetary spatial revolution."[24]

According to the ideology of the "scepter'd isle"[25] that was summarized for British readers in Shakespeare's *Richard II*, the English "nation" followed a foreign policy of "splendid isolation." That is what First Lord of the Admiralty Lord Goschen called British policy in 1896.[26] Goschen was echoing sentiments from colonial New Brunswick in the "New World," where people had long regarded Britain as the "Empress Island."[27] The inhabitants of other island-based empires such as Venice and Hormuz did likewise.

John Venn "mapped out" sometimes overlapping logical relations between finite collections of sets in much the same way that his contemporaries mapped out aggregations of imperial holdings. Leonhard Euler put forward the first planar graph theorem[28] and laid the groundwork for developments in topology when

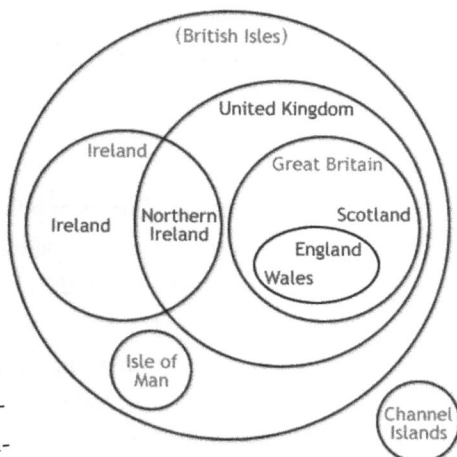

ILLUSTRATION 2
Islands Defined as Venn Diagrams. John Venn's island diagrams lend themselves to defining nations with many islands or large archipelagoes (e.g., Greece, Canada, and Greater Britain). Source: Collection Selechonek.

he solved a mathematical problem represented in terms of bridging islands in the Pregel River at Königsberg, the hometown of Immanuel Kant, who set the terms for modern philosophy.

EPISTEMOLOGY

An Enlightenment philosopher, Kant helped found the modern discipline of geography. From his chair on the island of Kneiphof in the Pregel, he offered lectures in geography and its human implications for forty years, more than he offered in any other subject.[29] Most of the times he taught this course, his *Diktattext* and announcement pamphlet listed the second module as "History of Lands and Islands."[30]

In the *Critique of Pure Reason* (1781), Kant writes that the realm of possible truth is a domain within which we can know something but which is surrounded by matters that we can never know. What follows next in the *Critique of Pure Reason* sounds like an island metaphor, so to speak, although islandology, in its consideration of the rhetoric of islands, eventually shows that it is more than that. "This domain [of possible truth]," writes Kant, "is an island [*Dieses Land aber ist eine Insel*] enclosed by nature itself within unalterable limits."[31]

Earlier in his career, when he wrote *Universal Natural History and Theory of the Heavens* (1755), Kant followed the view of Thomas Wright, in *An Original Theory or New Hypothesis of the Universe* (1750), and the view of William Herschel, in *Construction of the Heavens* (1785), that a planet or galaxy is like an island. In the *Critique of Pure Reason*, however, he is talking about one island—the territory whose boundaries define the limits of possible objectively knowable experience. (Invisible spirits, for example, necessarily fall outside those boundaries without necessarily being impossible as such, in the way that a square circle is impossible— that is, intrinsically contradictory.) The *Critique of Pure Reason* is, simply put, human reason's way of judging its own claims to knowledge with a view to laying out the limits, or horizons, of what it can legitimately claim to know. Everything that falls outside those limits is a kind of ocean filled with fog banks and melting icebergs[32] into which reason, prior to the *Critique of Pure Reason*, was inclined to stray. Kant calls this ocean "the region of illusion,"[33] a description that turns out to be fair warning to future generations.

A major goal of the *Critique of Pure Reason* is to reinterpret the indubitable tendency of human reason to overreach itself as an indication of our moral vocation. We cannot know whether (or not) human beings are free in an ultimate sense, for example. All our claims to knowledge, to the extent that they are justified, rest on the assumption that every event has a determinate cause. But because the "island," as Kant uses the term, merely defines the limits of things as they appear to us (and not as they are "in themselves"), the limits of objective knowledge do not preclude the possibility that we are ultimately free "in ourselves." Because morality requires that we think this (that we are free), we have permission to do so, given that there is no contradiction between saying that the world, as it appears to us, is necessarily determined and saying that the world, as it is, is free or might consist of free beings. The big questions—God, Freedom, Immortality—lie beyond the island of possible objective knowledge and thus define an ideal "noumenal" world,[34] which can be understood as the world as it really is (but which cannot be known by human beings), as the world that we project for our own moral purposes, or as both. One seeks not only to define the limits of the island, as Kant does. One seeks also to extend them, perhaps limitlessly, as Kant warns against, in such a way as to include the dark sea itself, as Nietzsche puts it in *The Gay Science* (1882): "The horizon seems clear" and then "the sea, our sea, lies open again."[35]

Some later thinkers ally Kant's use of the word *island* with his use of the term *nation-state*. (Just so, the politically definitive Treaty of Westphalia [1648] articulates the idea of single states.) Such thinkers tend to believe that the end of modern empires and nation-states will match up with the end of "island thinking" and hence, they hope, bring a certain political liberation. They call for "archipelagic thinking," as if an archipelago—with the various "bridges" between its islands, as Richard Rorty puts it[36]—does not already have precisely the same geographic limit or definition that an island has.

Kant, though, is also concerned with defining the boundaries of certain disciplines, mainly for moral reasons. If a rationally accessible morality should guide theology—rather than theology dictating what is morally right and wrong, for example—then it becomes necessary to lay out distinct disciplinary boundaries between "theology" and "philosophy." If we have to obey the law as it is and at the same time be guided by and strive to reform the law in light of a rationally established ideal of justice, then it is important to lay out the boundaries between "law" and "philosophy." That obligation, however, is a local matter primarily connected with the practices of the German university in the late eighteenth century. Here, too, many postmoderns, seeking out interdisciplinarity in "education," want archipelagic thinking, often in the antiuniversalist spirit of Jean-François Lyotard in *The Differend* (1983).[37]

The metaphorical use of *island* in the preceding paragraph follows Jean-Jacques Rousseau's *Emile, or On Education* (1762), the third book of which contains a famous reading of Daniel Defoe's *Robinson Crusoe* (1719), with its isolated hero. Because Crusoe is "on his island, alone, deprived of all the arts," he must figure out what he can know and what he cannot. Said Archimedes, "Give me a place to stand and I will move the Earth."[38] No wonder, then, that *Emile* was the one volume that Kant carried with him wherever he went in Königsberg.[39]

For Jacques Derrida, the Robinsonade query "Qu'est-ce qu'une île?" (What is an island?) turns easily into the political query "Qu'est une il?" (What is a he?).[40]

DIALECTIC
Thesis

The English term *island* includes two meanings in apparent disagreement with each other. The first, which I call a thesis, is the French-influenced meaning as something like "insulet."[41] This involves the separation, or "cutting" off, of land from water at the coast. The French definition, taken on its own as "land insulated by and defined against a surrounding terminus (*fin*)," even provides one model, albeit tendentious, for understanding the idea of *definition* itself in terms of how some words make for de-finable "islands of meaning" that intersect or overlap with the "shorelines" of other words—and perhaps how some words do not.[42] This meaning of *island* makes real sense, as we will see, only if we consider it together with its antithesis.

Antithesis

The other meaning of *island*, which I call an antithesis, is historically prior. It is of Norse origin: "water-land." This notion stresses that the noun *is* (the first part of the word) properly means "water"[43] and indicates the mixture of water and land at the limiting, or defining, "coast." Even as the biblical God separates water from earth (thesis), the Bible represents an earlier, even original, identity between earth and water (antithesis) that suggests *marshland, muck, mud, muskeg, bog, spunge*[44]—the sort of malleable, ever-changing humid material, or clay (*adam*), familiar to coastal cultures, from which the biblical God made the first human being (*adam*).[45]

Synthesis

These two meanings, taken together, suggest a complex definition of *island* variably at work in speaking about islands and in the logic of speaking about them.

Thesis and antithesis combined provide a third meaning more comprehensive than either has separately. Some persons might call this third meaning, which seesaws between identity and difference, "dialectical" in the Hegelian sense: a thesis and an antithesis, in apparent polar opposition to each other, are conjoined by a synthesis that cancels them out; at the same time, this synthesis incorporates and transcends them. It is not so much that Hegel was influenced by his vacations on the island of Rügen in the shattering way that his contemporaries were,[46] although he probably was. It is more that the dialecticians on whom he relied, Heraclitus and Plato, were influenced by the interactions of earth and water.

Hegel, who often visited Rügen, comes around to discussing islands in *Philosophy of History* (1821). Evincing a penchant for geographic determinism, he links coastlines with a particular stage in the development of the "World Spirit." Relevant here is the old philosophical controversy between Plato, for whom "both are two but each is one,"[47] and Heraclitus, for whom "water lives the death of earth and earth lives the death of water."[48] Whether or not the Heraclitean view is like that of modern physics, as Werner Heisenberg avers, it nevertheless links the physical universe with the human mind in much the same way.

How the older meaning of *island* (thesis) as "water-land" morphed in the Renaissance into the newer meaning (antithesis) as "water defined against land" is outside our present purview (but see Chapter 9). For the time being, it is sufficient to understand that, before the Earth was recognizably circumnavigated, the difference between island as "distant land" and as "water-isolated land" could be unreliable. In the fifteenth century (as the *OED* reminds us), the first part of the word *island*, which until then was largely associated with its original meaning of "water-land," began its association with the understanding of *island* as "land as defined against water and surrounded by water." Henceforth, it also was associated with the now "synonymous *ile*, *yle* (of Fr[ench] origin)." Sometimes the word was actually written *ile-land*, and when *ile* was spelled *isle*, *iland* "erroneously" followed as *isle-land*, *island*. The latter spelling became established during the decades before 1700.[49]

That meaning still obtains, however, in such toponyms for Baltic Sea places as *Eysysla* (land of island), which names the island at the mouth of the Gulf of Riga;[50] *Åland*, which names the island at the mouth of the Gulf of Bothnia; and *Zealand*, which names the large Danish island at the entrance to the Baltic. *Zealand*, variantly spelled *Sealand* and *Sjælland*, means "sea land" and names the island where Shakespeare's *Hamlet* takes place.

AMONG REGIONS that island studies examine is the Caribbean basin, an area that has made substantial progress in recent decades. One shortcoming of such studies, however, involves the logic of the critique of Hegelian dialectic that informs a dominant group of Caribbean scholars who seek to replace this dialectic (not a bad idea in itself) with a notion of natural "tidalectic." The instigator of this movement, the Barbadian poet Kamau Brathwaite, thus introduces the term *tidalectic* in his "Caribbean Culture: Two Paradigms" (1983):

For dialectics is another gun: a missile: a way of making progress: /farward [*sic*] / but in the culture of the circle "success" moves outward from the centre to circumference and back again: a tidal dialectic.[51]

In a 1991 interview with Nathaniel Mackey, Brathwaite says that tidalectics is "dialectics with my difference": "In other words, instead of the notion of one-two-three Hegelian, I am now more interested in the movement of the water backwards and forwards as a kind of cyclic, I suppose, motion, rather than linear."[52] In "New Gods of the Middle Passage" (2000), Brathwaite writes:

There're two ways of perceiving of experience in wester(n) [sic] philosophy. One is the one that we have been grown up on = dialectics (Hobbes, Hegel—great opp [sic] of Caliban & dialect—see my review of [the Guyanese historian and political activist] Walter Rodney's *How Europe Underdeveloped Africa* [1972] . . . —in which we assume there will always be a synthesis: the victorious leviathan. That there will always be a third factor of applied thought/action which gets us thru to success.[53]

That Brathwaite defines *tidalectic* as a "*natural* tidal procedure w/in a continuum rather than towards a fixed 'objective solution'"[54] may help to explain why the term has been influential in Caribbean island poetry, even as Brathwaite intended it to be. Thus he praises the Jamaican Rex Nettleford (1997) for "employ[ing] a tidalectic rather than a dialectic approach."[55]

Brathwaite's notion likewise informs much modern Caribbean island study. For example, Elizabeth DeLoughrey, in "Tidalectics: Navigating Repeating Islands" (2007),[56] and Paul A. Griffith, in *Afro-Caribbean Poetry and Ritual* (2010),[57] "use tidealectic as [the] controlling trope."[58] Silvio Torres-Saillant (1994) similarly writes:

A culturally centripetal Caribbean philosophy of history would separate itself from Hegelian dialectics with its teleological inexorability and would up-hold instead Brathwaite's concept of *tidalectics*, a far more humane notion. Contrary to dialectics, his notion permits us to see achievement in the Caribbean "success" of dialectics: synthesis.[59]

No matter its short-term utility for asserting regional autonomy ("European imperial Hegelianism versus indigenous tidalectic") and interdependencies of the national islands (French, Dutch, Spanish, and English), tidalectics remains unpacked if not, as we will see, unpackable: one heuristically useful model for what has historically dogged the more rhetorical of literary attempts to provide a consistent islandology.

NESOLOGY

Islandology is a hybrid word: the first part is English; the second, Greek. That doubleness, along with its neologistic aspect, helps explain its awkwardness, some of which an apparent synonym, *nesology*, might avoid. However, *islandology* better expresses the instability of the meaning of *island*. Even though it has come to indicate something like "island book"[60] or *isolario*, *nesology* means something other.

Many English-speaking people take the ancient Greek term *nēsos* to mean "island as insulet"—that is, land surrounded by water (thesis). However, *nēsos* is heuristically useful in coming to terms with "the difference between land and water" and with "the identity of land with water."[61] (To the packed geopolitical implications of the ancient Greek understandings of *nēsos* we will turn our attention in Chapter 8.) *Nēsos* actually includes meanings that avoid the usual Anglophone synthesis as outlined previously. For example, it can mean "peninsula" and "distant land"—as well as "swimming," as in "the land that swims in the sea."[62] There is, say, the ancient Greek toponym *Peloponnesus*, which includes the term *nēsos* and names a "peninsula," not an "insula."[63] There is also Mount Athos, likewise a peninsula called an island. For some time, in fact, the ancient Mount Athos actually had been separated from the mainland,[64] but even when all trace of the old canal had been lost, Mount Athos was separate in fiction, as in the Venetian manuscript book of Cristoforo Buondelmonti's *isolario* (island book) *Liber Insularum Archipelagi* (1430).[65]

Partly because of its variable ambiguities, the term *nēsos* came only recently into the English language, mainly as a toponym meaning "region of . . . islands."[66] The nesological idea characterizes islands as parts of *particular* archipelagoes and marks off all islands in general as part of a single "world" archipelago.[67] As Christian Depraetere puts it, "Islands are the rule and not the exception."[68] In this way, Herman Melville compares a group of artificial floating islands, or ships, with the archipelago of the world in his semiautobiographical *Redburn: His First Voyage: Being the Sailor-Boy Confessions and Reminiscences of the Son-of-a-Gentleman, in the Merchant Service* (1849):

Surrounded by its broad belt of masonry, each Liverpool dock is a walled town, full of life and commotion; or rather, it is a small archipelago, an epitome of the world, where all the nations of Christendom, and even those of Heathendom, are represented. For, in itself, each ship is an island, a floating colony of the tribe to which it belongs.[69]

The (universalist) idea is that there is only *one* archipelago in the world. The (particularist) notion is that the *only* archipelago or tribe in the world worth thinking of is one's own. Melville writes in his philosophical "travelogue" *Mardi, and a Voyage Thither* (1849) that

to the people of the [Mardi] Archipelago the map of Mardi was the map of the world. With the exception of certain islands out of sight and at an indefinite distance, they had no certain knowledge of any isles but their own.[70]

To see the world as one archipelago, then, is also to see only one archipelago in the world (see Illustration 3, Color Plate 1). The contradiction between the universal

ILLUSTRATION 3
Map of the World
as Archipelago.
From Muhammad
al-Idrisi of Ceuta,
*Nuzhat al-mushtaq
fi'khtiraq al-afaq*
(Tabula Rogeriana;
1325). Commissioned
by King Roger II of
Sicily, 1154. ("South"
is at the top.)
Source: Bibliothèque
Nationale de France,
Paris, MS Arabe 2221.

and the particular is here paramount, as for Gilles Deleuze in his archipelagic-endorsing essay on Melville's "Bartleby the Scrivener" (1853).[71]

The term *archipelago*, which raises such issues of particularism and universalism, is not ancient Greek, in spite of its two ancient Greek constitutive parts, *archē* (prime) and *pelagos* (sea). It was medieval writers who stitched together these two parts. At first, *archipelago* meant "the Mediterranean Sea" or, as the Roman historian Sallust called that inland sea, the *Mare Internum*.[72] Later on, the Venetians and their neighbors—island dwellers and city-state imperialists—gave the term its current meaning, "island region."[73]

Cristoforo Buondelmonti's *Liber Insularum Archipelagi*, the first widespread *isolario*, helped to introduce the particular notion of Aegean Archipelago that rekindled European interest in ancient Greece. The *Liber* became the model for Benedetto Bordone's Venetian-dialect *isolario*, the first to be printed. Likewise, Buondelmonti provided the exemplar for dozens of presumably global island books, including the *Islario general de todas las islas del mundo* (1545) by the Spanish cartographer Alonso de Santa Cruz[74] and *Le grand insulaire et pilotage* (1586) by the French Franciscan priest and explorer André Thévet. Such works came to define the entire world as if it were an Aegean-style archipelago of islands; they presented "a portolan of our entire globe"[75] along the lines one encounters in Umberto Eco's novel *The Island of the Day Before* (1995). More figural they were—like the cartographer Herman Moll's artful maps for Jonathan Swift's *Gulliver's Travels* and Daniel Defoe's *Robinson Crusoe*[76]—than geo-mathematical—like the astronomer Tycho Brahe's maps of the island of Hven, the site of his research observatory, Uraniborg, and his concomitant charts of planet Earth's location in the heavens.[77]

The island mapmaker Benedetto Bordone's own Republic of Venice had for its capital city an extraordinary metropolis of canals and islands, a floating urban environment with houses built on stilts and water-spaces between.[78] "Spaces between" is the literal meaning of *Nan Madol* in the Pohnpeian language.[79] Nan Madol is the island-city offshore from Temwen Island in the Federated States of Micronesia,[80] one of many places worldwide comparable with Venice or bearing variants of its name.[81] That the world as a whole is a Venetian archipelago is one gist of Melville's *Moby-Dick* (1851): "Through all the wide contrasting scenery . . . flows one continual stream of Venetianly corrupt and often lawless life."[82]

THE HEBREW WORD *ee* (or *ey*) suggests something like the archipelagic and distancing aspect of islandic thinking that the Greek *nēsos* conveys.[83] Etymologists connect *ee* with the Norse *ey* and hence with many British place-names.[84] (Comparable are the English-language terms *ait*, *eyot*, and *ait-land*.)[85]

Hebrew *ee* means "island" in the sense of "entirely water-surrounded dryland or terra-firma," but it also means "distant territories." The English-language Bible thus refers to the faraway "islands of the gentiles"[86] or "Islands of the Blest" (*makaron nēsoi*).[87] Shakespeare's Prince Hamlet fluidly refers to "the undiscovered country from whose bourn [meaning "horizon" or "limit"] / No traveller returns" (*Hamlet*, act 3, scene 1). The second meaning of *ee*, as distant "land with a seacoast," suggests Shakespeare's "Illyria," a setting for *Twelfth Night*, with its seacoast as "an island."[88]

TREATING ISLANDS as "cultural worlds" and coasts as permeable "cultural boundaries" has its intellectual costs. One such is the bias evinced by Greg Dening's *Islands and Beaches* (1980),[89] with its focus on the Marquesas, and his *Beach Crossings* (2005),[90] which is largely autobiographical. Because both works suffer from a definite narrowness, it is wise to investigate how comparative studies of particular cultures—even as they provide practicable case-style approaches to defining *island* (*nēsos*, *ee*) and isolating *definition*—also suggest potential or apparent universals cutting straight across human thinking in general.

In this regard, the Chinese American geographer Yi-fu Tuan writes in *Topophilia* (1974) that the idea of "the island seems to have a tenacious hold on the human . . . but it is in the imagination of the Western world that the island has taken strongest hold."[91] For the post-Renaissance world of Western exploration and empire, that obstinate hold was indeed complex and powerful. As thinking in terms of islands is nowadays variably widespread among places and peoples, islandology allows global access to contradictions within cultures and the long-term political and human implications of how human beings live, think, and speak.

HOW ONE THINKS about a water-surrounded land is sometimes linked with how one thinks about land-surrounded water. The result can be surprising. Consider the similar rhetorical and logical implication of a geographer's statement regarding worldwide water rheology (the logic of flow):

It is convenient to describe each of two hydrographic regions [those that are outflowing and those that are inflowing] by a single word: We may term through-flowing, or ocean, drainage "exoreism" (*ex-o-ré-ism*, from the Greek ἐξ, out, and ῥεῖν, to flow), [and] interior basin drainage "endoreism" (*en-do-ré-ism*, from ἐν, in, and ῥεῖν).[92]

On the one hand, this statement seems simple enough: the rainwater that flows *out* to the great ocean basin joins up with 82 percent of the water on Earth, and the rainwater that flows *in* to the interior basins (seas, lakes, and interior deltas)[93] joins up with the other 18 percent. On the other hand, the logical implication of *rheology* is that the way out and the way in are one and the same. First, an exorheic body of water can become endorheic and vice versa. The Black Sea, for example, was probably once an endorheic basin—an island of water surrounded by land. Now, however, it is part of the great exorheic basin—geological changes linked it with the Mediterranean by way of the Dardanelles and Bosporus straits.[94]

This historical observation, however, ignores the nesological and islandological factors at work that, in any case, level the difference between endo- and exo-rheologies and allow one to configure them in the same way. When one walks the shoreline of a body of water, one eventually returns to where one started. (The ability to be circumambulated is one definition of *island*.) But this happens when one circumambulates either an interior sea (following the path that hugs the shoreline of Walden Pond) or the exterior ocean (following a path that hugs the shoreline of the Americas). The tenth-century Muhammad Abu'l-Qasim Ibn Hawqal's book *The Image of the Earth* (Surat al-ard), sometimes translated as *Configuration of the Earth*, thus includes a wheel map of the endorheic Caspian Sea[95] that resembles his map of the world.[96] Where the former shows water surrounding land, the latter shows land surrounding water.

Unlike many European cartographers, Ibn Hawqal knew that the Caspian Sea—taken with its ambiguously marginal river tributaries—was landlocked. "This sea," he writes, "is not connected with any other; and if a person wishes to make a tour completely round [circumambulate] it, nothing will impede him but a few rivers which fall into it from various quarters."[97] Nesologically speaking, then, the difference in size between one body of water and another does not matter.[98] ("*Land locke* is when the land is about you," writes John Smith in *A Sea Grammar* [1627],[99] which he composed two decades after he founded the celebrated Virginia Colony at water-locked Jamestown Island.)[100]

How defining water-surrounded land is linked with defining land-surrounded water is suggested likewise by Herodotus's story of the Egyptian king Necho II (610–595 BC), in which Necho determined that Africa was an Asian peninsula or perhaps vice versa. His sailors started out at an Egyptian port in the Red Sea and ended at a Nile River port in the Mediterranean, thus almost circumnavigating the kingdom:

After two full years [they] rounded the Pillars of Heracles in the course of the third, and returned to Egypt. These men made a statement . . . to the effect that as they sailed on a westerly course round the southern end of Africa, they had the sun on their right—to northward of them. This is how Africa was first discovered by sea.[101]

Necho, according to Herodotus, concluded that the narrow region between the Pelusian distributary of the Nile River, which emptied into the Mediterranean and the upper reaches of the Red Sea, was actually an isthmus connecting Africa and Asia.[102] Egyptian rulers before Necho's time had a distaste for the sea and sea power.[103] However, during the rule of the ancient pharaoh Sesostris,[104] who had aimed to conquer the entire world,[105] there had been a plan to cut a navigable channel that would make Africa[106] an artificial island with the Egyptian navy at its helm.[107] Necho brought this plan back to life, and by the time that King Ptolemy II (283–246 BC) completed his work, Egypt was a great naval power operating with two navies, one based in Mediterranean ports, the other in Red Sea ports.[108]

When it comes to understanding the horizontal and vertical dimensions of landed islands (Africa) and islanded seas (Caspian), difference in size is inconsequential, whether between the one ocean basin and the many inland bodies of water or, as we will see, between the continental mainland and its attendant islands.

Horizontal and Vertical

HORIZONTALS
Continent

What is the difference between the dry land we call an "island" and the dry land we call a "continent" or the "mainland"?[1] Is it that we believe we can walk all around the shore of an "island" and get back to where we started, but, in the case of continents and mainlands, we are not so sure? The *Oxford English Dictionary* would seem to say so when it defines *terra firma* as "a mainland or continent, as distinct from portions of land partly or wholly isolated by water"[2] (See Illustration 4.) Yet nowadays, although we believe that *all* bodies of land on Earth are, in principle, subject to circumambulation, we still call some of these bodies "islands" and others "mainland." How come? Is it just a holdover from older times of relative ignorance?

Size would seem to matter. Ephraim Chambers's *Cyclopaedia* (1727–1741) states: "[The general term] *Terra firma* is sometimes used to mean 'a large continent,' in contradistinction to 'a small island.'"[3] D. H. Lawrence, in "The Man Who Loved Islands" (1926), writes: "An island, if it is big enough, is no better than a continent."[4] This notion may help to explain why we call both large islands and small continents "island-continents" (the toponyms *Australia* and *Greenland* are examples) even while we say that continents are not islands. Consider how Sebastian Münster, in the sixteenth century, conceived the nationally diverse continent of Europe in terms of a peninsular cartography that united its countries (see Illustration 23). In *On the Spirit of Hebrew Poetry* (1782–1783), even Johann Gottfried von Herder uses the terms *island* and *small continent* interchangeably when he points out that, although people generally think otherwise, diversity can develop more rapidly on small islands than on large continents.[5]

PEOPLE WHO LIVE on relatively small islands in archipelagoes often refer to a bigger island in their archipelago as the "mainland." Thus people from White Head Island in the Grand Manan Archipelago (in Canada's Bay of Fundy) call the archipelago's largest island, Grand Manan, "Mainland." Likewise, people from smaller islands in the Orkney Archipelago (off the northernmost part of Great

ILLUSTRATION 4
Terra Firma. Herman
Moll, 1701. Source:
Beinecke Rare Book
and Manuscript
Library, Yale
University.

Britain) call the archipelago's largest island "Mainland." Presumably this name is a corruption of the Old Norse term *Meginland*, which became *Mainland* partly in consequence of the tendency, among residents of small islands, to call relatively large islands "mainlands" and likewise actually to name them *Mainland*. Another example is Mainland Island in the Shetland Archipelago.[6]

Many people want to assert a simple "cutoff" between "islands" and "mainlands" based on absolute (versus relative) size. "Water-surrounded land up to such and such acreage is an island," they say; similarly, "Islands larger than that are mainlands." Walter Scott provides a relevant gloss on an English-language synonym for "island," *öe*, in his translation of the Danish-language poem "The Elfin Gray": "*öe*, an island of the second magnitude; an island of the first magnitude being called a land."[7] John Venn takes a similar position on the relationship between size and islandness in his *Symbolic Logic*:

The statement that St Helena [Island, where the Emperor Napoleon had been imprisoned] contains a large salt lake or sea, studded with islands, might call for some explanation as to what was meant. And yet, unless we insist upon some such limit as to the relative extent of the land or water assignable to compose an island, what definition could be laid down for either lake or island, on a closed surface like a sphere, which should not make such a statement strictly true?[8]

Privileging size in this way, no matter how attractive, rules out the realities of human experience. The distance between largeness and smallness, like that between proximity and distance, is ever more collapsed thanks to already ancient and always accelerating technologies of time and travel and theories for understanding calculus and the universe.[9]

Planet

It takes little imagination nowadays to consider that North America and the other continents are islands—along with the other hundred thousand "open-sea islands" on planet Earth. By almost inevitable logical extension, Earth itself is an island. Plutarch, a contemporary of Ptolemy, in his island-centered essay *On Exile* (ca. 100 AD), says,

Each planet, revolving in a single sphere, as on an island, preserves its station; for "the Sun [a planet, according to Ptolemy] will not transgress his bounds," says Heraclitus.[10]

In *An Original Theory or New Hypothesis of the Universe* (1750),[11] Thomas Wright bravely speculates that the Milky Way is a flattened islandic disk of stars and that the nebulae are separate Milky Ways.[12] Immanuel Kant, in *Universal Natural History and Theory of Heavens* (1755), follows the same logic when he suggests that "a planet is far less in relation to the totality of creation than is . . . an island in relation to the earth's surface."[13] Both the astronomer William Herschel, in *Construction of the Heavens* (1785), and the geographer Alexander von Humboldt, in *Kosmos* (1845–1862), say that Earth is an island in the solar system—and that likewise the solar system is an archipelago in the Milky Way (see Illustration 5).[14]

The ancient thinkers' understanding of Earth's earth as an island surrounded by water—cartographically expressed by Muhammad al-Idrisi of Ceuta (see Illustration 3, Color Plate 1) and many mapmakers of medieval Europe[15]—thus helped to ground the genuinely scientific hypothesis that the Earth was an island in the universe. In the medieval period, of course, the hypothesis that Earth was an infinitesimally small insular "point" was usually put into the mouth of the devil. One instance informs an English-language homiliary of the eleventh century, in which a devilish creature says,

that all this earth [*middaneard*] [is] no more of dry land [*dryges landes*] beside the great Ocean than a point pricked on a board [*prican aprycee on anum brede*].[16]

Much earlier (in the fifth century), Macrobius claimed that the Earth made up no more than a point.[17] Similarly, "Lady Philosophy" in Boethius's *Consolation of Philosophy* (ca. 524), which was translated by King Alfred the Great, says that Earth is a point (*ad caeli spatium puncti*) and, in that sense, has no size (*nihil spatii prorsus habere*).[18]

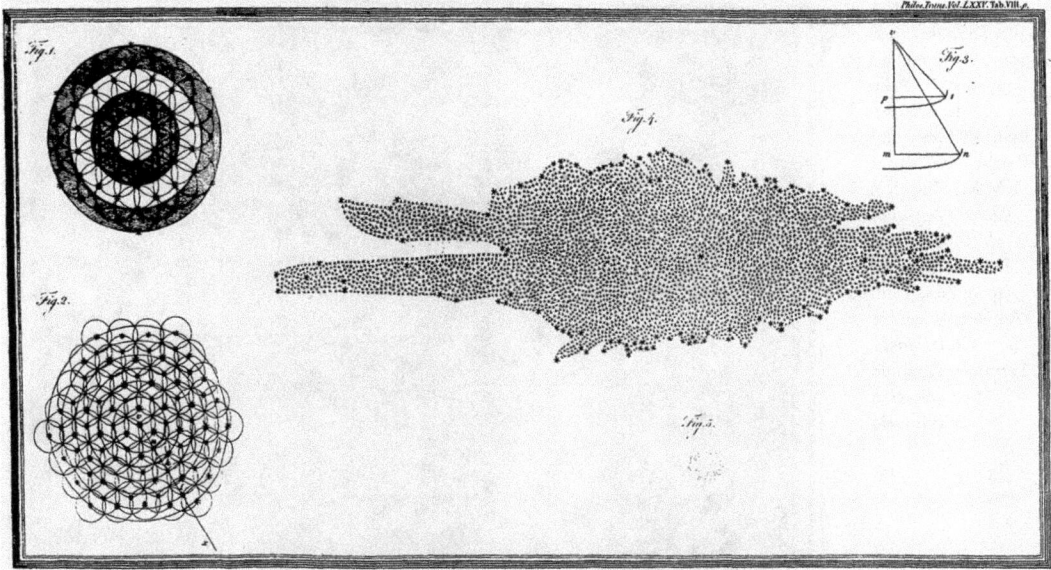

ILLUSTRATION 5
Planet Earth as
Island and Milky
Way as Archipelago.
Cartographic
representation of
planet Earth as
an island in the
Milky Way and the
Milky Way as an
archipelagic island, or
isolated archipelago,
in the universe. From
William Herschel,
"On the Construction
of the Heavens,"
*Philosophical
Transactions of the
Royal Society of
London* 75 (1785):
213–266. British
Library, 115.h.46,
p. 56. Reproduced by
permission.

This sort of islandological and Archimedean metaphor, which Kant explores in *Universal Natural History*, comes to play an important role in later astronomy. In *Other Worlds Than Ours* (1870), Richard Anthony Proctor writes, "Our earth is as a minute island placed within the ocean of space."[19] And Friedrich Engels, in *Feuerbach and the End of Classical German Philosophy* (1886), states that "not only does our group of planets [the solar system] move about the sun, and our sun within our *island universe [the Milky Way]*, but our whole *island universe* also moves in [universal] space."[20] By the early 1950s, such notions about the universe—that planet Earth is an island and that the Milky Way is an archipelago—were already science fiction commonplaces, as in Raymond F. Jones's novel *This Island Earth* (1952), whose cover shows overlapping planetary orbits in the style of a Venn diagram.[21]

The definition of islandness in terms of relative or absolute size finally involves a calculation about infinity. One recalls here the statement of Sicilian Archimedes (see Illustration 57):

Give me a place to stand and I will move the Earth.[22]

Mary Jones imagines such a place in *A Child's Geography of the World* (see Illustration 6).[23] The boy in her illustration sits nowhere—literally speaking, he is in *Utopia*.

In *Summer's Last Will and Testament* (1592), Thomas Nashe writes: "Every man cannot, with Archimedes, make a heaven of brass."[24] Here he is referring to Archimedes' construction—in brass—of an isolated artificial "globe" representing Earth.

VERTICALS

The way up and way down are one and the same.

—Heraclitus, *Fragments*

Tides

One often speaks about islands as if they existed only in a horizontal plane—as "land entirely surrounded by water—north, south, east, and west." Yet there is also a vertical dimension to islands—up and down—in relationship to sea level.

The rise and fall of the land in relation to the water, and the rise and fall of the water in relation to the land, are the waves and tides, which continually change the surface area of islands as well as their height and volume above the sea.[25] They change island into mainland and mainland into island. Wash-lands, bridge-islets, semi-islets, semi-islands,[26] and tidal islands appear and disappear.[27]

Land-tied islands likewise connect to and disconnect from the mainland.[28] Peninsulas when the tide is out and islands when the tide is in, they were the preferred dwelling places of the Veneti, inhabitants of the region called Armorica (meaning "place by the sea"), which included present-day Brittany. Longer-lasting drops in sea, lake, or river level can make an island part of the mainland by means of a new tombolo, ayre, or tidal flat. When the water drops enough—or the land rises enough—an island disappears beneath the surface, at least to the extent that the water is opaque. Such an island thus no longer has a horizontal or vertical dimension. "Now you see it, now you don't."[29]

Some islanders have cultures obviously informed by this vertical magic of the waters—the Marsh Arabs of Iraq, for example.[30] The coastal fishermen and -women of the Bay of Fundy likewise adapt to the extraordinarily high tides at Fox Point Beach near Parrsboro, Nova Scotia, on the other side of the Bay of Fundy, with its sixteen-meter tides.[31]

Islands sometimes sink permanently into the deep, or so we say. Ptolemy's *Aprositus Nēsos* comes to mind,[32] together with the long-standing *topos* of "lost" islands and island-continents: the lost Atlantis, the "Island of Women" in the Orkney Archipelago,[33] and many others.[34]

Subsidence and Uplift

The bottom of the adjoining sea is thickly covered by enormous brain stones.
—Darwin, *The Structure and Distribution of Coral Reefs*

One workable definition of island, first put forward by Bernhard Varen in the influential *Geographia Generalis* (1650), starts not from the land but from the sea. "The Earth," says Varen, "is covered in Water," but the part of that which stands out above the surface of the water, he says, are "Lands or Islands."[35] Varen takes into account less the constant shifts of sea level or the fact of floating islands and more the link between the geological or hydrological creation of islands and the zoological creatures, including beavers, earthworms, and coral, that help to make them.

Charles Darwin's last works consider such creatures. For example, in *Earthworms* (1881), he writes:

It may be doubted whether there are many other animals which have played so important a part in the history of the world, as have these lowly organised creatures [earthworms].

Some other animals, however, still more lowly organised, namely corals, have done far more conspicuous work in having constructed innumerable reefs and islands in the great oceans; but these are almost confined to the tropical zones.[36]

The casings cast up by worms help to wash islands away by loosening the soil, and they help to build them up by facilitating the growth of plants. In fact, Darwin's earliest works concern such creatures as are "living" and/or becoming "fossil" islands: the purview of *Coral Reefs* (1842)[37] joins the zoological history and the geology of Earth understood as a whole in the most brilliant islandology of modern times.

Wrote the satirical Thomas Seward about the theory that all life is explicable in terms of such shells as coral: "Great wizard he! By magic spells / Can raise all things from cockle shells."[38] Erasmus Darwin, Charles's influential grandfather and the evolution-theorizing author of *Zoonomia, or the Laws of Organic Life* (1794–1796), was himself a keen student of islands.[39] In 1771, he added to the family crest, which showed seashells, the motto *E conchis omnia* (everything from shells; see Illustration 7).

Charles Darwin's first paper, "On Certain Areas of Elevation and Subsidence in the Pacific and Indian Oceans, as Deduced from the Study of Coral Formations," was delivered to the Geological Society of London in 1837.[40] It set the stage for the great body of his work that followed. While his reputation was sealed with *On the Origin of Species* (1859), "Elevation and Subsidence" already included Darwin's extraordinary views on the evolution of the world as a whole, understood in terms of the interaction between the environment of the living (animal and plant species) and that of the nonliving (volcanic rock and seawater).

Coral Reefs and *Volcanic Islands* (1844)[41] deepen the insights of Darwin's first paper. For example, the theses of the distinctly islandological *Coral Reefs* point to the linkage between the gradual evolution of the Earth and the comparably unhurried evolution of its creatures. In the first

instance, *Coral Reefs* focuses our attention on a feature of the nonliving environment, volcanic explosions, some of which make for the sort of "sterile" places that
help elucidate species evolution and species extinction.[42] Such "nature's laboratories" come to mind as San Benedicto Island in the Revillagigedo Archipelago (in
Mexico) and Krakatoa (in Indonesia).[43] Just as important, *Coral Reefs* epitomizes
interest in the discovery of "fossils" (understood as the remains of animals preserved as rock).

In his study of the creation of islands, moreover, Darwin uses the placement
of coral reefs on the globe in such a way as to pinpoint the movement of the
Earth's tectonic plates. A reviewer in *Nature* (1874) properly writes about *Coral
Reefs* that

the natural history of a zoophyte was brought into connection with the grandest phenomena of the globe . . . with the progressive subsidence of more or less submerged mountains
and with the distribution of volcanic foci.[44]

As Darwin notes, a fringing coral reef that surrounds a volcanic island in the tropical sea grows upward as the island subsides (sinks), thus becoming a barrier reef
island. In time, subsidence carries the old volcano below the ocean surface and the
reef remains. At this point, the island is an atoll. Micronesia's Nukuoro Atoll is a
photogenic example (see Illustration 8).

ILLUSTRATION 8
Nukuoro Atoll,
Micronesia. The atoll,
photographed from
space, is 3.7 miles
in diameter. Source:
NASA.

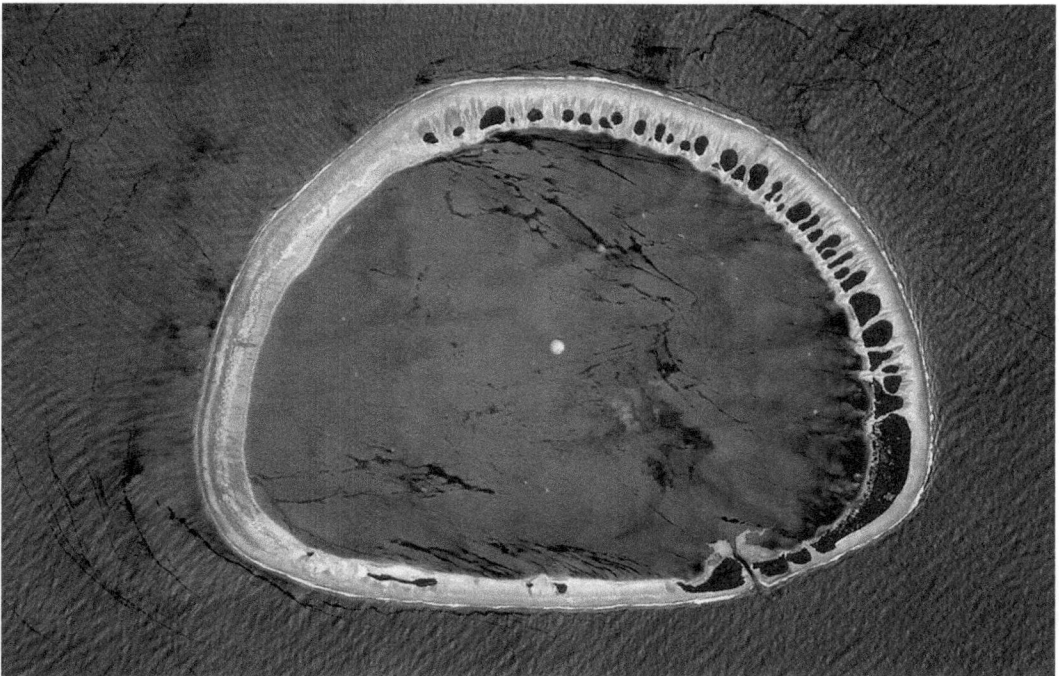

Darwin develops his insight about coral islands and subsidence and uplift into a methodology for charting the movement of the tectonic plates of the Earth as a whole. The physical interaction of (living and fossilized) animals with the human conceptualization of that interaction leads to a vision of the workings of the entire planet. Writes Darwin:

But now, viewing the appended map, it may, I think, be considered as almost established, that volcanoes are often (not necessarily always) present in those areas where the subterranean motive power has lately forced, or is now forcing outwards, the crust of the earth, but that they are invariably absent in those, where the surface has lately subsided or is still subsiding.[45]

The actual cause of subsidence and uplift along an apparently vertical axis we consider toward the end of the book, along with the view that the liquid ocean is itself an island floating on molten rock.

Animate Swimmers and Inanimate Floaters

One way to define an island is to say, "An island is solid matter of one substance, earth, surrounded on the horizontal plane by liquid matter of another substance, water." However, this says nothing about whether the earthen island is connected to the submarine earth beneath the liquid and thus part of it. The epigraph to Martinican poet Édouard Glissant's *Poetics of Relation* (1990)[1] is the Barbadian poet Kamau Brathwaite's dictum, in "Caribbean Man in Space and Time" (1975), that "the unity is sub-marine."[2]

Consider the rationale of John Donne writing, in *Meditation 17* (1624),[3] that "no man is an island." If Donne's meaning is that "every man is part of the continent, a part of the main,"[4] then the sentiment is puzzling, if not downright misleading. Almost no island is an island in the sense that Donne uses the term; all are part of the main insofar as they are of a piece with the solid submarine earth beneath the water. In this sense, islands are contiguous with the mainland in something like the way human beings are akin with the mud or clay (*adam*) out of which, the Bible says, God created them and to which they return after death.[5]

Landlubbers too—or, as Newfoundlanders sometimes call them, "land-crabs"[6]—often say, "All land is one land under the sea."[7] But it just ain't so. Landlubbers do not want to take into account those cases where floating islands, including ships, disconnect from the submarine earth.

Suppose, then, that an island is not "fast" to the submarine earth. Assume instead that, so far as one can tell, it is "loose" or "untethered" from it, or may as well be. Imagine now a floating inanimate creation (a ship), perhaps, or an animate creature (a whale).

BIG FISH

Islands can look like living creatures—as does Whaleback Island near Portsmouth, New Hampshire.[8] And living creatures can look like islands—as do the kraken[9] and leviathan.[10] So, too, do inanimate human creations look animate—such as ships when they are camouflaged as forests.[11] None of this means that islands are alive (animate). When Edmund Burke writes that "Spain is but a whale stranded on Europe's shores,"[12] *pace* Herman Melville in *Moby-Dick*[13] and Carl Schmitt in *Land und Meer*,[14] Spain stays inanimate.

And yet all islands—and coasts—already have a floating, liminal, aspect. An island, which sailors often confound with a whale, is at the limit a sort of "swimming land," apparently self-moving like an animate being. The etymology of *nēsos* links it with *natation*, meaning "swimming,"[15] and with problems of conceiving islands in terms of "flotation" and "animation." The strange sea monster, or *leviyatan*, named Leviathan, is a famous example. "Nothing on earth is his equal," says the book of Job.[16] The rabbis add that the monster was three hundred miles long,[17] which still leaves room on Earth for the land-dwelling Behemoth (see Illustration 9). Travel accounts and natural histories often describe a meeting with a giant fish that is also an island or with a small island that is also a giant fish.

We are now in the realm of "the big fish story" (a narrative that exaggerates the qualities of a fish that the teller has presumably caught) or the story of "the one that got away" (a narrative that describes a fish that the teller was unable to reel in).[18]

We read about the monstrous kraken of ancient lore in the *Natural History of Norway* (1752–1753) by the Danish theologian Erik Pontoppidan of Aarhus, whose native territory was on the Øresund just across the way from Zealand Island and Elsinore. Pontoppidan writes: "Amongst the many great things which are in the ocean, some have said in the past, is the Kraken. This creature is the largest and most surprising of all the animal creation."[19] He states that this huge animal looks like a group of small islands surrounded by seaweed.

In the Old English poem "The Whale," we read of the kraken's appearance:

[It is] like that of a rough boulder, as if there were tossing by the shore a great ocean-reedbank begirt with sand-dunes, so that seamen imagine they are gazing upon an island, and moor their high-prowed ships with cables to that false land, make fast the ocean-coursers at the sea's end, and, bold of heart, climb up.[20]

The Arabic-language *First Sea Voyage of Sinbad the Sailor* reports that Sinbad "landed" on what appears to be an island. In the animated movie *Popeye the Sailor Meets Sinbad the Sailor* (1936), Sinbad sings: "I'm Sinbad the Sailor, sing hearty and hail! / I live on an island on the back of a whale / It's a whale of an island— that's not a bad joke / And its lord and master is this handsome bloke."[21] Sinbad's "island" turns out to be a gigantic whale on which trees have been taking root "ever since the world was young."

According to the medieval bestiary *Physiologus*, the "aspidochelone [asp turtle] . . . is a great whale, that has what appear to be beaches on its hide. . . . This creature raises its back above the waves of the sea, so that sailors believe that it is

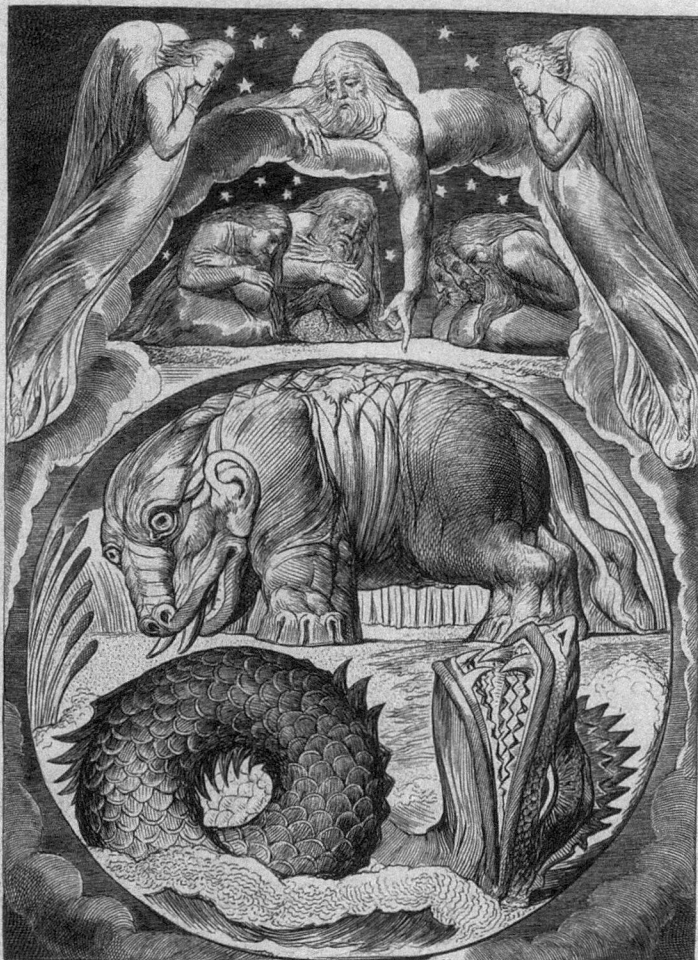

Within the illustration:

Can any understand the spreadings of the Clouds
the noise of his Tabernacle

15

Also by watering he wearieth the thick cloud
He scattereth the bright cloud also it is turned about by his counsels

Of Behemoth he saith. He is the chief of the ways of God
Of Leviathan he saith. He is King over all the Children of Pride

Behold now Behemoth which I made with thee

WBlake invenit & sculpt.

London Published as the Act directs March 8. 1825 by Will Blake N3 Fountain Court. Strand

Proof

ILLUSTRATION 9 Behemoth and Leviathan. William Blake's illustrations for the book of Job (1826). Reproduced by permission of the Morgan Library and Museum, New York.

just an island."[22] William of Normandy reports in his *Divine Bestiary* (1210) that "the upper part of [the asp turtle's] back looks like sand, and when it rises from the sea, the mariners think it is an island." The thirteenth-century Bartholomeus Anglicus reports that "the great fish seemeth an island. And if shipmen come unwarily thereby, unneth they scape without peril."[23] A thirteenth-century Middle English bestiary reports: "Cethegrande is a fis, / The moste that in water is. / That thu wuldes seien get, / Gef thu it soge wan it flet, / That it were a neilond / That sete one the se sond."[24] William Caxton has it thus in his *Mirror of the World* (1481): "In this se[a] of Ynde is another fysshe so huge and grete that on his backe groweth erth and grasse; and semeth proprely that it is a grete Ile."

According to the Norwegian *King's Mirror* (1250),

> There is a fish not yet mentioned which it is scarcely advisable to speak about on account of its size, which to most men will seem incredible. There are, moreover, but very few who can tell anything definite about it, inasmuch as it is rarely seen by men; for it almost never approaches the shore or appears where fishermen can see it. . . . In our language it is usually called the *kraken*. I can say nothing definite as to its length in ells, for on those occasions when men have seen it, it has appeared more like an island than a fish. Nor have I heard that one has ever been caught or found dead.[25]

A similar monster, called Jasconius, appears in the *Voyage of Saint Brendan the Navigator*, originally an Irish seagoing story, or *imram*, about Saint Brendan from the island of Fenit on the west coast. While searching for the infinitely distant "Western Isles," Brendan and his comrades come to feel an "island" swim under their feet. The entire "island" shudders.

> The island began to be in motion like a wave. The brothers rushed to the boat . . . the island moved out to sea.[26]

The island, like a wave, moves out to sea or becomes part of the sea (see Illustration 10).

SHIP

Some islands that swim are the artificial floats we call ships and boats. "A ship," writes Schmitt, "is as much a floating piece of land as it is a swimming dog."[27] In *White-Jacket; or, The World in a Man-of-War* (1850)—a precursor to *Billy Budd* (which was posthumously published in 1924)—Melville discusses, in political terms at once dystopian and utopian, the "floating island" that is a ship over which its captain rules. He compares the commodore of the navy, which is an "oaken archipelago," to the sultan of the Sulu Archipelago in the Philippines: "Upon these occasions, surrounded by his post-captain satraps—each of whom in

ILLUSTRATION 10
Saint Brendan on the Back of a Whale.
From Caspar Plautus, *Nova Typis Transacta Navigatio* (1621). Reproduced by permission of the collections of the Printing and Graphic Arts Department, Houghton Library, Harvard University, Typ. 620.22.697.

his own floating island is king—the Commodore domineers overall—emperor of the whole oaken archipelago; yea, magisterial and magnificent as the Sultan of the Isles of Sooloo."[28] (Airships—islands floating in the air—show the same characteristics, as we will see in Chapter 14.)

ON THE FLOATING ISLANDS that are man-made ships, some human beings live full-time. The regular life of the Moro Bajan "sea gypsies" of the southern Philippines[29] provides an example. In *Lalla Rookh* (1817), Thomas Moore writes about "that Eastern Ocean, where the sea-gipsies, who live forever on the water, enjoy a perpetual summer."[30] Melville considers himself one such ("stateless") gypsy when he identifies as an *omoo*—a rover among the islands of the Marquesas.[31]

Ships might be so big as to become floating "countries." Worth mentioning here is Jules Verne's 1895 *L'île à hélice* (Propeller Island),[32] a sequel to his 1871 *Une ville flottante* (A Floating City).[33] Verne's series of "Extraordinary Tales" (1863–1905) also includes islandic voyages under the surface of the water (e.g., *Twenty Thousand Leagues under the Sea*, 1872), beneath the surface of the land (e.g., *Journey to the Center of the Earth*, 1871), and up to the moon (e.g., *From the Earth to the Moon*, 1867). All such tales of man-made islands, Verne's clever publisher claimed, "outline all the geographical, geological, physical, and astronomical knowledge amassed by modern science and . . . recount, in an entertaining and picturesque format . . . the history of the universe."[34]

FLOATING ISLAND

Floating islands—disconnected vertically from the submarine *terra firma*—are not affected by the rise and fall of the water surrounding them in the same way that islands tethered to land under the water are. Francis Bacon, in his utopian novel *New Atlantis* (1624), hypothesizes a fictionally crucial sea island, Bensalem, that allows for a scientific "experiment solitary touching the super-natation of bodies";[35] yet natural floating sea islands are not so well known as artificial ones (ships). They should be better known. Their untethered condition often upsets preconceived ideas of ocean islands as linked to the earth below. In fact, there are floating islands all over the world: *floatons, sudds, tussocks,* and *hassocks.*[36] They can mark intranational and international borders[37] and carry refugee camps.[38] Naturally formed pumice rafts are huge,[39] float thousands of miles,[40] last for years, and distribute animals and plants across the world.[41] They appear to be regular sandbars, but they are not.[42] They "swim" like animate beings understood in nesological terms.[43]

The "sovereign territoriality" of floating islands is a matter for intense debate—much as it is for icebergs.[44] In Melville's terms from *Moby-Dick* (1851), one would call such islands not "fast fish" (an echo of a German-language term such as *Festland*) but instead "loose fish." Some call them "paradisiacal."[45] Others call them "fiery devils." Not surprisingly, the culture of people who live on floating

ILLUSTRATION 11
Floating Village near Nasiriyah, Iraq. "The whole country is under water, the villages, which are mainly not sedentary, but nomadic, are built on floating piles of reed mats, anchored to palm trees, and locomotion is entirely by boat." Gertrude Bell, 1916; Gertrude Bell Archive, Newcastle University. Photograph by Nik Wheeler, 1974. Reproduced by permission.

islands is a specific topic in anthropology. The floating islands of more northerly climes become a special concern for Europeans' fascination with pile dwellings (see Illustration 11, Color Plate 2) and icebergs.

In his remarkable book *On the Nature of Things* (50 BC), Lucretius seems to postulate that all land is, in this sense, islandic. He wonders why the seas do not overflow their bounds and comes up with an explanation supplementary to Aristotle's notion of evaporation. Lucretius holds that there is a pervasive underground connection between the oceans and the springs or sources of rivers. That connection renders the floating and the islandic all earth—mainland and island alike. There is, he says,

water distributed abroad through all the earth underneath, and all meets at the sources of each river, whence it returns over the earth in a column of sweet [fresh] water along the path which has once been cut for its liquid course.[46]

Wrote Henry David Thoreau in *Walden* (1854): "One value even of the smallest well is, that when you look into it you see that earth is not continent but insular."[47] Theories of continental drift and plate tectonics, as we will see, likewise suppose the liquid beneath the floating all.[48]

Material Substance and State of Matter

Some people claim that planet Earth is not an island because it is enveloped by something other than liquid water. The implication is that the substance (air instead of water) and the state of matter (gas instead of liquid) are crucial to how one thinks about islands and isolation. Corollary questions come to mind: Is (solid fresh or salt) water floating on (liquid fresh or salt) water an "island"? (Some etymologies of *insula* [island] derive the term from *salum* [sea] and/or *sal* [salt].)[1] Is the planet (solid) rock floating on (molten) rock? What about (solid) rock floating on (liquid) water? How about (solid) water floating on (liquid) rock? What about ships at sea?

If you do not want to think about islands in this way, you will escape the vertigo of it. Yes. You will also miss an essential linkage between geography and philosophy and a good part of human culture and language that focuses precisely on islands that float on water and in air.

ICE FLOE AND LAVA FLOW

They plied North-West among Ilands of Ice . . . some of them aground.
—Samuel Purchas, *Purchas His Pilgrimes*

We have seen that one way to define an island is to say that it is solid matter of substance A surrounded on the horizontal plane and perhaps also on the vertical plane by liquid matter of substance B. Generally we believe that A is earth and B is water. This belief deserves some scrutiny, both where the liquid and the solid are the same substances (icebergs that float in water and rocks that float in a pyroclastic flow) and where they are different substances (a glacier floating on solid rock).

The Far North of Canada is all about such places: watery permafrost, bog land, and *muskeg* (from the Cree word meaning "low-lying marsh").[2] Small islands of mushy peat crisscross invisibly with denser sphagnum tussocks; channels stretch invisibly underfoot. Stability comes, if at all, only from continuous firm ice or discontinuous permafrost—a meter beneath the "shell" or "crust" that is the visible surface. Walking here is death to the unwary.

Land's End is the toponym for the place where the polar pack ice begins and ends in the fall and spring of most years. For Canada, that is the place on the west coast of Prince Patrick Island, at the eastern opening of the Northwest Passage. Such places the world over give rise to conceptions like the German painter Caspar David Friedrich's *Sea of Ice* (1823–1824; see Illustration 12, Color Plate 8A),[3] inspired by William Edward Parry's *Journal of a Voyage to Discover a North-West Passage* (1821), as well as his *Wanderer above the Sea of Fog* (1818)[4] and *Chalk Cliffs on Rügen* (1818).

Places like these, where liquid and solid rock and water mingle, require a special vocabulary. Here *landbergs* are frozen freshwater glaciers sliding across the rock.[5] *Icebergs* are solid freshwater islands floating in a sea of liquid salt water. Some bergs are as large as Belgium (in the Southern Hemisphere half a century ago); some regularly threaten shipping lanes (Ayles Ice Island and Petermann Ice Island); and some are tabular, with steep sides and flat tops, running aground in the same spot year after year (Thwaites Iceberg Tongue in Antarctica and Pobeda Ice Island in the Mawson Sea).[6] *Land-fast ice* (saltwater) extends from the "land's edge" out to sea and generally forms shelf ice. *Drift ice* (saltwater) floats on the surface of liquid salt water and sometimes comes together to form "pack ice."

In unreliably frozen regions,[7] there are lakes and straits of liquid water surrounded by islands of solid water. An iceberg with a lake on it is a *meltie*; one with a strait is a *dry dock*. Relevant intermediate terms suggesting a permeable solid and liquid include *frazil ice*, *grease ice*, *slush*, *shuga*, and *pancake*, or *pan*, *ice*. In Varick Frissell's documentary-style film *The Viking* (1931), the sealers leap magnificently from pan to pan (see Illustration 13). In Newfoundland, these pans are straightforwardly called "island pans."[8]

ILLUSTRATION 12
The Sea of Ice. Caspar David Friedrich, 1823–1824. Bildarchiv Preussischer Kulturbesitz, Berlin / Hamburger Kunsthalle / Elke Walford / Art Resource, New York.

ILLUSTRATION 13
Sealers on Pan Ice.
Still from Varick
Frissell's movie *The
Viking* (1931). Photo
from Thomas Sweeny
Jr., "Filming an Arctic
Epic," in Bell and
Howell's monthly
magazine *Filmtopics*,
November 1930;
photo taken June 29,
1930. Source: Harvard
College Libraries.

The debate about whether Eskimo languages have more "basic" words or concepts for such water forms than other languages is an old one. Franz Boas, in *Handbook of American Indian Languages* (1911),[9] suggests that they do; other scholars argue that they do not.[10] This debate is not as important for us as an understanding of cultural variations[11] and modes of geographic metaphorization. After all, "land-surrounded water" (a lake) is like "water-surrounded land" (an island), and a "neck of water" (a strait) is like a "neck of land" (an isthmus).[12] Likewise, *strait* is a "foreland" (a place where two seas meet),[13] a *camel* is "a ship of the desert," and a *wave* is a "mountain of water." One might say that "a continent is 'an archipelago of insulated communities'"—as Charles Merivale puts it in *History of the Romans* (1865).[14] Newfoundlanders call the seashore between the high and low tides the "landwash,"[15] but it might just as well have been the other way around.

ILLUSTRATION 14 shows a lake (i.e., a body of liquid water) surrounded, or defined, by an apparent *terra firma* (a field of solid snow and firmly frozen ice). It does not show that the *terra firma* here is a glacier-become-iceberg—a freshwater floating ice island, or ice mountain, in the sea; an iceberg five times larger than

ILLUSTRATION 14
Island of Water on
Iceberg in Foreground
with Islands of Land
in Background.
Photograph by Nick
Cobbing, Greenpeace.
Reproduced by
permission.

Manhattan Island floating like a "loose fish" in the ocean of the Canadian Arctic. (What German speakers call an "ice mountain" [*Eisberg*], as if a floating mass of ice were connected to land below the sea, Newfoundlanders—who live on the sea's edge—more straightforwardly call "island of ice.")[16]

Many cultures also evince a sense that earth as well as water is infirm. In Iceland, solid water (i.e., ice) sometimes floats on liquid rock (magma). When, in 1996, an Iceland volcano exploded 2,500 feet below the surface of the Vatnajökull ice cap, the lava flowed beneath the ice, turning the glacier above into a floating island of solid water on a sea of molten rock.[17] Just so elsewhere, solid rock floats atop liquid rock—and atop liquid water—in the form of pumice rafts and lava balloons.

NORTHWEST PASSAGE

The [Northwest] Passage is a difficult piece of territory to categorize
because it is neither just land nor just water and the legal jurisprudence
for waters, let alone, remote, ice-infested, arctic waters, is not clear.
—Andrea Charron, "The Northwest Passage Shipping Channel: Is
Canada's Sovereignty Really Floating Away?"

The straits and islands of the Canadian Arctic Archipelago—36,000 islands extending 1,500 miles east–west and 1,200 miles south–north—are often the subject of border disputes with Russia, Europe, and the United States. Canada has more coastline than any other country: 55 percent of the world's total[18] is in Canadian territory. (Canada also has more islands than any other country, with the possible exception of Finland.)[19] Island quarrels here often center on tiny, uninhabited islets and skerries. Among them is the struggle over Hans Island—whose sovereignty is disputed by Denmark.[20] Hans has more "meaning" to both countries than just mineral ownership and sea passage.

Denmark has a national identity bound up with notions about land and water going back a thousand years and more.[21] Canada puts forward the same face. Canadian prime minister Stephen Harper puts it in the nationalist way: "Canada's Arctic is central to our national identity as a northern nation. It is part of our history. And it represents the tremendous potential of our future."[22] The Northwest Passage is Canada's "Holy Grail."[23] The phrase "Westward Ho!" helps to explain the successful expansionist ethic of nineteenth-century "United Statesian" culture; just so, the phrase "Up North" helps to explain the later countrywide nationalist "ethic" of Canada.

The ongoing search for a reliable northern passage began a year after Columbus "discovered" the Americas, in the wake of Pope Alexander VI's bull *Inter Caetera* (1493). The bull awarded the Southern Hemisphere of the New World to Catholic Spain and Portugal, thus denying easy circumnavigation to non-Catholic Europe and triggering the northern European search for a Northwest Passage. Most searches turned into "voyages of delusion."[24] An old belief, still common in the days of Captain James Cook in the late eighteenth century—that seawater does not freeze—helped ensure this failure.[25]

Explaining this apparently odd belief helps to unfold the difficulties of comprehending this part of the Earth. Explorers of the period were encountering *polynyas* in the cryospheric Arctic or Antarctic regions. A polynya is open seawater, a lake, surrounded by sea ice for most of the year. (The Russian word *polyj* means "hollow.") Thus a saltwater polynya is similar to a lake *lacuna*, or *lagoon*, except that it is surrounded by a *terra firma* of frozen water instead of one of solid earth. Some of these saltwater "ice-holes" are large (as much as 50,000 square miles). Some appear in the same general location year after year in the Weddell Sea (Antarctica) and the North Sea (in the Smith Sound between Ellesmere [Island], whose name recalls both a lake [*mere*] and a sea [*mere*],[26] and Greenland [Island], sometimes called a "continent"). Nowadays we better understand water upwellings and wind and water currents in the cryospheric regions than did Cook, but the unpredictability of most polynyas, especially in the Laptev Sea, still makes it difficult to accurately forecast when and if ice-holes will open—or close.

The early explorers' true tales of failure—and their false stories of success—have, over the centuries, come to constitute a repository of materials that informs Canadian culture, even the *imaginaire* of Canada. First, Anglophone Canadians sing the national anthem as "The True North Strong and Free." They recall the Victorian poet laureate Alfred Lord Tennyson's unswerving loyalty to England in the politically charged *Idylls of the King* (1856–1885). "That true North, whereof we lately heard" is how Tennyson puts it in the section of his poem entitled "To the Queen." Second, Francophone Canadians sing the national anthem instead as to "la croix" (the cross). By way of comparison, there is the relatively secular Québécois *chansonnier* Gilles Vigneault's celebrated French Canadian "national hymn,"[27] "Mon pays, ce n'est pas un pays, c'est l'hiver" (My Country, It Is Not a Country, It Is Winter). Vigneault wrote this song for Arthur Lamothe's movie *La neige a fondu sur la Manicouagan* (The Snow Has Fallen on Manicouagan; 1965), which the National Film Board of Canada had commissioned to mark the construction of the hydroelectric plant in the Manicouagan territory claimed by the nomadic Innu First Nation (or Naskapi, meaning "people beyond the horizon"). Third, the Inuktitut themselves, who are this region's "native people," chant "ᐊᖕᒋᔪᓘᑎᓪᓗ, ᐅ ᐸᓇᑕ" (*Sanngijulutillu*) and focus instead on the "snow compressed into ice." The Canadian *imaginaire* as such obtains whether one imagines the passage as going from west to east[28] or from east to west.[29]

No one European traversed the entire Northwest Passage until 1906. That year, the Norwegian explorer Roald Amundsen made the passage by *aqua liquida* (via ship). In 1921, Knud Rasmussen, the Danish Greenlander polar explorer, made the passage on *aqua firma* (via dogsled).[30] These days the Northwest Passage is "open" for part of the year thanks to "global warming." From that worry arises the Canadian political imperative to "recontinentalize Canada" in terms of "Arctic ice's liquid modernity" and "the imagining of [the] Canadian Archipelago."[31]

The Canadian government now claims that the waters of the Arctic Island region are "internal to Canada."[32] Prime Minister Stephen Harper has announced that "ships entering the North-West passage should first report to the federal government." Americans enjoy singing "This land is your land / This land is my land," from Woody Guthrie's classic song from 1944, but the Canadian Armed Forces now send out an annual military operation to the Arctic region, whose proper toponym is *Nunalivut*, meaning "the land that is ours."

Relevant here is Andrea Charron's scholarly paper "The Northwest Passage Shipping Channel: Is Canada's Sovereignty Really Floating Away?" (2004), published by the War Studies Programme in the Royal Military College of Canada and presented officially at a Canadian Defence and Foreign Affairs Institute meeting. Likewise relevant is a new contract ($39 billion) for building warships.

FEELING FOR SNOW

You can't win against the ice.

—Peter Høeg, *Frøken Smillas fornemmelse for sne*

Defining the term *terra firma* eventually involves distinguishing one material substance from another (earth from water) and/or one state of matter from another (solid from liquid). Hybrid material substances (muskeg) and hybrid states of matter (slush) fit easily into this definitional schema. Less easy to categorize, though, is "ice cap." By this term, I mean "a thick, solid body of water that caps, or covers, some large part of the surface of Earth, whether the part is dry, wet, or both."

There is the frozen water, including icebergs and ice shelves, that covers the liquid water. "At any given time," claims the Danish author Peter Høeg in his Greenlandic novel *Miss Smilla's Feeling for Snow* (1992), "floating ice covers a fourth of the earth's ocean area. The drift ice belt in Antarctica is 8 million square miles; between Greenland and Canada it's between 3 and 4 million square kilometers."[33]

This section focuses on the ice cap that rests on rock (earth) and the difference, in terms of human perception and institutional policy, that sea level can make. The part of this ice cap that rests on rock (earth) at sea level or above, most people think of as "glacier." However, a large part of the world's ice cap rests on earth that is below sea level. These areas people usually call "terra firma" (as if the ice there were covering firm and dry land), but this part of the ice cap actually is "iceberg" resting on submarine rock.

Danish Greenland has ice sheets that are sometimes five thousand feet below sea level. One example is the fast-moving glacial Jakobshavn Isbrae (in western Greenland), which terminates in the Ilulissat Icefjord.[34] Antarctica has its concomitant Bentley Subglacial Trench (in Marie Byrd Land), which extends downward into the trench a mile and half.

Some of these submarine-grounded ice sheets are fast moving: the Jakobshavn Isbrae sometimes moves 125 feet a day. Some are large: the Bentley Subglacial Trench is often called "the world's largest island that is not a continent."[35]

Is Greenland a single island landmass? The answer depends on how one thinks (or does not think) in terms of states of matter and sea level. If one thinks of an island in "Anglo-French" terms (a coastline that is "wet" [here liquid water] meets what is "dry" [here solid water] at sea level), then, all other things being equal, Greenland is an *insula*. Most traditional maps of Greenland thus represent the place as a large island with a few proximate smaller islands (e.g., Alluttoq) and archipelagoes (e.g., Upernavik). Representing Greenland in this way, though, masks the sense in which what we call Greenland is not all "land" understood in the sense of "earth or rock that stands at or above sea level." If Greenland's ice sheet were to disappear, it would look not like a large island but instead like a vast series of archipelagoes and islands. Where once there were solid mountains of water towering upward far above sea level, now there would be deep trenches of liquid water extending far below.

How would a traditional Anglo-French mapmaker draw Greenland if, feeling out "global warming," he were to represent all its water, both solid and liquid, in the same way? The question can have a heuristic purpose. Making such a map would reflect (1) knowledge of what lies above and below; (2) communicative means adequate to convey such factors as substance, state of matter, and sea level; and (3) expression of the tension between understanding an island as "water as opposed to land" and comprehension of it as "water as land."

The right feeling here for "ice ↔ land" (ice understood as both land and water) informs Bille August's English- and Greenlandic-language film version of *Miss Smilla's Feeling for Snow*.[36] The story begins with a "half-Greenlandic and half-Danish" hero, Miss Smilla, who resides on the canal-ridden, artificial island Christianshavn in Copenhagen—a city that is itself located on two islands (Amager and Zealand) and whose capital buildings house the Danish *Folketing*, or national assembly. The film ends in the general region of the previously mentioned Alluttoq Island, a site of an old Norse (Viking) settlement around glacier-covered Disko Bay, off Greenland.

Greenland is legislatively part of Denmark, and *Miss Smilla's Feeling for Snow* is, accordingly, a colonial and postcolonial story.[37] Settlement around Disko Bay was established by Erik the Red, who also had close links with the Danish settlement on Oxney (Island) in Britain. Erik set up Brattahlid and Gardar in the late tenth century, where the small town of Igaliku now stands, at an isthmus near Tunulliarfik Fjord. This spot was the location of the *ting*, the relatively democratic general assembly of the Danes' Eastern Settlement on Greenland.[38]

The movie *Miss Smilla's Sense of Snow* ends with a mainland Danish villain who cannot distinguish between water and ice; he drowns on the ice floes.[39] In fact, the movie's tipsy oil platform–island recalls the dangers of "Iceberg Alley."[40] *Iceberg Alley* names the dangerous marine region in the Davis Strait, separating Baffin Island (Canada/Nunavut) from Greenland (Denmark), where members of Greenpeace once scaled a 53,000-ton oil rig (at Atammik in Greenlandic waters) owned by Cairn Energy, a British corporation. The tides in the Davis Strait are fierce and range up to ninety feet. Especially in the winter, the weather there is hostile. Building any sort of platform city in those dire straits isn't easy, no matter what Cairn's feeling for profit, and not only for natural reasons, but also for political ones.

BERGMAN'S ISLAND

I want to see with my outer eye first and then with the inner eye.

—August Strindberg, letter to Torsten Hedlund

The Swedish filmmaker Ingmar Bergman (1918–2007) located his stories on islands and chose islands as his shooting locations even as he attempted to come
to terms with the psychological isolation of the human mind and with the political alienation of mainland life. The film critic Andrew Sarris notes that the
deeply neurotic Bergman, celebrated for his claustrophobic and islandic movies
and chamber scripts, generally confined himself to the "island of his mind."[41] This
is what Hamlet, whose bourn was the globe itself, calls the "mind's eye."

Bergman chose tiny Fårö (Island),[42] a remote islet with its own dialect, to be
his home. "Ever since my early childhood," he says, "I have felt rootless wherever
I've been. It is only since I came [to Fårö] (at around the time that I was making the movie *Through a Glass Darkly* [in 1961]), that I have felt at home in the
world."[43] Of Fårö, Bergman says, in the autobiographical *Magic Lantern*, "I had
found my landscape, my real home."[44] For him, it was the subject of such documentary movies as *Fårödokument* (1970) and *Fårödokument 1979* (1979)—a site
of presumed utopian potential.[45] It was as well the shooting location for such
films as *Through a Glass Darkly* (1961), *Persona* (1966), *The Passion of Anna* (1969),
Scenes from a Marriage (1973), and *Shame* (1968).[46] Yet Fårö was not Bergman's
only island. *Shame*, for example, was shot also on Gotland Island, as was *The
Touch* (1971);[47] *Summer Interlude* (1951) includes the tale of the instigating diary
written in the Stockholm Archipelago and was partly shot on Rosenön; *Summer
with Monika* (1953) had Ornö (Island) for its shooting location; and *Autumn Sonata* (1978) includes memories whose shooting location was Danish Bornholm
(Island).

Insular headlands likewise provide locations for Bergman's island movies. *Autumn Sonata* was shot mostly on Romsdal Peninsula (Norway); *Hour of the Wolf*
(1967) was filmed on the Bjäre Peninsula (Sweden); and *The Seventh Seal* (1957)
had Hovs Hallar, on the Bjäre Peninsula, for the *Hamlet*-like location of its first
scenes, in which the Knight plays chess with Death.[48] The never-completed movie
The People Eaters (mid-1960s) was scheduled to be filmed at Hallands Väderö (Island), just off the Bjäre Peninsula.[49]

Non-Scandinavian islands also figure importantly in Bergman's work: *Ship to
India* (1947), for example, as well as *Thirst* (1949) and *Waiting Women* (1952).

Noteworthy here is Bergman's lifelong affiliation with works by the island-
informed Swedish playwright August Strindberg,[50] chief among which being the
chamber play *The Ghost Sonata* (1907). Bergman directed *The Ghost Sonata* on
stage four times, twice in theaters in the Øresund region (1941, 1954). Significantly,

Strindberg's actual stage play ends with the projection of images of *Island of the Dead*, Arnold Böcklin's dark painting (1880s),[51] about which Strindberg had plans to write an entire play on its own.[52] Bergman, however, although he knew this painting and loved it as a child,[53] omitted the projection from all four productions. When a journalist asked why, Bergman answered that *Island of the Dead* was a bad painting. Perhaps so, but there are other reasons. Böcklin's work was a favorite of Adolf Hitler's,[54] and Bergman had enjoyed the Wagnerian aesthetics of Nazi rallies, including the island-modeled Nazi *Thingspiele*, during his time in Germany in the mid-1930s.[55] He alternately tried to explain and expiate these feelings. There is, for example, his 1943 stage production in neutral Sweden of the Danish playwright Kaj Munk's *Niels Ebbesen* (1942), about the Danish "tyrant slayer" who was killed by German troops in the fourteenth century. Munk's politics were already sufficiently conservative, however, so that what he disliked most about the Nazi occupation of Denmark was not that it was too dictatorial but that it was insufficiently "Nordic." At around the same time (1944), there was Bergman's anti-Nazi production of *Macbeth*, which he produced at the theater he was then directing in Helsingborg, Sweden. Decades later, Bergman returned to the Nazi period, in *The Serpent's Egg* (1977), which is set in Germany in the period of the Nazi rise to power during the 1920s. The Shakespearean question is that of *Julius Caesar*: Should one kill off a potentially tyrannical serpent in its egg? Even before it ventures outside its "shell"? Even before it moves beyond the "shoreline" of its own original space and seeks more *Lebensraum*?[56]

In all of these works, Bergman does not retreat from mainland to island simply to find a better politics or a more reliably humane construction of political assembly (*ting*), which he would have found in such places as Gotland Island.[57] Some film historians see his move to the island as a utopian gesture, but in most ways Bergman steps back from politics—even to the superficial extent of complaining about the legal requirement to pay taxes.

In this context, it is worth explaining why *Hamlet* should figure so importantly in many of Bergman's movies, including, most famously, *The Seventh Seal*, with its great shoreline scene, which recalls *Hamlet*'s opening. One explanation is that this Shakespeare play is a British islander's national history of Danish Helsingør (Shakespeare's "Elsinore"), a castle on the Scandinavian island of Zealand across the Øresund from Helsingborg, where Bergman was stage director. Helsingør, after all, is where Bergman staged his own play *Kamma noll* (Draw Blank, or Come Up Empty; 1948), with its island setting in the Stockholm Archipelago.

A second explanation involves Bergman's own attempts to get away from a more or less port city. Bergman was born (in 1918) in the port city of Uppsala and grew up in Stockholm (from 1920 to 1934); his childhood is partly the subject of his feature movie *Torment* (1944). He departed from this *hamn* (port), seeking to leave as much as Hamlet and Laertes sought to leave the "nutshell" (*Hamlet*) that was Elsinore.[58] The shooting location of his dockside movie *Hamnstad* (Port of Call; 1948) is thus the deepwater port of Gothenburg, across the Skagerrak from Danish Jutland, which is the birthplace of Hamlet, according to Saxo Grammaticus. When Bergman staged his version of *Hamlet* in Stockholm (1986), he had his actors use the Swedish-language translation by a Gotland islander,[59] who emphasized the port and portal aspects of Shakespeare's play.

A third explanation, likewise recalling the deeper identities and rivalries that inform the Scandinavian-British affiliation, involves a struggle over a patch of rock very much like the one, not large enough to bury the dead, for which the Norwegian Fortinbras is apparently sending his troops to do battle. This place would be Rockall Island—called "the most isolated small rock in the oceans of the world."[60] Bergman knew Rockall well. This tiny island has been variably claimed by the Faroe Islands (Denmark), Iceland, Ireland, and the United Kingdom.[61] Its geography was used to devastating effect during World War II, as Bergman helped to recall for others in his direction of the Swedish playwright Rudolf Värnlund's quasi-pacifist play *U-Boat 39* in Stockholm (1943). *U-39* was the first submarine (*Unterseeboot*) sunk by the British. After circumnavigating island Britain, it attempted to sink the aircraft carrier HMS *Ark Royal*—the floating island of Britain itself—near Rockall. Just so, as Todd Field suggests, Bergman provides a sort of aqueduct for projecting a submarine subconsciousness: "[Ingmar Bergman] was our tunnel man, building the aqueducts of our cinematic collective unconscious."[62]

Rockall is not the only "British" island that is arguably also "Scandinavian." Others include the various islands of the Scottish Orkney Archipelago. The Scandinavian aspect of *Macbeth* helps to explain Bergman's attraction to it. In his first stage production of the play in 1940, in which Bergman played the Scottish king Duncan, Macbeth's rival and/or co-conspirator is a Norseman—as had been suggested centuries earlier when the Icelandic Snorri Sturluson's *Heimskringla* and the *Orkneyinga Saga* reported the history of Thorfinn, Earl of Orkney. Bergman's second production of *Macbeth* in 1944 took place at the theater at Helsingborg, just across the Øresund from where Shakespeare set *Hamlet*. At around the time that Shakespeare was writing his Scandinavian play, King James I of England was negotiating with the Danes about sovereign ownership of the Orkneys, where the people still spoke a variant of Norse.[63]

No wonder Bergman wanted to shoot *Through a Glass Darkly* (1961). He says, "[*Through a Glass Darkly* is] set on an island, with four people who rise up from a twilight sea in the first scene and walk ashore to begin the drama. I had this idea that we would do the film in the Orkneys." But since they were "out of the question"[64] for Bergman, he sought out another shooting location. That is, he made his first trip to Fårö, where *Through a Glass Darkly* was finally shot.[65] In that way, a location off the island of Britain—a one-time player in the great Anglo-Scandinavian Empire of the North Sea—became Bergman's "Fårö." Says the First Gravedigger in *Hamlet*: "There [in England] the men are as mad as he [in Denmark]" (act 5, scene 1).

The Norwegian actress Liv Ullmann, who lived with Bergman for five years and had a daughter with him, made a television movie, *Faithless* (Trölosa), in 2000. She filmed it on Fårö. Erland Josephson, who had appeared in Bergman films beginning with *It Rains on Our Love* (1946) and ending with *Saraband* (2003), played Bergman as an elderly dramatist alone in an isolated coastal cottage. Bergman, for whom Fårö would seem to be both a real and a symbolic place,[66] appeared in *Faithless* as a variant of a "Shakespearean Prospero figure," but it is worth noting that, much as one might have expected, he did not do a *Tempest*. Not that he was not asked to. Michael Billington has it that John Gielgud, who redefined the role of Shakespeare's Prospero, the politician-scholar-artist in exile, in his famous performance at Stratford-upon-Avon in 1957, and who played Prospero in Peter Greenaway's movie *Prospero's Books* (1991), "never ceased wanting to make a movie of *The Tempest* directed by Ingmar Bergman."[67] But Bergman was no more likely to do that than put on a production of *Island of the Dead*. While Prospero sought to return to mainland Italy, Bergman chose to be buried on Fårö[68] in a coffin whose wood came from the local innkeeper.[69]

Part Two

Placemarks and
Cultures

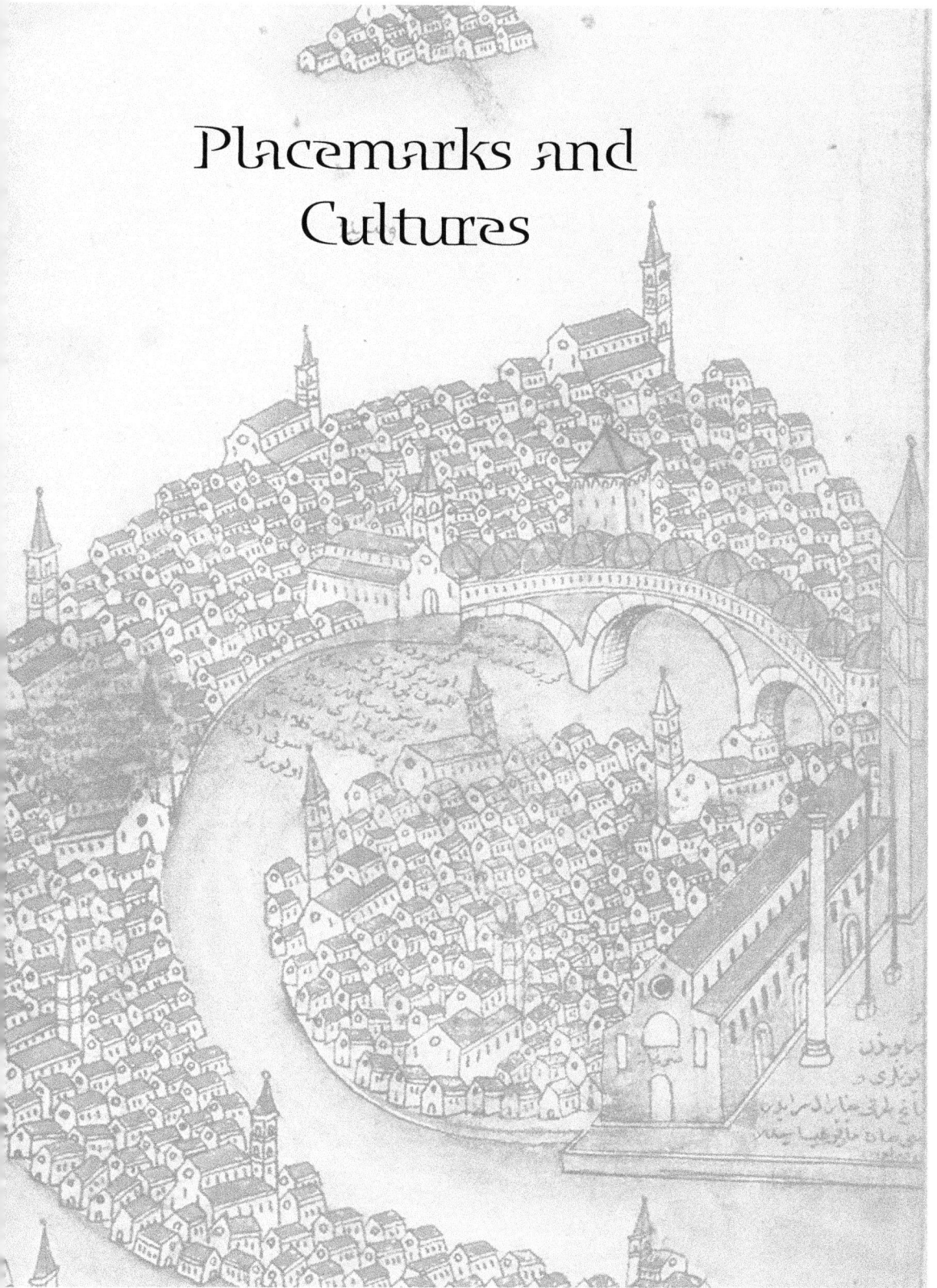

اولوقن
منزلى
باغ لمقه طار الدم یوم
نن حیاه مازوقیبا یحاه

Cities in Straits

STRAITS

This island [Hormuz] was formerly the greatest and most frequented
port on the ocean.

—Niccolao Manucci, *Storia do Mogor*

One way to begin categorizing the waterside cities in the world is to deal politically with those that are variably independent or autonomous city-states.[1] (That
method is one subject of Chapter 8.) Another way, less traveled, is to begin geographically with the list of island-cities located in "straits."

The word *strait* comes from the Old French *éstreit*, meaning "narrow" or "a
narrow isthmus of water, sometimes just a river, conjoining two larger bodies of
water." *Isthmus* itself derives from the ancient Greek word for "neck," the body part
that connects the head with the torso at a narrow "choke point."[2] Synonyms or
translations for *strait* include the English-language *hals* or *hawse* (literally "neck"),
which, in its Danish-language version, gives the town Elsinore—at the narrowing
of the Øresund—its name.[3]

Island-cities in straits constitute the main subject of this chapter. The list of
such places includes Montreal, which stands at the dangerous Lachine Rapids
of the Saint Lawrence River—whence the boatman can proceed westward, past
Detroit, whose name means "strait," into the interior of North America.[4] The list
also includes Manhattan Island in New York City—"The City," as Americans often call it—which stands at the "mouth" of the Hudson, a waterway that extends,
with limitations, from the island of Manhattan northward, via Lake Champlain
and the Richelieu River, to Montreal. Many other straits have "guard islands" that
"protect" the way through the waters,[5] some of them disputed[6] and a few shared
by two (or more) countries.[7]

Island-cities in straits also include those that border on variably endorheic (inland or inward-flowing) basins as well as exorheic basins. Astrakhan, for example,
has its Astrakhan Kremlin Island near the mouth of the Volga River, which flows
into the endorheic (closed) Caspian Sea. Mexico City was originally the Aztec

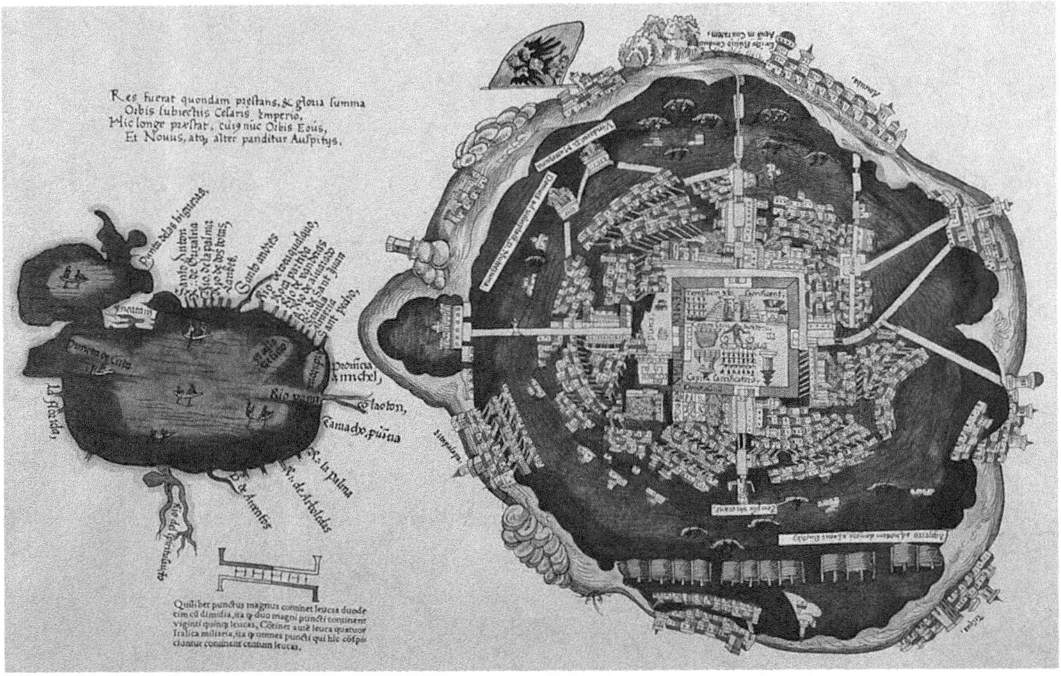

ILLUSTRATION 15
Tenochtitlán. From
Hernán Cortés (and
Pietro Savognanus),
*Praeclara Ferdinandi
Cortesii de Nova Maris
Oceani Hispania
Narratio* (Nuremberg:
Fridericum Peypus
Arthimesium,
1524). Bildarchiv
Preussischer
Kulturbesitz, Berlin /
Ibero-Amerikanisches
Institut, Stiftung
Preussischer
Kulturbesitz, Berlin /
Dietmar Katz.

"Tenochtitlán"—the sprawling water-city that the Spanish conquistadors found in the endorheic Valley of Mexico, a geological bowl into which forty rivers flow. Tenochtitlán was constructed in Lake Texcoco on interrelated natural and artificial islands together with massive dikes, as shown by Hernán Cortés in his 1524 cartographic representation (see Illustration 15) and in such subsequent works as Benedetto Bordone's 1528 Venetian-dialect *isolario*.[8]

Island-cities in straits often include port cities at deltas. There are, for example, the urban areas of the Niger Delta (Nigeria)[9] as well as those of the Golden Triangle on the Yangtze Delta (China).[10] Also on this list are Alexandria and Port Said on the Nile Delta (Egypt), Calcutta and Haldia on the Ganges Delta (India), Mongla and Chittagong likewise on the Ganges Delta (Bangladesh), Karachi on the Indus Delta (Pakistan), and Millingen aan de Rijn on the Rhine-Meuse Delta (Netherlands). Others include New Orleans on the Mississippi Delta (United States), Sulina on the Danube Delta (Romania), Tiksi on the Lena Delta (Russia), and Yangon on the Ayeyarwady Delta (Myanmar).

Insular Mumbai, which nowadays stands where the two estuarine branches of the Ulhas River flow into the Arabian Sea, is a telling example of such urban places. There once was a bay with an archipelago of islands: the Portuguese, who lorded over the area in the sixteenth century, called it Bombay (little bay in Portuguese); the Romans had known the area by the name *Heptanesia* ("cluster of seven islands" in Greek).[11] When the British arrived, they undertook massive earth-moving projects that changed the "lay of the land" as much as similar projects had changed the areas of Venice and Amsterdam. The eighteenth-century Hornby-Vellard Project, for example, aimed to make big islands out of small ones. The name *Hornby-Vellard* comes from *William Hornby*, who was then Bombay's governor, and from the local pronunciation of the Portuguese word *vallado* (fence, embankment).[12] The influential movie *Seven Islands and a Metro* (2006), says its director Madhusree Dutta, is precisely "about how Mumbai originated from seven small islands."

Rio de Janeiro, now mostly a mainland city on Guanabara Bay in southeast Brazil with its narrow mouth near Sugarloaf Mountain, began its European development when the French and then the victorious Portuguese (1565) founded the city on such islands as Villegagnon and Lajes. In 1502, Portuguese explorers assumed that the bay (*ria*) was a river (*rio*).[13]

Such island-cities as mentioned here are usually crucial to an understanding of the development of cosmopolitan trading cultures. Their locations provide imperial outposts or centers. The geographer Ellen Churchill Semple thus writes: "Islands located in enclosed seas . . . are the outposts of the surrounding shore, and become therefore the first objective of every expanding movement, whether commercial or political, setting out from the adjacent coasts."[14] Semple has many examples in mind,[15] and her position seems reasonable, but she fails to account for how some of these islands come to dominate the mainlands around them (rather than vice versa). Typical examples are Venice in the great *lacuna*, where modern capitalist culture early took hold, and Hormuz in the Persian Gulf, which controlled the trade route from Europe to India and China in the 1500s. These two island-cities in straits, Venice and Hormuz, we discuss in the following sections, together with a few ancillary examples.

ISLAND-CITIES

Thomas More's famous fictional land of "Utopia" is a definitively fabricated island. Its ruler, Utopus, conquered the original mainland place, called "Abraxas," and then decided that the sort of human culture he wanted to develop there required insulation from the mainland. Therefore, he ordered a giant earth-moving project by which the Utopians turned the mainland (that is, their part of it) into an island (entire of itself)—their mainland was "islanded."[16]

. . . this was no island at first, but a part of the continent. Utopus that conquered it (whose name it still carries, for Abraxa was its first name) brought the rude and uncivilized inhabitants into such a good government, and to that measure of politeness, that they now far excel all the rest of mankind; having soon subdued them, he designed to separate them from the continent, and to bring the sea quite round them. To accomplish this, he ordered a deep channel to be dug fifteen miles long.[17]

Turning mainlanders into islanders, Thomas More seems to expect his readers to know, has a military tradition. The ancient Persian emperor Xerxes, seeking to conquer Greece, made the peninsular mainland Mount Athos into an island by cutting a canal across it in 480–483 BC. He wanted thus to more easily invade.[18] Similarly, turning islanders into mainlanders has its tradition. The Macedonian emperor Alexander the Great, on his way to exploring and conquering a good part of the world, made the island-city of Tyre into part of mainland Phoenicia by building a mole, or causeway, that connected it with the mainland (332 BC). By means of this "disinsulation,"[19] he wanted to be better able to destroy the Persian naval base there.

The princely Alexander apparently wanted an island-city to mirror himself. The first plan to create such a place, presented to him by the islander Dinocrates of Rhodes, involved turning Mount Athos back into an island, as it had been during the time of the Persian tyrant Xerxes, and creating there a huge colossus of Alexander.[20] The plan was discarded. (Niccolò Machiavelli, in the chapter of his *Discourses on Livy* titled "Of the Beginnings of Cities" [1517], points out that the Mount Athos project was impractical.)[21] When he became an Egyptian pharaoh, Alexander was encouraged to start building a city on the island of Pharos, at the mouth of the Canobic distributary of the Nile River near a strip of land between the Mediterranean Sea and the then-navigable Lake Marcotis.[22] He ordered Dinocrates to connect island to mainland by a causeway.[23] Under subsequent Ptolemaic rule, this place, with its famous lighthouse, became a culturally syncretic center of great intellect and powerful empire. Philo of Alexandria writes that the island of Pharos was where scholars translated the Hebrew laws of Moses into common Greek[24] and where the city fathers convened the annual cosmopolitan Septuagint festival.[25]

The transformation of mainlanders into islanders (Utopia and Mount Athos) has counterparts when it comes to successful island–city-states like Venice and Hormuz. In both cases, people-moving and/or earth-moving projects turned mainlanders into islanders. There are two interrelated political and cultural factors here: a mainland people create an island or move to one, turning themselves into an island people; and the ongoing and ever-developing memory of that production or relocation informs a new culture and political economy.

Venice

Consider the early Venetians. They left the region that they called Terra Firma—mainland places like Padua, Aquileia, Treviso, Altino, and Concordia (Portogruaro) as well as the nearby undefended countryside—in order to avoid Germanic and Hun invasion.[26] They then immigrated to the great lagoon, where they joined the anglers already resident on the mudflats, people then called *incolae lacunae* (inhabitants of the lagoon). The new "settlers" found security on their islands, much as the Utopians found safety on theirs, and they gained the benefit of living in the midst of a great *lacuna* much as did the Utopians. As Thomas More reminds us,

The whole coast [of the artificial, or man-made, island of Utopia] is . . . one continued harbour, which gives all that live in the island great convenience for mutual commerce; but the entry into the bay, occasioned by rocks on the one hand, and shallows on the other, is very dangerous. . . . If these should be but a little shifted, any fleet that might come against them, how great soever it were, would be certainly lost.[27]

Venice was a "utopia" already realized when More wrote these words. More knew that the notion of island was an ambivalent figure of utopia, that the first printed *isolario* was authored in the Venetian dialect by Benedetto Bordone (1528),[28] and that all such books were part of the history of nautical cartography and imperial expansion.[29]

The process of making mainlander into islander and making land from water had a political and productive consequence in Venice like that for similarly marked economic locations. The Venetians began electing a duke in the eighth century. Within a few hundred years, the Venetian Archipelago became the world's great thalassocracy. With its naval power based in the Arsenal, Venice came to dominate a good part of the nearby mainland, Terra Firma. Soon it went on to rule a good part of the Mediterranean world as well. This "City of Canals" became the world's longest-enduring independent jurisdiction. With its republican "merchant princes," it lasted more than a millennium.

What made the Venetians powerful, in historical context, was *not* what they had found, namely, a swampland in a *lacuna*. Rather, it was their common wresting of land from water and of residing in a water-land (literally, *is-land*). Their swampland experience gave them a "recompensatory" edge. Like the canal city of Amsterdam in the "Lowlands," Venice developed a commercial empire from nothing by creating land where there had been none (the dikes) and wealth of which no one had dreamed (credit money). Out of the *lacuna* of real estate and unreal money, they made a new republican commercial dominion.

Out of nothing comes something: that is the great subject of the second part of Goethe's poetic drama *Faust*, one of whose principal subjects is political economy and the production of real estate. Shakespeare's island play *The Merchant of Venice* has, at its heart, the *Rialto*—an island toponym that, since Shakespeare's day, has come to mean something like "exchange" itself.

Twice in *The Merchant of Venice* we hear the same question: "What news on the Rialto?" What does *Rialto* convey? The toponym refers originally to an island in the Venetian lagoon, "Rivo Alto" (Ri'alto), one of the higher lagoon mudflats or sandbanks on which the formerly mainland immigrants settled. When the city of Venice was founded officially in the ninth century, the main urban office was located on that island, which was relatively secure from invasion insofar as it was both separate from the main island of Venice and close to it.[30] In the twelfth century, the Venetians built their first bridge across the Grand Canal, the pontoon-constructed *Ponte della Moneta* (Bridge of Money), which joined the Rialto to the main island.[31] By the mid-thirteenth century, in fact, the Venetians had built there a solid wooden bridge—called *Ponte di Rialto* (Rialto Bridge)—with commercial shops along the sides. In 1591, five years before Shakespeare began to write *The Merchant of Venice*, the Venetians built the present stone bridge, which actually remained the only way to cross the Grand Canal by foot until the 1850s. The Venetian painter Vittore Carpaccio's *Healing of the Madman* (1494) depicts an iteration of the bridge that is neither land nor water but both (see Illustration 16).

Rialto, the island where the archipelago-dwelling Venetians established their Exchange, gave its name to the bridge and thence to the world of European and

ILLUSTRATION 16
Rialto Bridge. Vittore
Carpaccio, *Healing
of the Madman*,
1494. From *The
Miracle of the True
Cross* series, Galleria
dell'Accademia,
Venice, 1494. Photo
courtesy of Scala /
Ministero per i Beni e
le Attività Culturali /
Art Resource,
New York.

other languages. The English traveler Thomas Coryate—the first Englishman to make the "grand tour of Europe"—writes in 1611: "The Rialto which is at the farther side of the bridge as you come from St. Marks, is . . . the Exchange of Venice, where the Venetian gentlemen and the Merchants do meet twice a day."[32]

The Merchant of Venice makes the further verbal leap from *Rialto* to *rialme*, meaning "realm" and/or "royal." Immediately after asking, "What news on the Rialto?" the Jewish moneylender Shylock asks, "Who comes there?" He who comes to the Rialto turns out to be Antonio, the *"royal* merchant"—a palpably oxymoronic phrase twice repeated in the play. Antonio is a dispossessed Christian aristocrat living in a commercial republic where the three "religions of the Book"—Islam, Judaism, and Christianity—coexist in relative harmony. The events that follow, however, invoke the tension between royal Christendom and republican merchantry in a way that destabilizes the juridically based political economy of Venice and sends us to a markedly utopian neverland—namely, Belmont, perhaps a veiled reference to the famous Marrano town of Belmonte in the mountainous borderlands of Portugal. From that region, Shylock's compatriot Tubal has likely fled—and Tubal is the ultimate source of the money that Shylock borrows in order then to lend it to Antonio.[33]

The forced conversion of non-Catholics by Catholic institutions is the underside of Shakespeare's comedy set in the "City of Water." The Turkish commander Piri Reis's maps of Venice show the location of the Rialto on the Grand Canal (see Illustrations 17 and 18).[34] Piri fought in the Venetian-Ottoman War (1499–1503) and helped to deliver many thousands of Marrano Jews and Morisco Muslims from the Inquisition.

Venice has given its name, if not also its conception, to countries and cities worldwide. The country of Venezuela is an example. One story goes that, when the Spanish visited the Venezuelan coast in 1499, Amerigo Vespucci named it Veneziola (Little Venice) because the pylon buildings of Lake Maracaibo reminded him of Venice.[35] There are dozens of cities and towns *named* Venice. Among them is Venise-en-Québec (Venice in Quebec), where I used to row as a child. There are also places that are *called* the "Venice of such and such a place." So, in the Northern Hemisphere, there is Saint Petersburg, called the "Venice of Russia," as well as Amsterdam, Bruges, Copenhagen, Hamburg, Manchester, and Stockholm. In the East and Middle East, there is Barisal (Bangladesh); Bandar Seri Begawan (Brunei); Lijiang, Suzhou, Tongli, Wuzhen, and Zhouzhuang (China); Nan Madol (Federated States of Micronesia); Alappuzha, Srinagar, and Udaipur (India); Jakarta and Palembang (Indonesia); Basra (Iraq); Osaka (Japan); Malacca (Malaysia); and Ayutthaya and Bangkok (Thailand). Worth recounting here are stories or histories involving Suzhou and Basra.

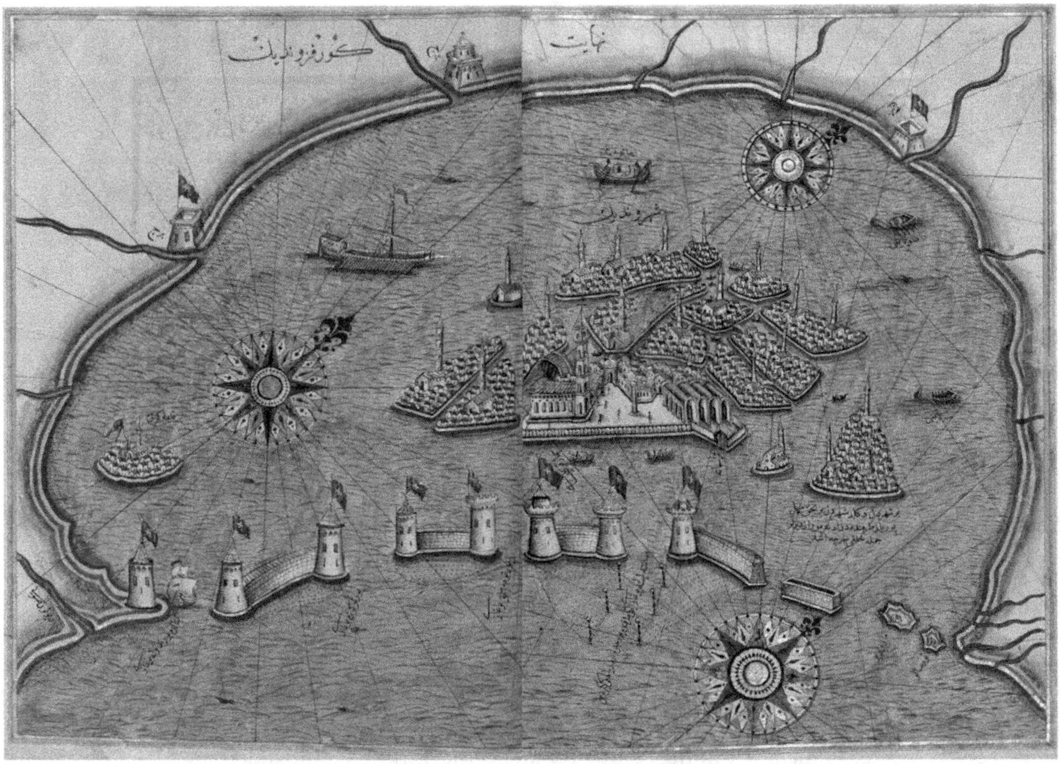

ILLUSTRATION 17
Map of Venice. From
Piri Reis, *Kitab-i
bahriye* (Book of
Navigation; 1521).
Reproduced by
permission of Walters
Art Museum.

Concerning Suzhou, called the "Venice of the East,"[36] the scholar might begin by discussing one of the influential Four Great Classical Novels of Chinese literature: the fourteenth-century *Shui Hu Zhuan* (The Water Margin), which is often attributed to Shi Nai'an from Suzhou, a city of canals on the shores and islands of Lake Taihu in the Yangtze River. "The marshes of Mount Liang" is how Alex and John Dent-Young, writing in 2002, translate the title. The Suzhou marshes were, at the time, the largest wetlands in northern China.[37] Righteous bandits, unhappy with governance from the mainland, use these water-lands as places of refuge and rebellion. They are the "outlaws of the marsh"—which is how Sidney Shapiro, writing in 1980, translates *Shui Hu Zhuan*—forming a gallant fraternity where "all men are brothers," which is how Pearl Buck, writing in 1933, renders the title.[38] The outlaws' headquarters are on Mount Liang, a precipitous "mountain-island" in the midst of the great Liangshan Marsh. A military general serving the unjust tyrant Tian Hu loses his temper with these raiders: "You marsh dwelling outlaws," he roars. "How dare you invade our territory?"[39]

Basra, the "Venice of the Middle East," presents a more complex situation, partly because this Iraqi city is located at a periphery of the water-land of the Marsh Arabs. Located at the Shatt al-Arab (Stream of the Arabs), Basra marks the

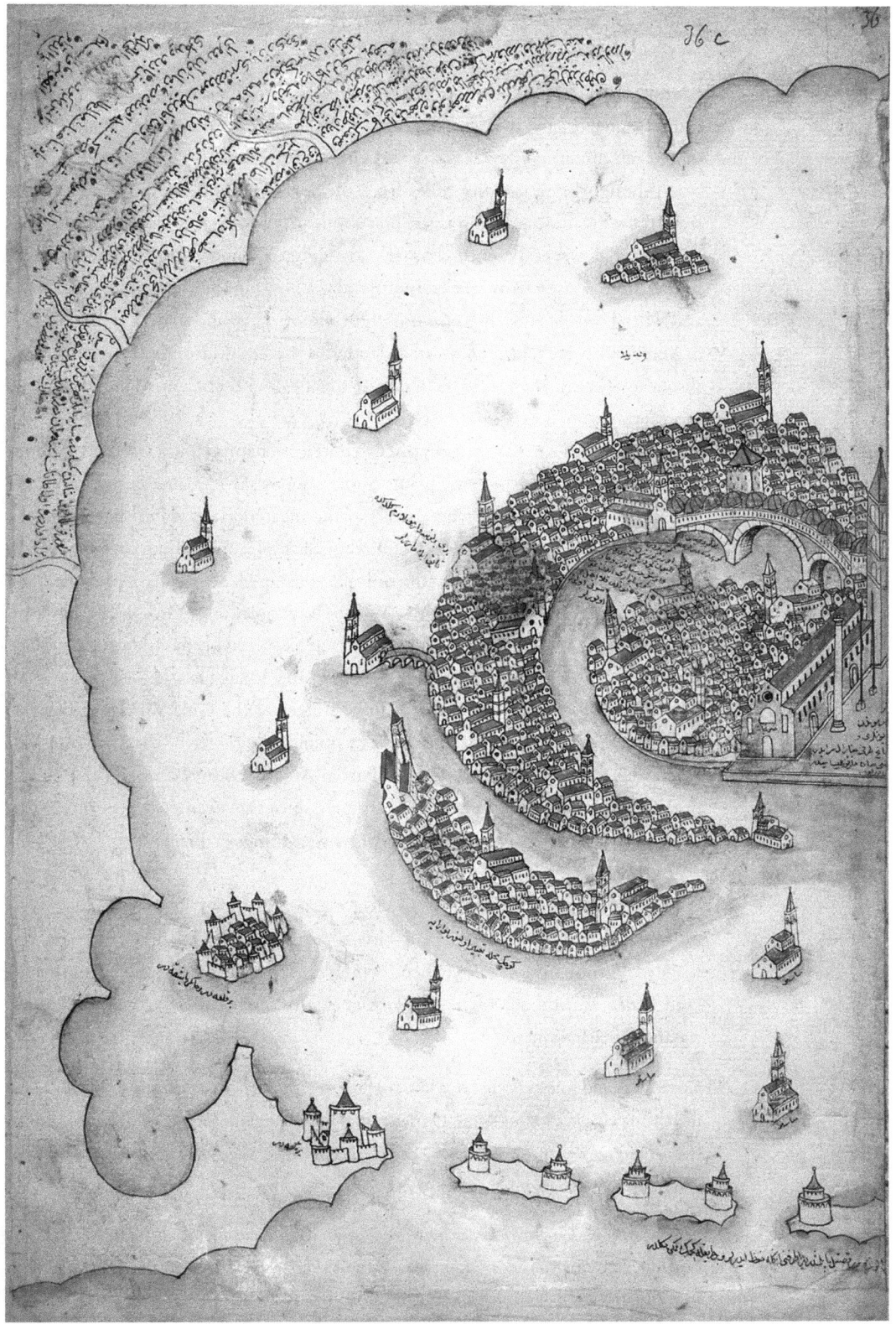

ILLUSTRATION 18 Map of Venice. From Piri Reis, *Kitab-i bahriye* (Book of Navigation; 1521). Bildarchiv Preussischer Kulturbesitz, Berlin / Staatsbibliothek zu Berlin, Stiftung Preussischer Kulturbesitz / Ruth Schacht / Art Resource, New York.

confluence of the Euphrates and Tigris rivers. Not only is it the regional homeland of the Marsh Arabs; it is also the home water-land of Sinbad, the great fictional sailor who discovers islands and island-cities worldwide.

Although Basra is thirty miles from the sea, its tides rise and fall about nine feet. The surrounding region is the first about which we have written records concerning tidal power. A tenth-century geographer, al-Muqaddasī the Jerusalemite, for example, writes: "The tide is a marvel and a blessing for the people of Basra . . . and when the tide ebbs, it is also useful for the working of the mills."[40] The islands of Venice likewise had tidal mills; so, too, did medieval Little Island, now a peninsula to the east of the city of Cork in Ireland,[41] and ancient Mahee Island, in Northern Ireland's Strangford Lough.[42]

Like most of the island-cities discussed in this chapter, Basra was a cosmopolitan, even universalist, meeting place: some scholars say its name derives from the Persian *bas-rāh*, meaning "where many paths meet"; others claim that it was the location of the Garden of Eden. Voltaire, in his philosophical novel *Zadig, or The Book of Fate* (1747), has it that "the universe was one large family which gathered together at Bassora [Basra]." Be that as it may, peoples and states of the Basra region have long had conflicts over border issues centered on the islands. Iraq quarreled with Kuwait over the Warbah, Bubiyan, and Failaka islands at the mouth of the Shatt al-Arab; this was a determinate trigger of the Gulf War (1990–1991).[43] The dispute over Abadan Island was the proximate trigger of the Iran-Iraq War (1980–1988). Qubban Island, in the Shatt al-Arab between Basra and the mouth of the Euphrates, has been a focus of struggle since the seventeenth century.[44] The tidal location that raises Basra up likewise often brings it down.

Hormuz

Hormuz, as an independent island–city-state ruled by the Muslim Hurmuzî, had its heyday from around 1300, when the Hurmuzî people moved there from mainland Persia, until 1507, when the Portuguese conquered it. In 1525, Piri Reis bemoans the situation:

Come now and listen to what it is like at Hormuz. Know then that this is an island. Many merchants visit it. . . . But know, O friend, that the Portuguese have come there and built a stronghold on its cape. They control the place and collect the customs—you see into what condition that province has sunk![45]

The island of Hormuz is located at a spot that is still globally crucial—in the bottleneck strait that connects the Arabian (Persian) Gulf with the Indian Ocean (see Illustration 19).[46] Not for nothing did the British Empire, in 1864, locate the

crucial articulation of its Indo-European telegraph cable on tiny Jazirat al Maqlab (Telegraph Island) in Khor Ash-Sham Fjord. There messages, crucial to British imperial dominion of the "Indian subcontinent," were sent between Karachi and London.[47]

Hormuz was also important to global trade and communications. Through this "eye of the needle" passes, even now, a good part of the world's oil. If the Rock of Gibraltar stands at the gates of the Greek divine hero Hercules, the island of Hormuz stands at the gate of the Persian god Hormuz (one etymology of the island's toponym).[48] Captain A. W. Stiffe, editor of *The Persian Gulf Pilot* (1870–1932), writes in his history of the island that "Hormúz became the capital of an empire which comprehended a considerable part of Arabia on one side, and Persia on the other."[49] "If the world were an egg," writes François Pyrard, then "Ormus would be the yolk"[50] (the *yolk* being "the core or best part").[51] Pyrard was on to something. He had been shipwrecked on Maalhosmadulu Island in the Maldives from 1602 to 1607 and published his *Voyage to the East Indies, the Maldives, the*

ILLUSTRATION 19
Map of Hormuz.
Johann Caspar
Arkstee and Henricus
Merkus, 1747.
Reproduced by
permission of the
National Library
of Israel, Eran
Laor Cartographic
Collection, Shapell
Family Digitization
Project and the
Hebrew University
of Jerusalem,
Department of
Geography Historic
Cities Research
Project.

Moluccas, and Brazil at the time that Shakespeare was completing his last play, *The Tempest*, in 1611.[52] He well knew his history.

Hormuz had been known to the Venetians at the time of the Hurmuzî population's emigration from mainland to island circa 1300. Marco Polo visited there at that time (see Illustration 20, Color Plate 8B). The Hurmuzî's removal westward to the island made a certain ethnic "sense," as they were part of an Arab population that had come over from the "Arabian" shore of the Gulf a century or two earlier and had founded, on the "Persian" shore, the settlement of "Old Hormuz."[53] The Hurmuzî never claimed status as "Persian,"[54] and Marco Polo reports that their ruler was the Arab leader Rukn al-Din Mahmud;[55] some historians say that the Hurmuzî dynasty was first established by the Yemeni chief Mohammed Dirhem Ko.[56]

Like the Venetians, the Hurmuzî left the mainland marshes for the offshore island for purposes of military defense and commercial security. Abd al-Razzaq Samarqandi, in his Arabic-language *Rise of the Two Auspicious Constellations and the Junction of the Two Seas* (written in the 1460s), says, "This city is also named *Dar al-Aman* [Abode of Security]."[57] Shakespeare's Venetian "merchant prince"

ILLUSTRATION 20
Marco Polo with Elephants and Camels Arriving at Hormuz on the Gulf of Persia from India. From *Livre des merveilles du monde* (ca. 1410–1412). Boucicaut Master (fl. 1390–1430) (and workshop) / Bibliothèque Nationale de France, Paris / The Bridgeman Art Library, Ms Fr 2810 f.14v.

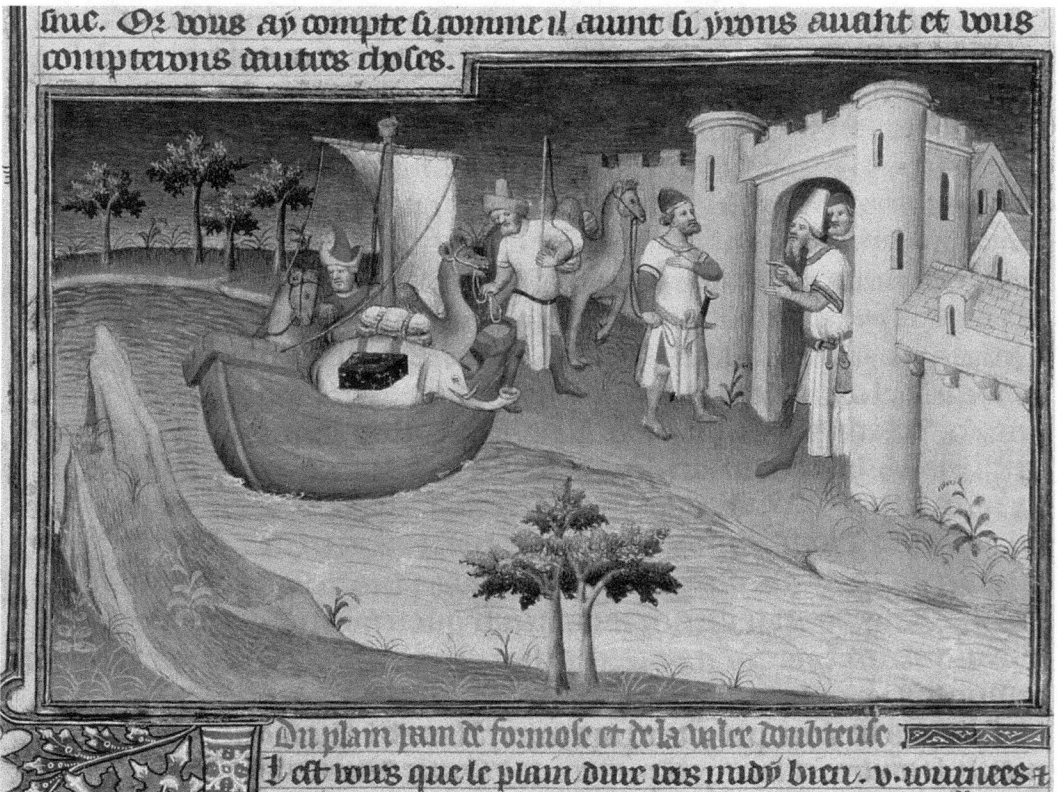

Antonio says the same about the importance of security for the prosperity and political stability of Venice:

> The duke cannot deny the course of law:
> For the commodity that strangers have
> With us in Venice, if it be denied,
> Will much impeach the justice of his state;
> Since that the trade and profit of the city
> Consisteth of all nations.
>
> (act 3, scene 3)

In this assessment of the basis of power in Venice, Shylock agrees with Antonio, as when the moneylender says, "If you [the Venetian court] deny me, fie upon your law! / There is no force in the decrees of Venice" (act 4, scene 1).

The wealth of Venice, such as it was, lay not in commodities but in the laws of trade. Like Venice, the island–city-state of New Hormuz had no "indigenous" wealth (other than fish), not even fresh water. Joseph Salbancke's *Voyage* (1609) says that Hormuz was "certainly the driest island in the world."[58] Most food and all water had to be imported.[59] The *Itinerary* (1510) by the Italian traveler and diarist Ludovico di Varthema, the first European to enter Mecca as a pilgrim, reports that New Hormuz "is an island and is the chief, that is, a maritime, place for trade. In the said island there is no sufficient water or food, but all comes from the mainland." Not commodities, therefore, but services made for commercial prosperity, and the particularly valued service was security together with a relative "openness" to strangers, sometimes called "tolerance," that is characteristic of certain seagoing and trading cultures. Samarqandi writes that "people of all religions, and even idolaters, meet in the city of Hormuz and nobody permits any hostile gesture or injustice against them." According to the Italian-Jewish explorer Jacob of Ancona, who traveled there in the thirteenth century, New Hormuz "is an open place for 'all the world.'" The Jewish population among the Hurmuzî then was 250.[60] (At that time, there were Jews living throughout the Gulf region.)[61]

Strait and lagoon island-cities like Hormuz and Venice often develop characteristically localized systems of taxation. Samarqandi noted that, "for all objects with the exception of gold and silver, a tenth of their value is paid by way of duty."[62] The Hurmuzî treated gold and silver differently from other commodities, partly as a way to effect monetary policy and encourage a reliable banking system. Pyrard notes disapprovingly that the governors at Hormuz "will for money let everything pass."

For English-language speakers, "Hormuz" became a byword for great wealth.[63] Fulke Greville set his tragedy *Alaham* there (posthumously published in 1633).

John Milton, who was fourteen years old when the Portuguese lost the island-city, refers to it in *Paradise Lost* (1667):

> High on a throne of royal state, which far
> Outshone the wealth of Ormus and of Ind,
> Or where the gorgeous East with richest hand
> Show'rs on her kings barbaric pearl and gold,
> Satan exalted sat.[64]

And Samuel Purchas writes: "And so the inhabiters of Hormuz doe say, that 'all the world' is a ring, and Hormuz is the stone of it."[65]

The fifteenth-century Chinese traveler Zheng He visited Hormuz many times between 1413 and 1433. He reports that there were no poor people there.[66] Samarqandi, around 1442, writes that Hormuz "has not its equal on the surface of the globe." In 1469, the Russian merchant Afanasy Nikitin went to Hormuz and remarked on its cosmopolitan aspect: "Hormuz is a vast emporium of 'all the world'; you find there people and goods of every description and whatever is produced on earth you find it in Hormuz."[67]

A definite political and commercial expansionism accompanied and materially supported the foundation of the Hurmuzî state. The Hurmuzî even established a few towns back on the Arabian side of the Gulf: "The Hurmuzî rulers developed Qalhât on the 'Umânî coast in order to control both sides of the entrance to the Gulf."[68]

What happened to Hormuz? The island is still there, but the cosmopolitan and imperial trading culture is gone. There are several factors to consider here, some with implications for the same region in the present day and some with implications for all island–city-states in straits.

First, Hormuz, along with many other islands in the variably enclosed seas of this region, was a border island of the disputed sort. Nowadays, among such are also Bubiyan and Warbah (Kuwait versus Iraq); Qaruh and Umm al Maradim (Kuwait versus Saudi Arabia); and the two Tunbs and Abu Musa (Iran versus Sharjah, United Arab Emirates). Elsewhere around the Arabian Peninsula, there are similar disagreements about island sovereignty: the Hanish Islands at the Bab-el-Mandeb Strait (Yemen versus Eritrea); an archipelago of four islands, including Duwamish, at the Bab-el-Mandeb Strait (Yemen versus Saudi Arabia); and Ras Doumeira and Doumeira, also at the Bab-el-Mandeb Strait (Djibouti versus Eritrea). Nearby are disputes about Diua Damasciaca, Ras Kamboni, and Badmadow (Somalia versus Kenya), and about Migingo and the lake islands of Lolwe, Oyasi, Remba, Ringiti, and Sigulu (Kenya versus Uganda). Political powers often highly prize such islands (including Hormuz), whether those powers are nearby (e.g., Iran) or far away (e.g., Portugal and England).

The second factor to consider is that such places are often ethnically complex in such a way that their domestic politics are related to that of surrounding mainlands. The Iranian director Babak Payami's movie *Secret Ballot* (2001) shows that this is true even today on the island of Kish in the Gulf. *The Merchant of Venice* concerns similar problems in Venice—problems that eventually destroy the Venetian Republic.[69] In the case of Hormuz, the decisive moment was the arrival of the Portuguese in the Gulf in 1507, which meant the expulsion of the Jews and other such groups.[70] At the time, Catholic missionaries from Iberia complained about the religious and linguistic tolerance of Hormuz. "Ormuz [is] a Babel for its confusion of tongues" was the view of the Spanish missionary Francis Xavier.[71] He had in mind not only the Arabic-language groups from the eastern mainland and the Persian-language groups from the western mainland but even the unique Hurmuzî dialect, which went by the name *Psa*. (The existence of this language is noted by the brilliant linguist Duarte Barbosa, brother-in-law and traveling companion of world circumnavigator Ferdinand Magellan.)[72] There were other languages at Hormuz: Judeo-Arabic and Gujarati as well as Chinese, Spanish, and the idiolects of the prominent Baghdadi and Fali tribes. Saint Francis Xavier did not much like this diversity, especially when it came to Jews. He established a Jesuit missionary in Hormuz, under the administration of the Flemish Jesuit missionary Caspar Baertz.[73] Soon thereafter all Hurmuzî Jews, if they did not convert or feign conversion, fled the island.

The conquest of Hormuz by the Portuguese in 1507[74] was followed by an invasion carried out by the joint forces of Abbas the Great (Persia) and King James I (England) in 1622. The Anglo-Persian capture of Hormuz changed the balance of power and trade in the entire world, opening up both Persia and India to English traders and closing off that trade to the Portuguese and Spanish. Although the Persians sent most Hurmuzî[75] back to the Persian mainland, a few made it to the Musandam Peninsula, a modern-day enclave of the Sultanate of Oman.[76] The people there maintain something of the old heritage.[77]

In his *Storia do Mogor; or, Mogul India*, written in the seventeenth century, the Venetian-born writer and traveler Niccolao Manucci—who spent most of his life in India after age fourteen—wrote the following about the demise of Hormuz:

This island was formerly the greatest and most frequented port on the ocean, where dwelt traders to every region in India men of great wealth so that a merchant possessing more than a million of patacas [unit of currency] was not a man of very great account. Shah 'Abbas considered that by making himself master of Hormuz, and transferring the port to the mainland, lying not over a league from the island, he would be able to draw all this wealth into Persia. But [Abbas] was frustrated in his object because the [Hurmuzî] traders were afraid of his interference.[78]

That was that, then.

These days, there is talk of officially reviving the sort of "trade zone" around Hormuz that once existed in these waters.[79] Islands in straits such as those in the Gulf can just as soon promote peace and trade as encourage war and piracy, just as soon bridge differences between peoples as isolate them. Elsewhere around the Arabian Peninsula—at the strait passages of the Red Sea, the Gulf of Suez, and the Gulf of Aqaba—there are similarly idealist plans afoot to construct real bridges at strait islands, even when those islands are disputed by several national powers. One such plan involves building a bridge from Djibouti to Yemen across the Bab-el-Mandeb, a strait at the mouth of the Red Sea. This bridge would pass through Perim Island. A second plan involves building a causeway from Egypt to Saudi Arabia, across the Straits of Tiran at the mouth of the Gulf of Aqaba. This bridge would pass through Tiran Island, an Egyptian-administered jurisdiction claimed by Saudi Arabia.[80] Along with Sanafir Island (Saudi Arabia) in the Straits of Tiran, Tiran Island plays a militarily and symbolically important role. Israel occupied both islands, Sanafir and Tiran, during the Six-Day War (1967)—even as it occupied man-made Green Island (Egypt) at the southern opening of the Suez Canal (1969) and Shadwan Island in the mouth of the Gulf of Suez (1970).

Bridges and tunnels as well as dikes are, one hopes, ideas whose time has come.

Seoul

Hermit Kingdom

The region we call "South Korea" is a peninsula, or an "almost island." It juts out into the Yellow Sea (west), the Sea of Japan (east), and the Korea Strait (south). The political state, the "Republic of Korea," that currently occupies this peninsular region, is less a peninsula than an "island." That is, South Korea as such is *entirely* surrounded. On the one hand, it is circumscribed *naturally* by the three seas. On the other hand, it is landlocked *politically*: on the north by a closed border—the presumably demilitarized zone that defines South Korea's land region against that of the Democratic People's Republic of Korea, the state that currently occupies the land region we call "North Korea." To the extent that one understands geography in terms of the "general relationships between nature and politics," South Korea is therefore an island.

The way that political and natural borders work in tandem is also already suggested by the Korean term for island, *do*, which can be interpreted as a phonetic transcription of *hanja*, or Chinese characters, meaning both natural "island" (島) and political "province" (道).[81] Just so, South Korea is naturally "isolated" from foreign incursion in the sense that one sometimes speaks of a "hermit kingdom"— an old-style Western imperial pejorative for Korea.

The Western world's knowledge of Korea came during the later periods of the Joseon dynasty, almost entirely by way of the water-surrounded islands, of which there are around 3,500, that surround mainland and/or peninsular Korea. They are the means by which the West came to understand this region—and, concomitantly, the means by which Korea came to a new understanding of itself. In the eighteenth century came such travelers as Louis-Antoine de Bougainville (1766–1769), Jean-François de La Pérouse (1785–1788), Étienne Marchand (1790–1792), Antoine Raymond Joseph de Bruni d'Entrecasteaux (1791–1793), Samuel Wallis (1767–1768), Philip Carteret (1767–1768), and James Cook (1768–1771, 1772–1775, and 1776–1780).[82] Certain countries on the Western rim did not wholeheartedly welcome Western visitors. Korea was called a "hermit kingdom" because of its self-chosen "isolation."[83]

Here one analogy to the Korean tradition of isolation is the Japanese insulating policy that the Nagasaki-based Japanese translator Shizuki Tadao in 1801 called *sakoku* (locked country).[84] The German Engelbert Kaempfer had articulated this policy in his *History of Japan* (1729),[85] which Tadao knew in its Dutch translation as aiming to create a closed (or insulated) country in an open sea. The Dutch had established their post for trade with Japan in 1609 on Hirado near mainland Nagasaki, and soon enough Japanese rulers corralled Western visitors on Dejima (protruding island), which in 1634 was constructed offshore from Nagasaki for the dual purpose of encouraging and limiting foreign trade and intellectual exchange.

Dutch sailors were exploring the Korean islands later in the century. One of these was shipwrecked in 1653 on Korea's volcanic Jeju Island. Some shipwrecked sailors escaped: Hendrick Hamel,[86] whose 1668 work about Jeju was translated into several languages, was the major European source of information about Korea for two full centuries![87] (That Jeju is so different from mainland Korea neither he nor his readers could then know. These days, it is administratively a "special self-governing province," and much of its language or dialect remains incomprehensible to mainlanders. For example, the word for "change"—that is, money over the price given back to the customer—is *juri* in the island language[88] and *geoseureumdon* in mainland Korean.)

The arrival of Westerners on Jeju is still an important plot line in the expression of Korean culture, as in *Tamra, the Island* (2009), directed by Yoon Sang-ho and Hong Jong-chan.[89] In this Korean television series with a seventeenth-century historical aspect, a Korean woman free diver (*haenyo*) from Jeju encounters a British sailor who washes up on the beach.[90]

Later exploration of Korea by Westerners likewise involved the outlying islands. In 1787, for example, Jean-François de Galaup, comte de La Pérouse, was in the vicinity of Jeju and Dagelet islands;[91] Jean-Baptiste Cécille was in the region of Hongdo.[92] In 1851, Captain Rivalan and Louis Charles Nicholas Maximilian de Montigny were on Bigeumdo, and Louis Helot and Admiral LaPierre were on Sinsido.[93] In 1856, Nicolas-François Guérin was on Geomundo (1856). It is not surprising that most explorers came to refer to Korea as an archipelago[94]—as if "mainland Korea" did not exist or the mainland were itself an island. This is the proposition with which I began this section. Demoulins, in the mid-nineteenth century, explored Wido and the Ko-Koun-to Archipelago[95] in this spirit. James MacDonald's firsthand account of a European expedition from China to Korea to rescue the crew of the *Narwal* in April 1851, first printed as series of articles in the *North China Herald* (May–August 1851), is a tale of island hopping merely around the peninsula.[96]

Ganghwa

The history of Ganghwa Island, which is of special interest in Korean islandology and archipelagic nesology, dates back to the period of Dangun, who was the partly mythical "founding father" of ancient Korea. In fact, Ganghwa marks a decisive location in Korea as the delta estuary at the mouth of the Han River, which with its treacherous delta currents and shallows flows up to the modern capital of Seoul. During the thirteenth-century Sambyeolcho Rebellion, Ganghwa served as the capital city. It is thus not surprising that the French campaign of 1866, which aimed to force reparations from Korea—a decisive turning point in the history of Korean isolation—focused principally on Ganghwa Island. In 1866, Admiral Pierre-Gustave Roze, who led the French forces, rightly called Ganghwa the "key to Seoul";[97] scholars and journalists still consider it "a prism for viewing Korean history."[98]

When Admiral Roze pillaged the royal city on Ganghwa, he entered the Joseon Royal Books and Manuscripts Collection and sent the material housed there, much of it six hundred years old,[99] back to Paris. (There it remained hidden in the National Library of France until 2012.)[100] This material mattered as much to the domination of Korea as the island itself. Jean Henri Zuber, an officer and artist who accompanied Roze, wrote, "The fact that books can be found even in the house of a poor peasant is something that we can only look upon with admiration and humbled pride."[101] Zuber did not know about Ganghwa's association, as old as the late thirteenth century,[102] with movable type, but he did rightly pinpoint the island's geopolitical significance and national import.

The French invasion was followed by other incursions into Ganghwa: American in 1871, for example, and Japanese in 1875. By the end of the nineteenth century, Korean "isolation" was ended.

Or so it seemed. In fact, isolationism returned: after the defeat of the Japanese imperial government that had ruled Korea for decades (1910–1945; see Illustration 21), after the Potsdam Conference that divided Korea (1945), after the uprising on Jeju Island (and the related massacres of its residents by mainland South

Koreans),[103] and after the Korean
Civil War (1950–1953). Nowadays,
because the channel of water to the
north of Ganghwa Island (in South
Korea) is what separates South from
North, the region is dangerous thanks
both to political circumstance (the
"intranational" border or "demilitarized zone") and to natural condition
(the delta). Ganghwa now thus plays
little overt role in the country's commercial life because Seoul cannot use
the Han River as a main thoroughfare.

The Han River creates other difficulties for South Korea's capital city
Seoul. North Korea, for example, has
had political reasons to flood Seoul by
opening the sluices of its dams, as it
did during the Water Panic of 1986.[104]

ILLUSTRATION 21
Japanese Postage
Stamp, 1931. The
Greater Empire of
Japan is depicted
as a stringlike
quasi-archipelago
that includes the
formerly independent
Korean Peninsula
(with attendant
islands), previously
Russian islands (with
part of Sakhalin),
and a hitherto
Chinese peninsula
(Liaodong) and
island (Liaodong).
Permission granted
by Collection
Selechonek.

One might compare here the border region where North Korea meets Russia and
China at the Tumen River in the insulated Rajin-Sonbong region.[105]

The bombardment of Yeonpyeong Island (2010) near the Han Delta is one of
many incidents arising from the disputed island-border region that also includes
Daecheong and Baengnyeong islands.

One can discern here the interaction of two aspects of a single conflict typical
of island-states. The defining *civil* (intranational) war aspect of insular South Korea's situation bears first on international disagreements with other nation-states.
With China, for example, South Korea has a dispute over ownership of the usually
submerged Socotra Rock in the Yellow Sea. With Japan, it disputes ownership of
Tsushima Island in the Korea Strait and Dokdo in the Sea of Japan, whose name
means "isolated, or sole, island." Both the civil and the international aspects of
Korean islandology recall the domestic struggle that figured in a recent Korean
film, *Castaway on the Moon* (2009), in which a soulful Mr. Kim is marooned on
the isolated Bam Island in the Han River—smack in the middle of Seoul, the
Korean capital (*seo'ul*).

Bamseom

Seoul is bisected into northern and southern sectors by the Han River and lies
thirty miles upstream from Ganghwa Island and thirty miles from Ganghwa's

island neighbor to the south—Incheon International Airport, which was built on "landfill" in the natural sea channel between Yeongjong and Yongyu islands. The shallow areas here have been a center of Korean history since the Three Kingdoms strove, during the first millennium, to control major trade routes from Seoul to the Han Delta and thence to the Yellow Sea and China. Nowadays, the marine route is barred: the delta is too close to North Korea. That is one reason that Busan, located at the southern end of the Korean peninsula and facing the Korea Strait and Japan, is now the center of isolated Korea's shipbuilding.

That Seoul is the country's cultural capital and a major industrial player is the main reason that the director Lee Hae-jun chose an island in the Han River in Seoul as the focus for *Castaway on the Moon* (*Kimssi pyoryugi*), known in English as *The Wandering Mr. Kim* and as *Castaway on the Moon*. In the film, a young man who is isolated from a society to which he is in financial debt, tries to commit suicide by jumping off Seoul's Grand Seogang Bridge. (Korea has the highest suicide rate in the world.) He is a refugee from the gold-clad 63 Building on nearby Yeouido, an island in the Han, which is Seoul's main district for capital development and credit banking. The bridge where he chooses to end his life joins the Han's two shores, passing over Bamseom (Bam Island), which is actually two uninhabited islets. His suicide attempt fails, and he is washed ashore on Bamseom like flotsam or jetsam. (*Pyoryugi* actually means "adrift" or "afloat" more than "castaway.")[106] At first, the young man's dilemma seems like that of Daniel Defoe's Robinson Crusoe. In fact, the word *pyoryugi* conjures up tales of the Robinsonade variety, as in Hendrick Hamel's internationally influential *pyoryugi* about being washed up in 1653 on the Korean island of Quelpart,[107] also known as Doi (barbarian) and Jeju. In the movie, the young castaway is within sight of the urban mainland of Seoul itself: he sees overhead the bridge to the mainland, but he cannot reach it.[108]

Parallel to this young man is a young woman who is herself isolated in a room in a high-rise apartment building near the Han River. An agoraphobic, she has not left her home for three years, preferring instead to take nighttime pictures of the three-dimensional island in the sky that is the moon. A "Cyholic,"[109] she spends her daylight hours playing *Cyworld*—the Korean "virtual money" extravaganza. *Cy* here indicates both "cyber" (digital) and the Korean for "relationship." (When discussing isolation as it includes psychological alienation in Korea, it is worth keeping in mind that the suicide rate among South Korean women is twice that of women in any other nation-state on Earth.) Paralyzed psychologically as she is, the female hero of *Castaway on the Moon* resembles Jimmy Stewart's character in Alfred Hitchcock's *Rear Window* (1953),[110] never more obviously so than when she trains her telephoto lens not on the moon, where she usually looks, but on

Bam Island and discovers the castaway. What follows is a curiously "salvational" romance. From the deserted island, there is opportunity for rebirth and maybe even peace.

Rebirth from a previous state is suggested also by the history of Seoul's Seonyudo (Angel's Resort Island),[111] which begins at a time when the islet was a beautiful mountain peninsula, Seonyubong, with a beach. That is the place depicted in the artwork of *Seonyubong Peak* (1742) by the eighteenth-century, "true view" Joseon landscape painter Jeong Seon. There followed the destruction of the island by the imperialist Japanese, who cut it down in 1925. Seonyudo afterward became the site of a "water purification" plant (1978) and finally an "ecological showplace" (2002). The historical trajectory suggests an idealist islandological aspiration as much as does the existence in present-day Seoul of the Seoul Floating Island—three tethered artificial islets, Vista, Viva, and Terra,[112] floating in the Han.

Two cinematic works attempt to explain the cultural place of Korean islandology differently from *Castaway on the Moon*. First is Kim Dae-seung's *Emperor of the Sea* (2004), a fifty-one-episode historical epic[113] focusing on the great Korean naval commander Jang Bogo (787–846) from the island of Wando (Cheonghae). Jang controlled the seas between southwestern Korea and China's Shandong Peninsula during several decades of the powerful Silla dynasty (57–935). The second work is Choi Young-hwan's *Blood Rain* (2005), a partly fictional film whose action takes place on "Donghwa Island" (a fictional amalgam) during the first decade of the nineteenth century. At that time, the neo-Confucian Joseon dynasty (1392–1897) was experiencing a decline partly because Catholic missionaries were making island-based incursions into Korea. The historical context of the film is at once colonial and international: Donghwa Island both pays tribute to mainland Korea and trades with China. The political struggle obtains between two forces: equality, suggested by the arrival on Donghwa of western Catholicism and a Korean middle class, and inequality, represented by an unreliable Joseon ruling class in relationship with a fearful, ignorant peasant class. The struggle transforms the once-peaceful island culture of Donghwa into a sort of "hell on earth," an island condition that forecasts the fall of the Joseon dynasty itself at the end of the nineteenth century. By the same token, the historical condition predicts the metamorphosis of peninsular mainland Korea into a geopolitical crossroads, partly based on shipbuilding. It foretells of the Korean sort of worldwide superhighway that *Blood Rain*'s twenty-first-century audience would know well.

Shipping Lanes

That South Korea has developed a large industry producing ships, or floating islands, is partly because island nations, defined as such by both natural barriers and political boundaries, often look to shipping lanes. In fact, South Korea now produces more than half of the world's advanced-technology container vessels. It also produces military and icebreaking vessels, partly for geopolitical purposes. Korea has its sights on the northern and southern world passages that have become realistic in the wake of climate change. The South Korea icebreaker RV *Araon*, commissioned in 2009, operates from the Korean King Sejong Station on King George Island (in Antarctic waters). At the other pole, Korea operates out of the Norwegian island of Spitsbergen (in Arctic waters). It also shows interest in Canada's Northwest Passage and Russia's Northeast Passage, its ships often entering both by way of the narrow Bering Strait, which slips between Big Diomede Island (Russia) and Little Diomede Island (United States). These northern routes lop thousands of miles off the older shipping routes.[114] Korea ships goods from Ulsan/Busan (home to the world's largest shipyard, operated by Hyundai Heavy Industries) to Rotterdam in the Netherlands, even as it buys up inshore mineral rights and real estate around the Mackenzie River delta in Canada's Yukon.[115]

The old dream of a "global superhighway," with its main link at the Bering Sea, seems almost realized by such Korean travels. Long ago, world geopolitical thinkers envisioned that highway in terms of the telegraph. Consider, for example, the planned European "North America Link" that was to connect with the "Russian-American Telegraph Plan" in 1864.[116] At the same time, there was discussion worldwide about a Bering Strait railroad bridge. Consider William Gilpin's idea (1860, 1890) for a Cosmopolitan Railway, "compacting and fusing together all the world's continents."[117] The engineer and poet Joseph Baermann Strauss put forward the first proposal for just such a bridge in his senior thesis at the University of Cincinnati in 1892.[118] Cincinnati was then famous for its Hegelian, world-historical constructions seeking to overcome both geographically distant shorelines and logically opposite poles.[119] (Strauss was no idealist visionary; he went on to design and actually build the Golden Gate Bridge.)

A concomitant transportation plan would be a Japan–Korea undersea tunnel or bridge. This project was under discussion during Japanese rule of Korea, beginning in 1917. Had it been carried out, it would have united Japan with Korea by going under or over the Korea Strait by way of Iki Island or Tsushima Island. These days, Tsushima is claimed by both Koreans and Japanese. The difficulties in macro-engineering seawater projects are often political ones: the dam macro-project at the Strait of Hormuz, which would join Iran with Oman (or with the United Arab Emirates), is a case in point.[120] The geopolitical world attracts and repels both globalization and insularity even as geography often seems to call out for them.

If "Ganghwa Island is the key to Seoul," as Pierre-Gustave Roze wrote in a famous letter to the French naval minister in October 1866,[121] then the meaning of insulation and isolation is likewise a key to understanding Korea. In this chapter, the major geopolitical arguments are mirror images of each other.

First, the insular status of South Korea radically affects its self-definition. The movie *Castaway on the Moon* hypothesizes and expresses a cultural and economic malaise. Its sign is a high rate of alienated withdrawal and isolated suicide and apparent escapism into a figurative *Cyworld*. An experiment like that of South Korea, in what might be called the "sociology of rapid change and commercial expansion," is taking place in the United Arab Emirates, where, as we will see, there are also other "worlds." Second, Korean islandology, insofar as it is both natural and political, is linked with South Korea's understandable success as a sort of new Venice, albeit without the military superiority and expansionism of an imperial power such as the elderly Great Britain.

In this light, one sees at once the smaller and larger aspects of a general island-ology. On the one hand, there is the isolated individual. On the other hand, there is the circumferentially surrounded "desert island" in the heart of a city and the globe-encompassing ships that traverse the world's new and old passages and thus seek to transcend an older insular status.

The World

In the city of Dubai, in the United Arab Emirates, is an archipelago called The World, or World Islands, constructed in the Persian/Arabian Gulf (see Illustration 22). This real estate development, which is an arrangement of islands that resembles a map of the world, has hydrological and cultural consequences[122] as well as economic penalties. Sheikh Mohammed bin Rashid Al Maktoum, the current ruler, plans to sell various "countries" to a wide assortment of private purchasers. (His fellow and rival ruler in the contiguous emirate—Sheikh Khalifa bin Zayed Al Nahyan—holds dominion over the island-city of Abu Dhabi.) So far, Sheikh Mohammed has run into financial difficulties thanks both to the current world "recession"[123] and to natural declensions, among which is the islands' gradual sinking into the waters of the Gulf.[124] One might recall the old story about King Canute the Great of Denmark commanding the tides to halt. Likewise, one might remember Herodotus's wry gloss on the tyrant Xerxes building both a canal through Mount Athos and a bridge across the Hellespont.

Maktoum's World is a giant water/earthwork, visible from space, that rivals any of the great earthworks projects we discuss in *Islandology*. (Whether it rivals the hypothesis of Athens as an island is another matter.) It is, after all, pleasant to count all of Earth's particular islands and continents as part of a single or universal "world archipelago" where people live peacefully—a union of insular nations in real estate heaven. One might almost call this utopian idea "The World," as

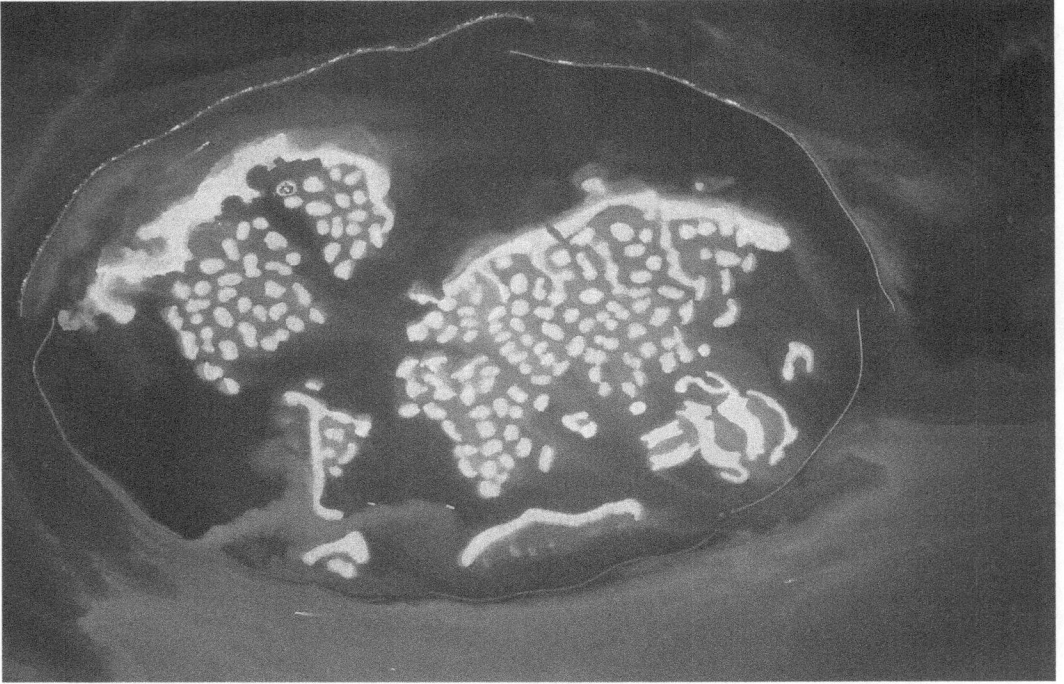

ILLUSTRATION 22
The World. The
photograph shows
a man-made real
estate archipelago
in the United
Arab Emirates.
Photographed from
the International
Space Station,
Expedition 16,
April 3, 2008.
Source: NASA.

configured by Sheikh Mohammed's insular real estate venture, were it not for the unrealistic requirement, both political and logical, that it eventually pass between the maelstrom of Charybdian particularism (the sinking islands) and the tempestuous beauty of Scyllan universalism (the rising oceans). Maktoum's World, as we will see, cannot rival either the political hypothesis of ancient Athens as an island republic or the democratic postulate of Westminster as a *ting*, or insular parliament.[125]

Naming and Sovereignty

WHAT'S IN A NAME

O, be some other name!

—Shakespeare, *Romeo and Juliet*

The global purview of *Islandology* means that the vexed question of how best to name places (on Earth), both on land and on water, has practical consequence as well as theoretical import. In *Islandology*, this question seems unavoidable: we name more than eight hundred islands and two hundred bodies of water.

Consider the case of the island that in the Preamble I call *Guanahani*. Some scholars who use this name follow the reported toponymy of the Lucayan natives in 1492. Other scholars, though, may call the place *San Salvador* Island, following the toponymy of Christopher Columbus, who claimed the place for Spain. Yet others want to call it *Watling* Island, following the nomenclature of the English privateer John Watling, who made it his base of piratic operations.[1] Enough has been said already to suggest the complex issues that come to bear in naming this place.

At the Chute

One island can serve as well as another for illustrating the naming problem, so let me take, for a second example, an island I knew well when I was a young child. Its short toponymical history simply expresses the political and technical issues at hand.

The spot is a tiny islet, no more than thirty-five square feet, in the Rivière du Nord (North River). It rises from the surface of the water at a gentle widening of the river just below the rapids, or *chute* (waterfall), where a meandering trout-filled brook enters the river near a road intersection, in the village of Val-David, Quebec, where three paths meet—Rue Sainte Olive, Rue Frenette, and Chemin de la Rivière. Up until the 1940s, this islet had no name to speak of—so far as anyone I am familiar with knows. (Most of the world's hundred thousand un-inhabited islands likewise have no name of which we speak.) In the early 1950s, four Yiddish-speaking friends and I gave this islet a toponym: *Île Petite*. The etymology is relatively straightforward. The French-language phrase *île petite* (little island) became the place-name *Île Petite*. Substantive phrases often become proper names,[2] and this one was ours.

When an island has one name only, however, another name is sure to follow. We children assumed that our name was more or less "private"—idiolectal. That was not to be. As soon as the "private" name *Île Petite* went "public," there were squabbles both about the name and about the island's mythic ownership, as it were. Within a month, a once-friendly gang of six-year-old Francophone children from the other side of Rue Frenette insisted on calling the place *Petite Île*. As soon a gang of Anglophone children from Rue Sainte Olive heard that the island at "The Shoot" (for so Anglophones spelled out the rapids there that Francophones called "La Chute") had two apparently Francophone toponyms (*Île Petite* and *Petite Île*), they responded in three ways. First, they called the island, which until then we *all* had figured was somehow in the public realm, their "property." Then they gave the island a name of their own: *Petty Place*. Third, they sent out a couple of nine-year-old thugs to beat us all up—which they did.

We children of Val-David now had a war or two on our hands. One was a civil war between two apparently pro-Francophone groups: some Francophones wanted the toponym *Petite Île*; others wanted *Île Petite*. The other was an international war: the Francophones who wanted the name *Petite Île* and the Francophones who wanted the name *Île Petite* united, or ganged up, against the Anglophone outsiders, who all wanted *Petty Place*. This struggle for an island and its name lasted three months until we hit upon a scheme for ending it that we had read about in a book of Norse sagas retold for children: single combat.

On the island itself, Hébert, the best fighter from our Francophone union, would fight Foster, the best fighter from the Anglophone group. The winner, as representative of the whole, would get to choose the name. It was a David and Goliath fight, and Hébert won. Foster's gang still claimed to have some rights of memory, however, and kept on calling the island *Petty Place*. The *holmgang*, or island duel, did not work out the problem in the way we had hoped. (Later in this book we will see that the same ploy did not work for the Scandinavians whose history Shakespeare tells in *Hamlet*.)

There were, from time to time in Val-David, other attempts to make peace. The Yiddish-speaking parents—guardians of us children on the Francophone side holding out for the name *Île Petite*—were afraid that someone would get hurt. They were frightened that someone would be "*shot* at the *chute*," as one trilingual schoolteacher in Val-David put it. The grownups suggested a fourth name: קליין אינדזל (*klayn indzl*), meaning "small island." None of us children liked the idea. Switching alphabets seemed especially disturbing (almost as if we were becoming Serbs and Croats scrapping over pieces of territorial nomenclature). In any case, we had gotten used to the conflict and could hardly imagine speaking without the differences.

Island toponymy is always politically tendentious and linguistically multifarious. No matter how *apparently* inconsequential the island, its toponym requires more care and attention than any toponymic committee can afford to give to the matter—and perhaps more wisdom than human beings can ever muster. There are, however, shorthand methods to deal with the problem. In *Islandology*, I usually provide only one name for each island, using nomenclature from standard toponym dictionaries, gazetteers, and geographical indices.[3]

Another method would be to give islands no verbal name at all but instead match each place with a number. The center point of our once nameless island in Val-David's North River would be N 46°01′46.87″, W 74°12′46.07″. Global positioning systems (GPS) and "geographic information systems (GIS) present themselves as quick and "value-free" ways to pinpoint spots on the globe. Numbering is an attractive solution to the problem of how to name places because it seems "value neutral." *Seems*, though, is the right word here. There are definite problems. Global positioning systems, for all their supposed "neutrality," require the hypothetical positing of something real outside the globe that they seek to measure. We consider this limitation when we imagine a human being sitting, as if he could, in outer space. Archimedes, the first person to make a globe representing planet Earth as a spatially measurable island, thus isolates himself from Earth, saying, from some such soapbox as Utopia,

Give me a place to stand and I will move the Earth.[4]

What is wrong with this idea, if we do not already get the joke of it, we explore throughout *Islandology*.

The geography of speaking about islands is concerned with the humanly inescapable intersection between the natural (or physical) environment and human (or political) culture. Mathematical geomatics, like geospatial technology, has for its object the measurement and location of the physical environment. The study of speaking about islands, however, needs to consider as well the significance of linguistic naming. The multiplicity of toponyms sometimes reveals, sometimes conceals, deeply consequential issues. By way of example, consider the problem of naming the tiny island just north of the modern-day Kingdom of Morocco whose coordinates are N 35°55′, W 5°25′. The island's Spanish-language name is *Isla de Perejil* (Parsley Island). Its Arabic-language name is *Jazirat al-Maʿdanus* (also Parsley Island). Newspapers in the kingdom used to call it *Leila* (the island), which is a "mispronunciation" of the Spanish *la isla* (the island).[5] These days, Spain and

Morocco still dispute sovereign ownership of the island; they recently threatened to go to war over it. The Moroccan king, Mohammed VI, now uses the French toponym *Toura* to refer to the island. This is the French pronunciation of the Berber *Tura* (empty). In this, the king wants to recall Morocco's non-Spanish and non-Arabic history.[6]

Whatever one calls a particular island, the sound that issues from one's mouth, even the alphabet that one uses to write the word, often becomes a shibboleth—a linguistic "password" used to identify you as a member of one cultural group as against another.[7] A person can lose his or her life if he or she chooses the wrong place-name; likewise, a person can lose his or her life if he or she pronounces the right place-name in the wrong way. Island toponymy does not differ qualitatively in this regard from mainland toponymy: everywhere languages differ, alphabets vary (קליין אינדזל), and pronunciations fluctuate. Empires change, orthographies morph, civil wars happen.[8] Those islands that are the broken and traded fragments of past empires perhaps serve well to present the typical political case for all.[9]

The Ancestry of Hydronyms

> I know too well there are some who have such an aversion to their
> [Celtic] mother tongue that they profess a hearty desire of seeing it
> entirely abolished, that no remains of it may be left in this Island
> [Britain], so great an eyesore is the language of their forefathers become
> unto them.
>
> —Thomas Richards of Coychurch, *A British or Welsh Dictionary*

Whether islands differ in regard to such matters from other places depends on the isolation of definition and the relative age of some kinds of place-names, especially those that are hydronyms, or water based. Worth considering here, in the context of the British Isles, is the fact that, with the striking exception of hydronyms, there are few Celtic words remaining in the English language. The Anglo-Saxon and Anglo-Norman dialects of the invading peoples did not easily absorb the then-indigenous tongues of the Irish, Manx, Scots-Gaelic, Breton, Cornish, Welsh, Cumbric, and Pictish peoples. However, because hydronyms generally change more slowly than most other kinds of words and names—a fact surprisingly widespread across the world[10]—there remains the Pictish *aber* (river mouth), the Cumbric *ince*, and the Welsh *abona* (river), whence Shakespeare's own *Avon*.[11] The name of the River Ouse, in Yorkshire, contains *usa*, which means water.[12]

The main words for "island" in most of these relatively "indigenous" British languages are variants of *innis*.[13] The contested etymology of this word includes John Strachan's position: "Celtic *eni-stî*, in-standing." For Strachan, *innis* etymologically means "standing or being in the sea."[14] Insofar as "to stand in" differs from "to float in" and "to be surrounded by," the Celtic notion of island differs from the Norse and Latin notions. Relevant here is Shakespeare's "Welsh play," *Cymbeline*, which is set precisely in the Milford Haven *ria* (drowned river valley) of Wales. Variants of *innis* point also to the nesological aspects of "island" we have considered: for example, the Middle Welsh *ynys* means "land" (in the sense of "realm") as well as "island,"[15] and many of the insular (Britain-based) Celts conceived of "otherworlds" as being actual islands.[16] Also relevant here are the Irish westward-looking *immrama* (navigational stories), like the wonderfully nonhistorical *Voyage of Máel Dúin* (ca. 1000) and the *Voyage of Saint Brendan the Navigator*. No one I knew in Val-David cared a whit for geography.

As a child, I once went around the village of Val-David to ask people, "Where does the North River empty out?" and to inquire of them, "What did the Indians, *les autochthones*, call our river?" I always received the same response: a blank stare. It was, for my interlocutors, as if the Algonquian and Iroquois Nations of Quebec had never been and all that mattered now was the road and the rail line that linked the village—call it what you will—with the city of Montreal. Not every place has preserved its hydronymy. All Canadian canoeists, however, know well enough that the Indians paddled the North River, or the Rivière du Nord, from Lac Brûlé (north of Val-David) southward to the Saint Lawrence River near Carillon Island. And from there who knew? Maybe they went eastward, down to the Gulf of the Saint Lawrence and the Atlantic. Perhaps they went westward, up to the Great Lakes and the Chicago area, and then southward, down the Mississippi River to the Gulf of Mexico—thus marking the eastern half of North America as the single island it was and is.

THE TROUBLE WITH ISLANDS

A Fast-Fish belongs to the party fast to it. . . . A Loose-Fish is fair game
for anybody who can soonest catch it.

—Melville, *Moby-Dick; or, The Whale*

Melville calls these two statements "a system which for terse comprehensiveness surpasses Justinian's *Pandects*."[17] The Roman emperor Justinian's sixth-century compendium of Roman law was intended to present everything. (The Greek term *pandektes* means "all-containing.") On the one hand, Melville essays to boil down the universal (catholic) principle to regulations involving two kinds of islands: the floating, artificial creations that are the whalers of human beings, and the floating, animate creatures—the whales—that the human beings seek.[18]

Here the Roman doctrine of "a thing that is not *owned*" (*res nullius*, thing of no one) interacts with the doctrine of "a creature that is not *known kin*" (*filius nullius*, son of no one). Justinian's *Pandects*, for example, discusses the ownership of newly generated islands: "When an island rises in the sea, though this rarely happens, it belongs to the first occupant; for, until occupied, it is held to belong to no one."[19] Following suit, an ironic Melville suggests in *Moby-Dick* that an island as such is a *res nullius* in the same way that a bastard is, according to Justinian's code, a *filius nullius*.

Melville's two statements are brilliantly terse: "A Fast-Fish belongs to the party fast to it. . . . A Loose-Fish is fair game for anybody who can soonest catch it." The reader of *Moby-Dick* soon understands how the statements relate to the ownership of individual human beings (slaves or spouses) as well as to the sovereignty or autonomy of islands and mainlands. The harpoon line, broadly understood, can turn a loose whaler into a fast one (owned, as it were, by a powerful whale) as easily as it can turn a loose whale into a fast one (belonging now to the mariners). On Newfoundland, in fact, the noun *shore-fast* refers to "line and mooring attaching cod-trap or seal-net to the land."[20] Goethe's *Faust*, by way of comparison, seeks to hold "fast" to the land he has wrested from the sea.[21] A decisive series of questions arises here, at once natural and political: Is it the whale that owns the whaler? Or is it the whaler who owns the whale? Which whaler gets to keep the whale? Thus the question, "Whose 'land' or 'island' is it?"—like the question, "Who gets to name it?"—often gives rise to debate and to war.

One of the main ideas that the geographer and philosopher Immanuel Kant presents in his *Metaphysics of Morals* is "the unity of all places [*Plätze*] on the face of the Earth [*Erdfläche*] as a spherical surface [*Kugelfläche*]."[22] This idea can sometimes militate against notions of private ownership (Fast-Fish), as the global island traveler Herman Melville ironizes it. Melville recalls in *Moby-Dick* his own place as global sailor on the floating island that was the whaling ship *Acushnet* (1841–1842),[23] on which Melville set sail from New Bedford, Massachusetts—at the mouth of the Acushnet River—the center of the world's whaling industry.[24] In 1886, moreover, Melville recalled a similar issue involving property and empire when he was an international customs inspector at the New York Custom House on the island of Manhattan at the mouth of the Hudson River. Melville was then, in 1866 or earlier, beginning to write *Billy Budd*, his manifest gloss on the Nore mutiny that took place where the Solent meets the sea at Spithead in the channel between the Isle of Wight and Portsea Island. Spithead was the foremost site of the Royal Navy and the backbone of the British Empire.[25]

A few lists and case studies, of the sort that follow, can give a hint merely of the full panoply of issues that arise, especially when it comes to uninhabited islands, commodity-rich islands, or strategically located islands in straits.

Autonomy

If a state were "essentially" a more or less insular island, then an island-state, whose borders match its circumferential coastline, would seem to have an advantage when it comes to self-representation and a claim for autonomy.[26] Yet the same forces that make for cohesion (the shore) sometimes also make for what some theorists of the "built environment" call *discohesion*.[27] After all, most "islands are detached areas physically and readily detached politically," as Ellen Churchill Semple argues in *Influences of Geographic Environment* (1911).[28] (Semple inherits her understanding of the nation-state in islandological terms from Friedrich Ratzel, whose *Politische Geographie*, published in 1897, and various essays on *Lebensraum*[29] deemphasized "happenstance" circumferential borders.)

Apposite to the notion of autonomous island, whose natural shores match its political walls, are the "free" places that individuals establish as their own and the "prison" places that nation-states likewise establish.

On many isolate islands, an individual can get away with hoisting up a flag and calling the place his sovereign nation-state (e.g., Swains Island in the Tokelau Archipelago). Some such island claimants call themselves "micronations." Examples include residents of such man-made islands as the Principality of Sealand (off the coast of Suffolk, England),[30] the Republic of Minerva (in the Minerva Reefs),[31] and the Esperantist Republic of Rose Island (in the Adriatic).[32] The rusted-out ship hulk in the movie *Iron Island* (2006), directed by the Iranian filmmaker Mohammad Rasoulof, becomes a state of its own. The usually short-lived political experiments of island "seasteaders" play a role when defining the political sovereignty and borders of relatively longer-lived states. Thus islands whose "political" borders match up to their "natural" coastlines often provide disturbing case studies of geography defined in terms of the intersection between natural and human factors (see Illustration 23).

The idea of the relatively autonomous—islandic—prison is venerable. In his chapter on punishments and rewards in *Leviathan*, Thomas Hobbes thus defines *imprisonment* itself in terms of its islandic expression:

Under this word imprisonment, I comprehend all restraint of motion caused by an external obstacle, be it a house, which is called by the general name of a prison; or an island, as when men are said to be confined to it.

The catalogue of actual prison islands is long,[33] and their story a staple of literature.[34] Some islands become, moreover, quarantine spaces for foreigners (e.g., Dejima in Nagasaki) and for would-be immigrants (e.g., Ellis Island in the United States). Hobbes speaks of islands as prisons.[35] Human beings often create

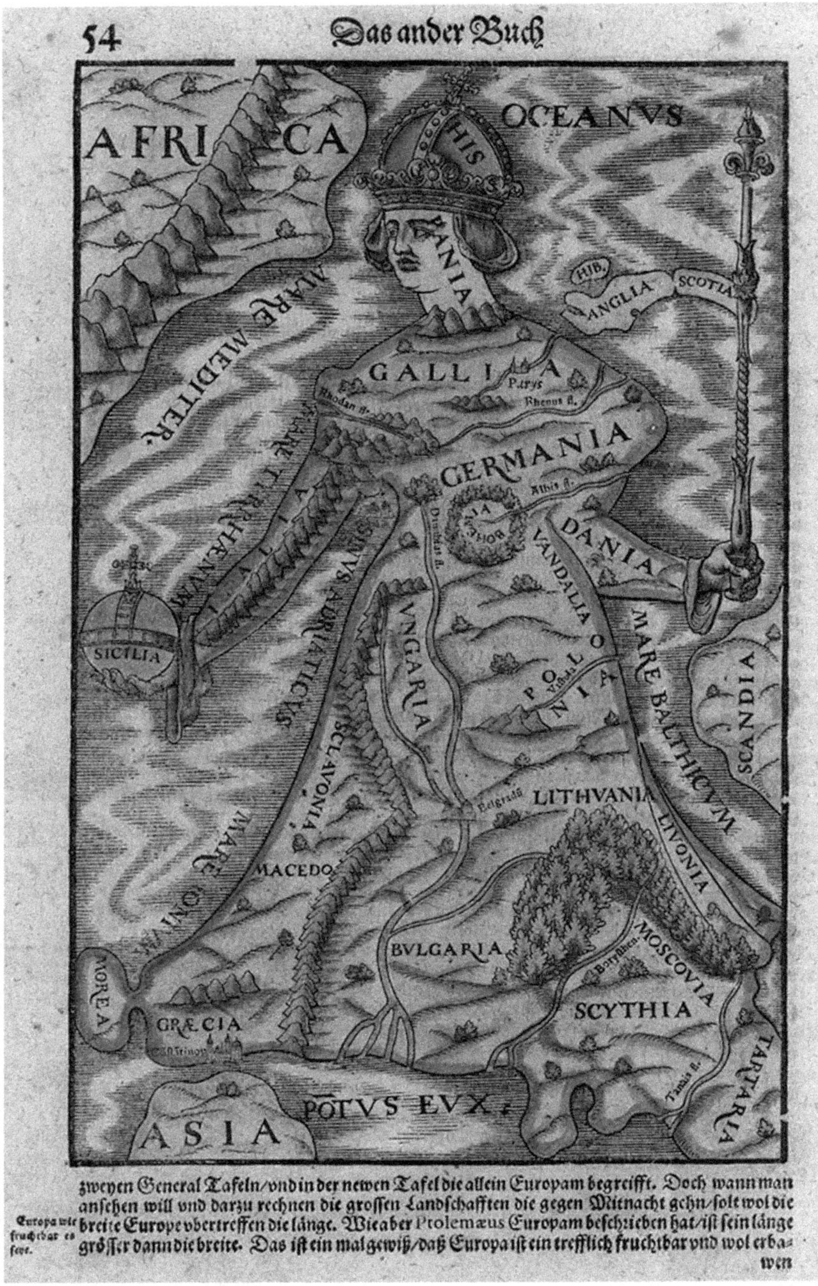

ILLUSTRATION 23 *Europa Regina*. The map shows Europe as a personified island, or *presqu'île*, in the form of its sovereign royal person. Included in Sebastian Münster's best-selling *Cosmographia* (1544) beginning in the 1570s. The cartographic type of "Europa Regina," first drawn by Putsch (Johannes Bucius Aenicola) in 1537, pretends to unify Europe as an "island-continent" whose political borders match its natural shoreline. (Rotate the map 90° counterclockwise to see "north" at the top.) Heinrich Bunting's map *Asia Secunda Pars Terrae in Forma Pegasir*, in *Itinerarium Sacra Scripturae* (Travels According to the Scriptures; 1581), similarly personifies Asia as an island in the form of the horse Pegasus; Bunting also includes a personified map of Europe. Harvard Widener Library, *Cosmographia*, v.1=HWoBA6, p. 55. Reproduced by permission.

"artificial" islands—ships—for political purposes: among "prison hulks" are the German SS *Deutschland* and SS *Cap Arcona*[36] and the island ships of the American Civil War.[37]

Other islands become storing-away places for diseased persons.[38] In literature and history, for example, there is the Greek warrior Philoctetes stranded on the island of Lemnos.[39] Famous such places include the maritime quarantine station (*lazaretto*) on Santa Maria di Nazareth Island in the Venetian Lagoon (1403) and New York City's quarantine anchorage off Bedloe's Island, which was a way station for contagious mariners before it became the foundation of the Statue of Liberty. The leper-colony island Spinalonga off the Greek island of Crete was originally a peninsula; the Venetians eventually transformed it into an island.[40] Robben Island (South Africa) was a leper colony before it became a prison.

Condominium

Many islands are split "officially," in terms of space, between two or more nation-states that have "shared sovereignty" (*condominium*) in more or less friendly or hostile ways. Germany and Poland share Usedom in the Baltic. Borneo is divided among Malaysia, Indonesia, and Brunei. (Malaysia and Indonesia share a 2,000-kilometer border.)[41] Cyprus is divided among the sovereign state of Cyprus, the Turkish Republic of Northern Cyprus, the United Kingdom's sovereign military bases, and the United Nations buffer zone. On mainland Europe, Turkey and Greece share an island, known as "Q," in the Maritsa River.[42] Condominium divisions also characterize Hispaniola (Haiti and the Dominican Republic), Ireland (United Kingdom and the Republic of Ireland), New Guinea (Indonesia and Papua New Guinea), and Sebatik (Indonesia and Malaysia). There are also Saint Martin (France and the Netherlands), Tierra del Fuego (Argentina and Chile), Timor (Indonesia and Timor Leste), and Bolshoi Ussuriysky (China and Russia). Canada shares, with the United States, Province Island in Lake Memphremagog (between Quebec and Vermont) and two islands in Boundary Lake (between North Dakota and Manitoba). From the time of Canadian Confederation (1867) until that of Newfoundland's union with Canada (1949), Killiniq Island, at the northernmost eastern tip of the Labrador Peninsula, was divided between the two; Killiniq (meaning "ice floes") is split nowadays between the province of Newfoundland and Labrador and Canada's semiautonomous territory of Nunavut.

In terms of time, as well as space, political states can share island dominion. France and Spain administer Pheasant Island in the Bidasoa River during alternating periods of six months. (More conventionally time- and space-shared islands would include Cuba, where the area around the Guantánamo Bay Naval Base has been leased "indefinitely" by the United States from the Republic of Cuba.)

Treriksröset is the time-shared artificial-island boundary cairn at the tripoint where the border of Finland meets the borders of Sweden and Norway (ten meters from the shore of Lake Goldajärvi). Saudi Arabia and island-state of Bahrain similarly share the King Fahd Causeway Embankment "40:66."

War by Any Means

> The unit of land which fits within the retina of the approaching eye is a token of desire.
>
> —James Hamilton-Paterson, *Seven Tenths: The Sea and Its Thresholds*

For millennia, human beings have made war for island sovereignty and have taken islands as "staging grounds" for making war. The trend is clear from the Persian invasion of Greece in 480 BC to American island hopping across the Pacific Ocean after the bombing of Pearl Harbor, the lagoon on Hawaii's Oahu Island, in 1941.[43] One group occupies an island even as another claims the right of occupation. Stories of trans-Pacific hopscotching include the voyages of native islanders[44] and the European and Asian colonization of the Pacific islands.[45] There are similar examples the world over: Africa,[46] Asia,[47] South America,[48] and Central America,[49] as well as the Asian and Pacific marine regions.[50] Many island sovereignty issues center on old imperialist "holdings" of the United Kingdom[51] and mainland Europe.[52] The United States has had a similar role to play.[53]

Islands that have "disappeared" are as likely to be fought over these days as those that are visible above the surface of the water.[54] Brazil disputes with Paraguay the river islands submerged by the reservoir of Itaipu. Sundarbans—a huge network of extraordinarily complex tidal waterways, mudflats, and islands of mangrove forest—would seem to belong to both India and Bangladesh. Islands that have only partly disappeared are likewise grounds for warfare. Vozrozhdeniya in the enclosed Aral Sea once marked the international border between Uzbekistan and Kazakhstan; when the water level changed in 1991, the island become a mainland peninsula shared between the two states.[55]

Kawehnoke and Hans

> Hänschen klein geht allein   Little Hans goes alone
> In die weite Welt hinein.   Out into the wide world.
>
>   —German folk song

Each of the potential island disputes just listed is different from the others and deserves a case study of its own, but there is not the space here to provide that. Worth mentioning, though, are two islands, Kawehnoke and Hans, claimed by a country, Canada, with the world's largest archipelago and with a global reputation for relative peacefulness. Yet Canada has territorial differences with Russia, which planted a flag under Canadian-claimed waters, and with the United States, which disputes parts of the Beaufort Sea and parts of the Grand Manan Archipelago. The situations of Kawehnoke and Hans, also disputed, are more complex in such a way as to suggest the direction of further case studies.

Kawehnoke (Cornwall Island) in the Saint Lawrence River is part of tiny Ak-wesasne (the Saint Regis Mohawk Reservation). This First Nation's territory strad-dles the borders of *two* countries, the United States and Canada, and includes two linguistically distinct Canadian provinces, Anglophone Ontario and Francophone Quebec, at the geographic spot where the Saint Lawrence River is joined by its tributaries, the Raquette and the Saint Regis. Kawehnoke is a pas-sageway for the movement of goods and people between Canada and the United States. American and Cana-dian officials occasionally call this traffic "smuggling" and "human traf-ficking"; the Mohawks, who claim limited sovereignty in the region, regard it as "legitimate exchange." In summer, the goods move back and forth between Canada and the United States by way of a land high-way that passes through Kawehnoke; in winter, nature's "frozen river" car-ries both goods and people with-out much American or Canadian supervision.[56]

Two countries with reputations for being relatively "peace loving," Canada and Denmark, dispute own-ership of Hans—a tiny uninhabited island of only a half square mile, on

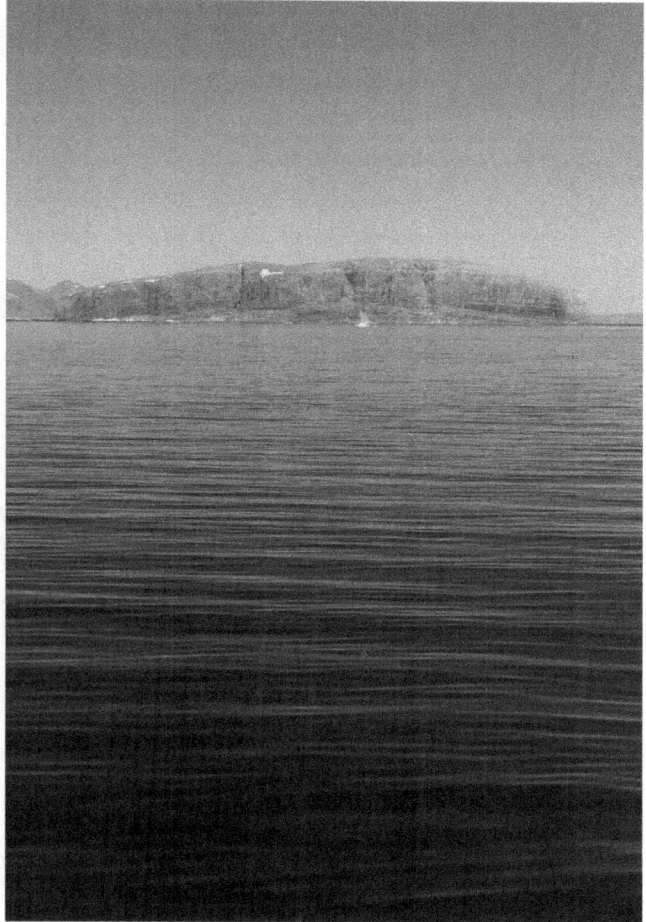

ILLUSTRATION 24
Hans Island Viewed from the South. The Inuktitut call the island Tartupaluk. Photo by Pappychou. © Dave Walsh. Reproduced by permission.

the northernmost periphery of the Canadian Arctic Archipelago (see Illustra-tion 24). (The archipelago includes 36,000 islands and extends 1,500 miles east–west and 1,200 miles south–north. Russia and the United States make various claims. Russia, for example, has sent submarines to plant flags on underwater ter-ritory that Canada claims.[57] The United States makes claims on the Beaufort Sea border area.)[58] Hans is a guard island in the Nares Strait. The Nares, named for the navigator who was seeking "open polar seas there,"[59] joins two seas. To the south is the Atlantic Ocean; to the north is the Lincoln Sea. The strait also provides a potential link between North America and Europe. To the west of the Nares is Canada's Ellesmere Island, which is part of the (Canadian) autonomous territory of Nunavut and is the fifth-largest island in the world, although with a human population of only 146. To the east is Danish-held Greenland—the world's largest noncontinental island and least densely populated dependency or country. (Even Antarctica is not immune from island disputes.)[60]

Whose island is it? The rules of ownership sometimes seem deceptively simple, and they hardly boil down to legal niceties about open and closed seas.[61] Islands are more like creatures on the high seas. As Melville says, "A Fast-Fish belongs to the party fast to it. A Loose-Fish is fair game." His sly dictum suggests even that a harpooned whale might have an ownership claim on the ship it drags around. The question of ownership here thus pertains partly to a literary issue: who, speaking of islands, gets to name ships at sea and the land, utopian or not, they come up against.

Utopias and Laboratory Hypotheses

A map of the world that does not include Utopia is not even worth
glancing at, for it leaves out the one country at which Humanity is
always landing. And when Humanity lands there, it looks out, and
seeing a better country, sets sail. Progress is the realization of Utopias.
                    —Oscar Wilde, "The Soul of Man under Socialism"

ONE PLACE AS ANOTHER

Islands, both as a part of, and apart from, the world, allow development of the
idea of a place sufficient for understanding other places. A particular island's
"biography" is, after all, both circumscribed and typical.[1] For many people, the
Orkney Islands are the "sufficient place,"[2] say, but in the last analysis, the particular place (per se) does not matter much. Writes Aldous Huxley in his utopian novel *Island* (1962)—a counterpart to his dystopian *Brave New World* (1932):
"Nobody needs to go anywhere else. We are all, if we only knew it, already there."[3]
One finds the same sentiment in William Dean Howells's utopian island novel
A Traveler from Altruria (1894).

Herman Melville's island travelogues and novels are the narratological apex
of this method of inquiry, whereby the islandic part becomes the true whole.
Melville writes in *Moby-Dick; or, The Whale*: "Queequeg was a native of Rokovoko, an island far away to the West and South. It is not down in any map; true
places never are."[4] His works go some way toward explaining the general interest
of early documentary filmmakers and anthropologists in particular islands and
"island people." Melville's island "ethnography" was a combination of fact and
fiction that provided a foretaste of exotic island cinematography. In 1918, Martin
and Osa Johnson released *Among the Cannibal Islands of the South Seas*, which they
filmed in the Vanuatu area. Such documentary-style filmmaking, writes Robert
Flaherty, inspired his own quasi-ethnographic cinematographic work, especially
Nanook of the North (1922), whose most famous scene Flaherty shot on Walrus
Island in Hudson Bay.[5] *Nanook of the North*, often called "the first feature-length

'documentary,'" takes its cue from Melville's "Great White." Flaherty and his backers advertised it both as a generalizing documentary, calling it on a promotional poster "the truest and most human story of the Great White Snows," and as a particularly thrilling adventure story, calling it "a picture with more drama, greater thrill, and stronger action than any picture you ever saw."[6]

So far as philosophy goes, one particular place on the globe is as good as another.[7] All the philosopher needs is the ability to set up a "fulcrum." As the ancient Greek Archimedes might have it:

Give me a place to stand and I will move the Earth.[8]

It is only the general, islandological hypothesis: the architectonic "what if" of human thinking. So surveyor Henry David Thoreau writes in *Walden*: "One value even of the smallest well is, that when you look into it you see that earth is not continent but insular. This is as important as that it keeps butter cool."[9]

Charles Darwin, as we see in *Islandology*, treats the experimental status of islands in several ways. One, which we consider elsewhere in this book, involves inanimate volcanic islands and animate coral islands. This doubly islandological focus—both inanimate and animate—allows for a comprehension of geological and marine subsidence and uplift together with a theory of worldwide plate tectonics.[10] A second way in which Darwin uses islands as laboratories involves worldwide animal speciation, meaning "the formation of new and distinct species in the course of evolution."[11] In his informal and private notebooks of 1838–1839, for example, he was already noting that, if the passage of *time* is an agent of speciation, then isolation in *space*, of the sort that islands provide, makes a good laboratory for his theories:

If [geographical] separation in horizontal direction is far more efficient in making species, then time, (as cause of change) which can hardly be believed, then, uniformity in "geological" formation intelligible.

No, but the wandering & separation of a few probably would be most efficient in producing new species: also one being reduced in numbers, but not so much then, because circumstances.[12]

No one before Darwin's time—or since—has understood islandology so comprehensively (see Illustration 25).

HYPOTHESES

Writes Thomas More in the epigraph to his islandic *Utopia*: "I alone of all nations [*terrarum*], without philosophy, have portrayed for mortals the philosophical city."[13] Islands are both a world apart and a part of the world. (Put another way,

ILLUSTRATION 25
Coral Island and
Encircling Coral Reef
Creating a Lagoon.
From Charles
Darwin, *The Structure
and Distribution of
Coral Reefs* (London,
1842). Reproduced
with permission from
John van Wyhe, ed.,
*The Complete Work
of Charles Darwin
Online*, http://
darwin-online.org
.uk/. Harvard College
Libraries.

an island's biography is both circumscribed and typical.)[14] Thus thinking about "small" or "deserted" (i.e., heretofore "uninhabited") islands can make for representative case studies of understanding the "large" world. Thanks to their typicality, size, and isolation, many islands have laboratory status in both the natural and the political sciences.[15] Without the whole "world as laboratory,"[16] one might have, in this way, at least a whole part of it.

Islands figure in the social sciences. For the specialized medical and public health statistician, there is Saint Kilda (the farthest part of the United Kingdom from the "mainland"), whose residents had an unknown disease that causes infant deaths. For the Malthusian demographer, Tikopia (Solomon Islands) has a presumably unusual cultural "acceptance" of infanticide. Was this a policy "meant" to control population growth? For the sexual pathologist, there is Clipperton Atoll (France), whose residents have a possibly "irregular" tradition of rape; the Colombian author Fanny Buitrago depicts such a place in her novel *Los Pañamanes* (1979), about an island where the culture is informed by conditions of rape introduced by Spanish conquerors. There is also Pitcairn Island (United Kingdom), where forced incest is sometimes discussed in terms of an experimental "breeding ground."[17] The literature from such islands often borders on the incestuous. In *Christina, the Maid of the South Seas: A Poem* (1811), which concerns Pitcairn, Mary Russell Mitford writes:

> Her mother was my mother too;
> My father hers, alike we grew;
> She call'd me brother, but such love
> For sister ne'er did brother prove.[18]

For those who seek a laboratory to study the fact or fiction of "full sexual liberty," there is, presumably, Pukapuka (Cook Islands).[19] For the military scientist, there is Vieques (Puerto Rico), where the United States practiced aerial bombing for decades. Likewise, there is Fangataufa Atoll and Moruroa Atoll (Tuamotu Archipelago in French Polynesia in the South Pacific). There the French government detonated nuclear test weapons from 1966 through 1996 after Algerian independence, when the French could no longer easily detonate them on the "island in the desert" that was Algeria's Reggane.

The search for "clean" islands of the celebrated Galápagos sort goes on,[20] but geographically inclined social scientists sometimes regard state political expansion and contraction somewhat as zoologists and geologists regard natural evolution.[21] For them, there is the never-ending spectacle of European states and individuals establishing themselves as royal or imperial rulers of small, presumably previously "uninhabited" or "outpost" islands. Examples of this sort include the island regime of Ramón de Arnaud on Clipperton Atoll and that of Eloise Wagner de Bosquet on Floreana Island (Galápagos Islands, Ecuador).

The hypotheses that populate literary works about islands are legion.

What if a man and his daughter were shipwrecked alone on an island, the only other inhabitants there being variously "indigenous" natives? Shakespeare's *Tempest*.

What if a bourgeois family (husband, wife, and offspring) were shipwrecked on an island? Johann David Wyss's *Swiss Family Robinson*.

What if a wealthy woman and her servant were stranded on a desert island? Would sexual desire change economic class bias? What would happen when they returned to mainland "civilization"? Lina Wertmüller's movie *Swept Away* (1974),[22] set mostly on Comino Island in the Maltese Archipelago. The social "class analysis" informing Wertmüller's work responds to that in Michelangelo Antonioni's movie *L'avventura* (1960), which is set partly on the tiny, almost wholly deserted volcanic island of Lisca Bianca, near Panarea, in the Aeolian Archipelago north of Sicily (see Illustration 26).[23]

What if a sailor were marooned on a desert island? Daniel Defoe's *Robinson Crusoe*, the literary *locus classicus*, which is also based partly on a "true-life" experience.[24]

What if a child were marooned (alone) on an island? Would he develop language on his own? Ibn Tufail's *Philosophus Autodidactus* (twelfth century) and Ibn al-Nafis's *Theologus Autodidactus* (thirteenth century). In those works feral children live in seclusion on deserted islands and eventually encounter castaways from the outside world.[25] The parallels here are many. About Caliban in *The Tempest*, Stephano asks where this "savage" learned his particular language: "This is some monster of the isle. . . . Where the devil / should he learn our language?"[26]

What if two or more children were abandoned on an island? What sort of society would they develop? William Golding's *Lord of the Flies* (1954) and

ILLUSTRATION 26
Lisca Bianca. This
isolated island
was the setting
of Michelangelo
Antonioni's movie
L'avventura.
Walterfilm.
Permission granted
by Collection
Selechonek.

R. M. Ballantyne's *Coral Island* (1857), the latter a fictional response, on the level of social anthropology, to the already massively influential studies of the animate and inanimate evolution of the Earth in Charles Darwin's influential *Coral Reefs* (1842).

What if, in the midst of war, a group of sailors were to land accidently on an island where, despite the fact that the island was "enemy territory," they were treated kindly by the local residents? The movie *Tanzen*, set on Deer Isle in Canada's Bay of Fundy, which takes this hypothesis as its premise. Christine Marion Fraser's novel *Rhanna at War*, set on a fictitious island in the (Scottish) Hebrides, does much the same: a Nazi bomber crew crashes on an island where they come to believe that they are in "paradise."[27] John Boorman's movie *Hell in the Pacific* (1968) likewise tells how two soldiers from enemy camps who do not know each other's language are marooned on the same small island.

THREE ISLAND UTOPIAS

You'ld be king o' the isle, sirrah? . . .

Had I plantation of this isle— . . .

—Shakespeare, *The Tempest*

The genre of "the island 'utopia'" includes well-known works of ancient literature and philosophy. The gamut ranges from the lost island of Atlantis in Plato's *Timaeus* and *Critias* (fifth century BC) to Euhemerus of Messene's Panchaea in the *Sacred Description* (fourth century BC).[28] The work of Theopompus (fourth century BC) includes the fictional island Meropis, beyond "the world ocean,"[29] with its regional *anostos* (place of no return). Lucian of Samosata's description of the peoples of the Moon, in *True History* (second century AD), focuses on warfare between the denizens of insular planetary islands: the people of the Sun and the Moon struggle over who will colonize the morning star. It includes a description of the lives of those who live inside an animate floating island (a giant whale) and a panorama of the "Isles of the Blessed."[30] "I chant copious the islands beyond," intones Walt Whitman in *Leaves of Grass* (1855).

The political bent of such works usually involves more of a critical description of the author's own political place (*topos*), or state, than a prescription of what one should do to improve one's own place. Diskin Clay and Andrea Purvis thus write:

> A work of [classical] utopian literature does not present an ideal society as a model for social reform: it works rather by indirection in directing the reader's critical gaze away from the society created by the utopian author onto his or her own society. Its focus is not the ideal but the real.[31]

For many moderns, on the other hand, the role of utopian island literature often changes to prescription. One example is Francis Bacon's *New Atlantis*, which is less a presentation of a *utopia* (nowhere place) than of a *eutopia* (excellent place) to be emulated.

Utopian island literature often gives rise to humor. Lucian and Theopompus, for example, have passages that mock the islandic travels of Homer's Odysseus and his tales of heroic *nostos* (homecoming).[32] For most utopian writers, there is also the tension between the somewhere—Archimedes' *pou*—being realizable and it being "merely" ideal. The historically irresolvable opposition here—between what cannot and what can successfully be emulated—gives rise to such straightforward sentiments as Plato's warning that, as Voltaire puts it in "The Prude" (1752), "the best is the enemy of the good."[33] Likewise, that tension provides fodder for myriad good laughs and reminds one to be silent or to be esoteric when articulating viewpoints of which the head of state might disapprove, at the risk of losing one's own head.

The first two works that we discuss are distinctly "comic." The author of the third, Thomas More, ended by losing his head. More actually invented the word *utopia* to represent an esoteric critique of the English state. When at trial he refused to give up the doctrine of papal supremacy, relying instead on the sort of esotericism that informed *Utopia* and promoting the legal maxim that "silence presumes consent,"[34] the king had him decapitated. His own good humor, as we will see, comes out much later in the partly "Shakespearean" play *Sir Thomas More* (1592).

Atoll K

What if a couple of bumpkins were, out of the blue, to inherit (as if one could) a heretofore unknown island extraordinarily rich in uranium? This is the hypothesis of the last movie made by the comedy team of Stan Laurel and Oliver Hardy, *Atoll K* (1951), where *K* refers to nuclear reactions.[35] Other titles were the German-language *Dick und Doof erben eine Insel* (Laurel and Hardy Inherit an Island), the working title *Robinson Crusoeland, Utopia, Entente cordiale*,[36] and the Danish-language *Gøg og Gokke På Atom-øen* (Laurel and Hardy on Atomic Island).

The story of *Atoll K* begins in London, with a view of Parliament at Westminster. Laurel and Hardy learn from a solicitor that they have "inherited" property—a desert island in the South Pacific. On the way there, the duo and their friends encounter a tempest. Shipwrecked, they find themselves on an atoll that, having just emerged from the sea, is not yet anyone's "property." Laurel and Hardy decide to set up a new country there, "Robinson Crusoeland," that has "utopian" aspirations. All is well for a while, but soon visitors arrive and discover massive amounts of uranium. People now come from all over the world, many with unsavory intentions and dystopian means. Usurpers soon condemn Laurel and Hardy to be hanged. As luck would have it, it is a tempest that suddenly sinks the island back into the Pacific that saves them. Our two heroes, set adrift, hold on to the hangman's scaffolding, now a flotation device, from which they were to hang. Thus reborn from the salty waters, Stan and Ollie are cured of their dream of a "heaven on earth." They then wend their way to the desert island, the taxable "property," that they had inherited.

The first spoken line of *Atoll K* is "Who has never dreamed of knowing Heaven on Earth, of sailing away to a desert island? Well, here's a story of such a dream come true." The "truth" is Heaven on Earth is no place. It is merely utopian.

Political contexts of the postwar period help to explain the ironic purposes and commercial weaknesses of *Atoll K*. First, there was the proliferation of international islandic nuclear weapons testing. In 1947–1948, the United States began testing on Bikini and Enewetak atolls in the Ralik Chain of the Marshall Islands. In 1952, Great Britain began testing on Trimouille, one of the Montebello Islands of Australia. Also in 1952, the Soviet Union was planning to move the testing of major nuclear weapons to the Novaya Zemlya Archipelago in the Arctic Ocean.

Second, the United States—and Hollywood in particular—threatened writers and cinematographers who did not keep to a narrow and right-wing nationalist path. In that context, *Atoll K* had a worrisome subject as well as a then-controversial crew (John Berry, Orson Welles, and John Houseman). Berry's *Hollywood Ten*—a documentary exposé of the House Un-American Activities Committee

(HUAC)—came out in movie theaters in 1951, the same year as *Atoll K*. Predictably, *Atoll K* was "mutilated" both in the production phase and in distribution.

Old New Land

One does well, no matter in what political context, to ask about any utopian adventure whether its organizing idea is realizable. By way of illustration, consider Theodor Herzl's 1902 utopian island novel *Altneuland* (Old New Land). What if two nineteenth-century men from urbane central Europe took a twenty-year trip to a distant island, passing through the same place on their going and coming? In this parable-like story, the Viennese heroes find Palestine destitute on their way to Rarotonga Island. On their way back, they find it transformed into a cosmopolitan modern society.

The novel's first Hebrew-language translator rendered its German title, *Altneuland*, as *Tel Aviv* (ancient hill or spring). That is how, in 1909, the city of Tel Aviv in present-day Israel got its toponym.[37] "Wenn ihr wollt, ist es kein Märchen" is the epilogue to Herzl's *Altneuland*.[38] A standard translation is "But if you do not wish it, all this that I have related to you is, and will remain, a fable." Relevant to this matter of wish and will in the case of a Jewish homeland are two North American ideas for a refuge island for the Jews. The first, called "Ararat" (after the islandic mountaintop where Noah's ark came to rest), was to be located on Grand Island in the Niagara River between Canada and the United States (1824). The second, promulgated by Franklin Delano Roosevelt's secretary of the interior in the wake of Kristallnacht, was Baranof Island in the Alexander Archipelago, part of the territory of Alaska (1939).[39]

Thomas More's Skull

> The commander Utopus made me, who was once not an island,
> into an island.

> —More, *Utopia*

In *Utopia*, Thomas More generally pretends to hope for a true incarnation on Earth of Plato's ideal republic.[40] That incarnation begins, in Utopia, with King Utopus's turning a peninsula into an island. Scholars debate whether More was influenced by the fourteenth-century Byzantine thinker Nicephorus Gregoras's view of Mount Athos—"a self-sufficient agricultural institution modeled on Plato's polis"[41]—as both natural peninsula and ideal island; Ernst Bloch, in *The Principle of Hope*, considers possible implications for the history of utopian thinking.[42]

Two decades after More published *Utopia*, Henry VIII had More's head removed from his body (1535) and ordered that the head be placed on a pike for a month, for all passersby to see. The skull eventually showed through the rotten flesh. (Jester Yorick's smelly skull in *Hamlet* comes to mind.)

We will never know whether Henry VIII had in mind one irony of the punishment as understood in the context of the Western philosophical tradition: that his skull was to Thomas More's brain what the artificial coastline of More's *Utopia* was to his island polity. In his "science fiction" dialogue *Timaeus*, in fact, Plato thus associates the "lost" utopian island of Atlantis with the human brain. "Out of this," he writes, "[the god] fashioned, as in a lathe, a globe made of bone, which he placed around the brain."[43]

In fact, Ambrosius Holbein's widely circulating fictional maps of More's Utopia usually show it resembling a skull or human cerebrum.[44] As presented in Illustration 27, Utopia is depicted as a head joined at the neck to a continental body by an *isthmus* (neck)—or separated from it by a canal.[45] Here Holbein both follows the presentation of the *nēsos* Mount Athos in a 1484 *isolario* (island book) by the German cartographer Henricus Martellus[46] and contributes to the tradition that "an island is an [artificial] figure."[47] The human skull, so considered, is both the "coastline" of an insular nation-state and the "shell" of an individual human brain. When Shakespeare named his theater "The Globe," he referred not only to the globe that is the Earth and to the brain that is Hamlet's "distracted globe"

ILLUSTRATION 27 *Utopiae Insulae Figura*. Woodcut by Ambrosius Holbein showing the island and the facing page of text. From Sir Thomas More, "De optimo reip. statu, deque noua insula Utopia," *Apud inclytam Basileam* (1518). New York Public Library / Art Resource, New York.

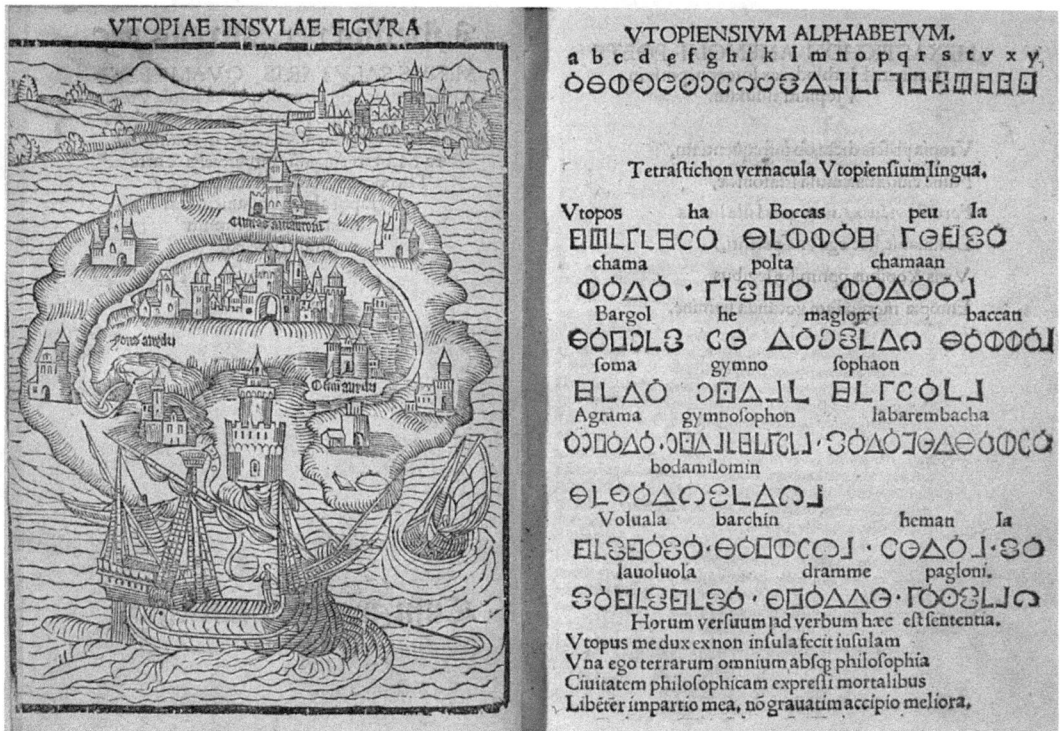

Quand ce corps fut plus rapproché de l'endroit où j'étais, il me parut être d'une substance solide, aplati à la base, uni, et qu'il réfléchissait très-clairement la mer sur laquelle il planait.

Je m'arrêtai sur une hauteur à deux cents pas environ du rivage, et je vis ce même corps s'abaisser presque en ligne parallèle avec moi à un mille de distance. Je pris mon téles-cope, et je découvris un grand nombre de personnes qui allaient

but also to the fabricated peninsular stage where the fabrication of his own brain, *Hamlet*, was to be performed.

The association of the realm with the head that bears it, or of the island country with the brain that gives birth to it, informs the language of the anonymous play *Sir Thomas More* (1592), one of whose authors was Shakespeare. Here Lord Surrey, speaking with Erasmus himself, says, "This little isle holds not a truer friend / Unto the arts" than Thomas More. More himself dryly remarks to his jailer, "More rest enjoys the subject meanly bred / Than he that bears the kingdom in his head."[48]

Utopia is the *locus classicus* for dozens of works of utopian island "fiction," ranging from Francis Bacon's *New Atlantis* (1624) to Samuel Butler's *Erewhon, or, Over the Range* (1872).[49] Its most trenchant imitation, though, is the work of the Irishman Jonathan Swift, who much admired More's *nova insula Utopia*. He said about the brainish More that he was "the person of the greatest virtue these islands ever produced."[50] Swift's work itself set a new standard for utopian and dystopian conceptual island adventures, which appear in his anonymously published *Gulliver's Travels* (also known as *Travels into Several Remote Nations of the World*). The cartographer Herman Moll, an émigré from either Holland or Germany, produced the maps for *Gulliver's Travels* (see Illustration 28); he also produced the fictional charts for Daniel Defoe's *Robinson Crusoe* (1719),[51] considered earlier in this chapter, and the factual map of South America that shows that continent's shores with the toponym *Terra Firma* (1700).[52] In the same long-lived tradition is Arnold Bennett's satirical play *The Bright Island* (1924),[53] in which members of the "English governing class" are marooned on an undiscovered island (recalling also *The Tempest*).

Moll, who had fled the Scanian War, was more carefully esoteric than More in presenting political hypotheses in fictional garb. That did not stop him, however, from repeating the claim, first made in Garci Rodríguez de Montalvo's 1510 fictional work *Las sergas de Esplandián* (The Exploits of Esplandian), that "California is an Island." Montalvo writes:

Know, that on the right hand of the Indies there is an island called California very close to the side of the Terrestrial Paradise.[54]

Moll went further, saying that he "had in [his] office mariners who have sailed round it."[55] He also displayed California as an island in his well-known *Codfish Map* (1720).[56] Taking this old sixteenth-century island fiction for cartographic fact, Hernán Cortés and Francisco de Ulloa went off to find the "island of California." Its actual "discovery" turned a Californian utopia from an island into a mainland.

Politics, Philosophy, Epic Drama

WAR OF CIVILIZATIONS

The ancient Greek toponym *Euripus* has become a general term in European languages. It means "a strait that has strong and dangerous tides and currents flowing between an island and the mainland."[1] The actual Euripus Strait separates island Euboea from mainland Boeotia.[2] Its narrowest part (forty meters) is at the town of Chalcis. Whereas Mediterranean Sea tides are regularly small and gentle, at Chalcis they are large and violent. The Euripus current races—like a riptide—in one direction, and shortly afterward, with equal velocity, runs the other way. Here this Heraclitean up-and-downness seemed, to scientists for millennia, to make impossible ever understanding the mysteries of tides. Here, scarily, "the way up and the way down are one and the same."[3] (See Illustration 29.)

Before becoming the subject of scientific writing, the Euripus was already a subject for the early epic writers Homer and Hesiod, as well as for the tragedians Aeschylus (*Agamemnon*) and Sophocles (*Antigone*).[4]

The tragedian Euripides' proper name memorializes the sea battle between Greeks and Persians that took place in 480 BC at the Euripus Strait, during the same days that the famous Battle of Thermopylae took place.

For those who fought by sea the whole aim of the fighting was concerned with the channel of Euripos, just as the aim of Leonidas and of his band was to guard the pass: the Hellenes accordingly exhorted one another not to let the Barbarians go by into Hellas; while these cheered one another on to destroy the fleet of the Hellenes and to get possession of the straits.[5]

The Sicilian historian Diodorus, in *Library of History* (first century BC), makes the same analogy between the *ravine* at Thermopylae, site of the land battle of Thermopylae in 480 BC, and the simultaneous water battle at the *strait* of Euripus.

Euripides' war plays, *Iphigenia in Tauris* (ca. 413 BC) and *Iphigenia at Aulis* (ca. 405 BC), both involve the Euripus Strait.[6] The latter tells how the Greek forces, gathering before their journey to Troy in order to get Helen back, were "caught"

ILLUSTRATION 29
Bridge of the Euripus.
Engraving from the
*Penny Magazine of
the Society for the
Diffusion of
Useful Knowledge*,
May 4, 1833. Harvard
University Libraries.
Permission granted
by Collection
Selechonek.

in the narrow Euripus by a lack of wind. They overcame their situation only when Agamemnon sacrificed his own daughter Iphigenia. She herself reports:

Only then could the Trojan War begin in earnest. Where Euripus rolls about its whirlpools in the frequent winds and twists the darkening waves, my father [Agamemnon] sacrificed me to Artemis for Helen's sake, or so he thought, in the famous clefts of Aulis.

For there [at the Euripus Strait] lord Agamemnon mustered his expedition of a thousand ships of Hellas, wanting to take the crown of Troy in glorious victory and avenge the outrage to Helen's marriage, doing this favor for Menelaus.

But when he met with dreadful winds that would not let him sail, he went to burnt sacrifices, and Calchas had this to say: "Lord and general of Hellas, Agamemnon, you will not set free your ships from land until Artemis has your daughter Iphigenia as a victim."

Euripus is where the epic contest begins—and ends. Pindar asks, "Was it Iphigenia, slaughtered at the Euripus far from her fatherland[,] that provoked . . . the heavy hand of . . . anger?"[7] Was Iphigenia perhaps spirited off to yet another island? Perhaps even a far-off Asian one like Tauris?[8] The question importantly suffuses ancient Greek tragedy. Agamemnon in *Iphigenia at Aulis* settles the issue. "The birds are still at any rate and the sea is calm, hushed are the winds, and

ILLUSTRATION 30
Iphigenia in Aulis
Commemorative
Stamp. Penny
Nickels, Wells Tuthill,
2010. Reproduced by
permission.

silence broods over the Euripus."[9] Aulis was the site of sacrifice. (See Illustration 30.)

POETS' CONTEST

One history of Greek epic literature has it that the rival poets Homer and Hesiod, taken together, make up a full picture of the origin of national European epic literature. Homer set the first standard for island adventure, and, not surprisingly, legend says that he was born on an island. Which island has been a matter of long-standing debate. Simonides says that Homer's birthplace was the island of Chios.[10] Aristotle, whose mother came from the Euripus region, suggests that Homer was from the island of Ios. Others argue for the island of Salamis. A few say the island of Cyprus. Some accounts say that Homer was from Cumae—an urban colony on the west coast of Italy established by Greek travelers from the island of Euboea (in particular the city of Chalcis) on the Euripus Strait. The author of *The Contest of Hesiod and Homer*, written in the second century AD but based on a fourth-century BC tradition about the poets' rivalry, has it that Homer was from the island of Ithaca.[11]

Hesiod's home, in any case, was Boeotia on mainland Attica, which borders the western side of the Euripus Strait. According to *The Contest*, following a lead from Hesiod's own *Works and Days*, the landlubberly Hesiod sailed on a ship only once in his life:[12]

For never yet have I sailed by ship over the wide sea, but only to Euboea from Aulis, where the Achaeans once stayed through much storm when they had gathered a great host from divine Hellas for Troy, the land of fair women. Then I crossed over to Chalcis, to the [funeral] games [in honor] of wise [King] Amphidamas where the sons of the great-hearted hero proclaimed and appointed prizes.

In *The Contest*, Hesiod's poetry presents "peace and domestic agricultural industry." He wins over marine islander Homer, whose poetry presents war and destruction. "The philosophic landlubber often wonders at the eternal restlessness of his naval brother-man," writes Richard Francis Burton in *Two Trips to Gorilla Land* (1876).[13]

Hesiod's single *fer*ry ride of forty yards across the Euripus, from mainland to island, matches up, as meta*phor*, to all the Odyssean voyages, between the Asian and European mainlands and among the Greek islands that Homer reports in the *Iliad* and the *Odyssey*: Aeaea—home of the enchanting Circe. Aeolia—the floating island home of Aeolus, Keeper of the Winds. Capri—home of the Sirens. Ogygia—home of the nymph Calypso. Scheria (Drepane)—home of the Phaeacians and King Alcinous. Thrinacia—home of the cattle of Helios.

ISLANDS RAZED AND RAISED

The Hellespont (i.e., Dardanelles) is the strait that connects the North Aegean Sea with the Propontis (Sea of Marmara), thence the Bosporus Strait and thence the Euxine (Black Sea). Thus it separates "Europe" from "Asia" or, if you like, defines Europe and Asia against each other (see Illustration 31).[14] It does so in much the same way that Homer suggests in the *Iliad*, a work that describes the Trojan War between the European Greeks and the Asian Persians. Herodotus especially understands the Trojan War this way. The Hellespont for him is the great cultural, commercial, and political "crossroads," back and forth, of human history.

The actual waters of the Hellespont flow in two directions, east and west, at the same time. One stream flows above the other: the upper stream, via a surface current, from the Sea of Marmara to the Aegean; the lower stream, via a deeper current, from the Aegean to the Sea of Marmara.[15] This double flow, back and forth, amazed the ancients and inspired them. "The way up and the way down is one and the same," writes the Greek Heraclitus from nearby Asian Ephesus.[16]

Herodotus, who lived farther down the coast at Halicarnassus, explains to the Asian Persians, or "translates," the ways of the European Greeks (Hellenes) regarding the giving and taking of women (Helen). Doing so in his *Inquiries*, he reports what it meant, in rheological terms, when in the early fifth century BC the Persian tyrant Xerxes I built bridges across the Hellespont. He did that, says Herodotus, in order to allow his army to "ferry across" (*metaphorein*) the Hellespont and conquer Greece. When a sea storm destroyed the bridges, Xerxes was so enraged that he sought to punish the strait, giving the water three hundred lashes of the whip and then branding it with red-hot irons. Herodotus says, wryly, that this was a "highly presumptuous way to address the Hellespont."

A little later in the *Inquiries*, Herodotus reports that, in 492 BC, after a tempest destroyed the Persian navy while it was sailing around the peninsula of Mount Athos, Xerxes ordered his engineers to build a canal across the isthmus that linked Mount Athos to the mainland. The channel, completed in 480 BC, turned the *peninsula* at Athos into an *insula*. "On this isthmus" which joins Mount Athos to the mainland, writes Herodotus, "stand . . . [mainland] towns . . . the inhabitants of which Xerxes now proposed to make islanders." The canal completed, Xerxes' navy was able to sail safely directly to the region of Eretria, on the island of Euboea

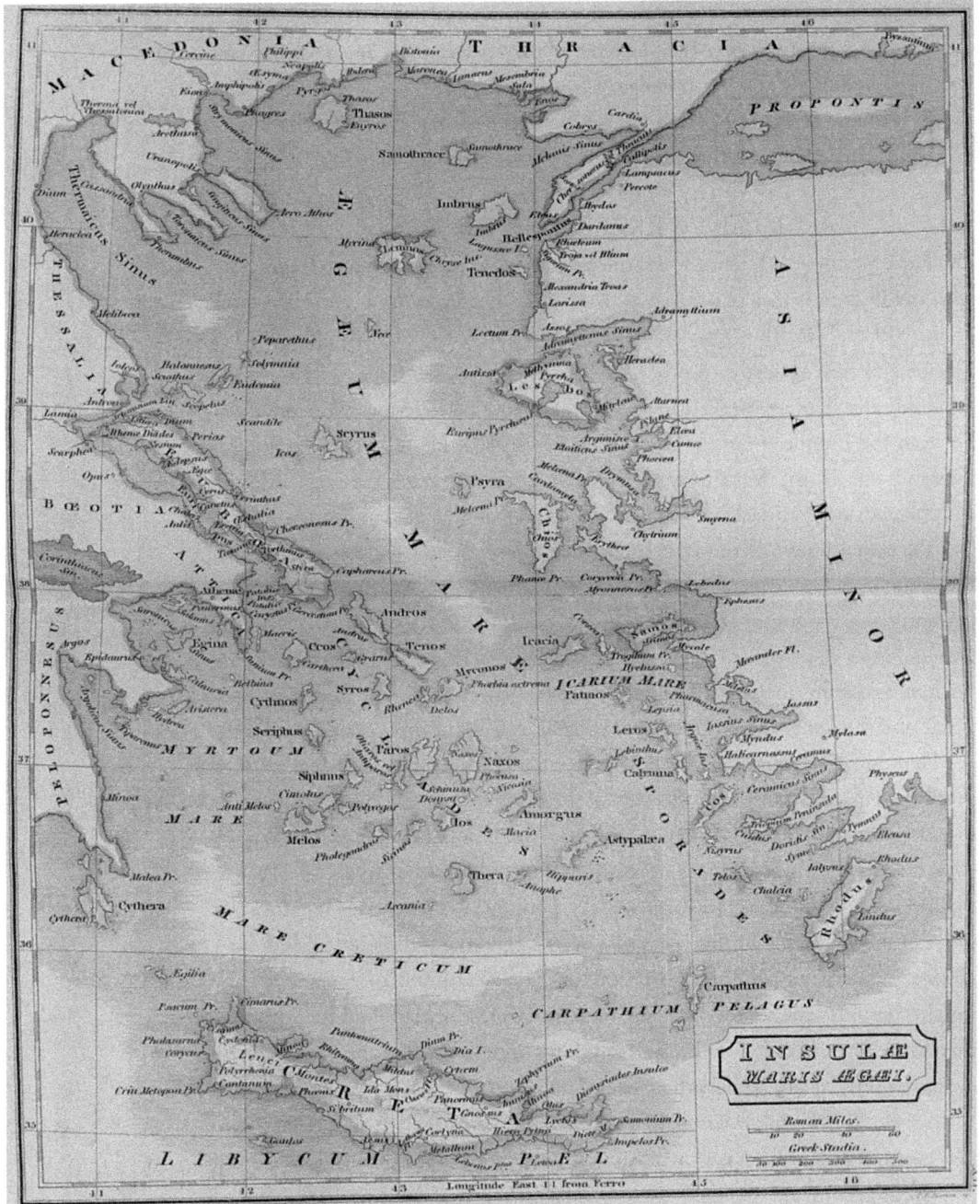

ILLUSTRATION 31 *Insulae Maris Aegaei*. From Samuel Butler, *An Atlas of Ancient Geography* (ca. 1840). Butler's grandson (who shared his grandfather's name) wrote the utopian satire *Erewhon* (1872) and put forward the notion that Homer's *Odyssey* was written by a young woman from the island of Sicily. Courtesy of the Harvard Map Collection.

near the Euripus Strait. At Euripus, the Persian tyrant hoped he would be able to conquer European Greece.

Herodotus thus twice emphasizes the prideful aspect of the Eastern tyrant. At a neck of water (Hellespont), Xerxes builds a bridge that joins one continent with another. At a neck of land (Athos), he builds a canal that joins one sea with another. "I cannot but conclude," writes Herodotus,[17] "that it was mere ostentation that made Xerxes have the canal dug—he wanted to show his power and to leave something to be remembered by."[18] (See Illustration 32.)

IF ATHENS WERE AN ISLAND

Zeus would have made [Knidos] an island, had he wished it.

—Herodotus, *Inquiries*

If we [Athenian mainlanders] were islanders, who could be more secure from attack?

—Thucydides, *History of the Peloponnesian War*

The ancient Greeks did not group islands according to whether they were parts of the same natural or geological archipelago (which is not even an ancient Greek word, as we have seen). It was the Athenians who came to do that, more according to overtly political linkages involving marine commerce and thalassocratic politics as well as variably thalassic religions:[19] the late fourth- and third-century *koion tōn nēsiotōn* (League of the Islanders),[20] the Delian League of Cyclades,[21] and the like amphictyony centered on the island of Calauria.[22] These conditions helped partly to model the rise of the Athenian navy as well as its empire and democracy.[23] Above all, as we will see, was the remaking of Athens into an island and the abstracting reformation of the notion of island from a natural to a partly political and fully geopolitical idea. The making of an apparently natural island into an apparently human-made one, whether by earthworks (bridges and channels) and/or by politics, and especially the consciousness of the making, is part of the principal understanding of Athenian political culture itself. Ancient Athens, a mainland *polis*, was refashioned into an island.[24] One might almost say that, for the Greek as for the Norse, democracy is "an island thing."

Many ancient Greeks wrote about islands. There is, for example, *The Islands* (Aristophanes)[25] and *Greece, or The Islands* (Plato Comicus),[26] as well as writings by Aretades of Knidos[27] and later Diodorus of Sicily, who in the first century AD made *nēsiōtika* (concerning islands)[28] the entire subject of his *History*. The deeper politics of nesology, however, is to be found in the works of two men: Herodotus (484–425 BC), the "father of history" and author of the *Inquiries* into the Greco-Persian wars, who hailed from the general region of Halicarnassus in Asia Minor; and Thucydides (460–395 BC), an Athenian general and author of the *History of the Peloponnesian War*. These two put forward implicit theories of nesology and politics in their works, many of which need teasing out.

Fernand Braudel's notion that, for the Greeks and some others in the Mediterranean region, there are "islands that the sea does not surround,"[29] is accurate so far as it goes and likewise helpful insofar as it seems to take the edge off the sometimes surprising meaning of *Peloponnesus*. After all, the "island (*nēsos*) of Pelops" is actually a peninsula.[30] These days, the Peloponnesus is the dominant part of Greece, artificially connected to or disconnected from the rest of the mainland in two ways. (Roadwise, it is connected to the mainland by the Rio–Antirrio Bridge, completed in 2004, where the Strait of Rion meets the Gulf of Corinth. Shipwise, it is disconnected from the mainland by the Corinth Canal, completed in 1893, where the Saronic Gulf meets the Gulf of Corinth.) Mani, one of the peninsulas of the Peloponnesus itself, is similarly "an almost-island (*nēsos*)." Émile Kolodny says the same, on a demographic level, about peninsular Mount Athos.[31]

Useful as Braudel's reminder might be, it does no more than remind us of the apparent ambiguity also present in the meaning of many English-language terms. These include *chersonese*, from *chersos* (dry land) and *nēsos* (island); *byland*, as in William Harrison's chronicle of British history[32] and Tristram Risdon's study of the regions of Devon;[33] *promontory*, which can refer both to a landmass overlooking water and to one overlooking land; and *hoo*. (Sutton Hoo in Suffolk is an escarpment overlooking the Deben River and the site of an important ship burial site two miles from Rendlesham, the seat of the Anglo-Saxon kings of East Anglia.)[34]

What is required further in understanding ancient Greek peninsular nesology has to do with natural and political engineering as well as hydrology and rheology. Kolodny, when he calls Mount Athos an island, has in mind the geographic demographic conditions (cultural insularity) rather than conditions of physical geography such as natural or man-made isthmuses/bridges and canals/moles. In ancient times, already, there was considerable debate about physical geography and what it might mean for human beings to purposefully change it. For example, there are the varying descriptions of what it meant to turn a peninsula into an island (as Xerxes did to Mount Athos) or a natural canal into an isthmus (as Xerxes did to the Hellespont). (Did not the sailor and island explorer Odysseus take up a ship oar and wander into the dry hinterland seeking a "land that knows nothing of the sea" and someone who would mistake an oar for a winnowing fan?)[35]

On the one hand, there were reasons not to turn dry mainland or ambiguous byland into water-island: for example, it would transform the mainland into a prison. ("Don't Fence Me In" [1934] is how Cole Porter and Robert Fletcher might have put it.) On the other hand, it would transform the mainland into a secure fortress. (Another group on the Mediterranean, the Jews, would make an "island" around the Torah: "Put a protective fence [Aramaic סייג (*syag*)] around the Torah.")[36] Federico Borca allies here the Roman idea that the Semitic *polis* Carthage, stretched between salty lagoons, was a "peninsula surrounded by the sea."[37]

Herodotus, in the nesology that pervades his *Inquiries*, takes a particular interest in the debate about whether to turn the Greek colony of Knidos into an island.[38] (This place, a Carian neighbor of Herodotus's own Halicarnassus in Asia Minor—the next large peninsula to the south—was the homeland of Aretades of Knidos, who wrote a book about islands.)[39] Knidos was built partly on the mainland and partly on the island of Triopion (Cape Krio), where the Mediterranean meets the Aegean. In olden times, the land on which it stood was connected to the mainland by a causeway and bridge. Its citizens sought advice from the oracle at Delphi as to whether to protect themselves from attack by cutting a canal through the isthmus. The oracle replied: "Dig not the isthmus through, nor build a tower! Zeus would have made an island had he wished it."[40] (No wonder the poet Pindar says that the island is a *tower* of Kronos!)[41] Making the urban center (*asty*) of Knidos an island fortress would disconnect it from its mainland area (*chōra*), with political and military consequences. For Herodotus, as Paola Ceccarelli puts it, "fortifying an isthmus and digging a trench are put on the same plane. This leveling makes for an evolution [in thought] towards the abstraction of the concept of island."[42] (Opposite to the view that a mainland ought not be made into an island is, as we have seen, the view of Thomas More: the first act of King Utopus is to build a channel, making an island out of his realm.)[43]

Corinth, like Knidos, provided a case study—one closer to Athens. The oracle at Delphi warned the Corinthian tyrant Periander, around the year 600 AD, against cutting a canal at the isthmus of Corinth. While there may have been a commercial reason or hydrological rationale for such advice,[44] the warning, when digested by Athenians, had more to do with the oracle's own situation. In any case, the plan was not put into effect until 1893, when the canal was built. The Diolkos—a sort of railway "land bridge"[45] or stone carriageway that provided portage from the Gulf of Corinth to the Saronic Gulf[46]—had been constructed in ancient times; it was arguably the first large railroad and worth an interesting description by Strabo and many others.

Just how these ideas influenced subsequent Athenian political theory and geopolitics is important here. The story might begin with a brief discussion of the ancient Greeks' understanding of the international wars, especially the Second Persian Invasion (480–479 BC), which provided Herodotus with subject matter, and the intranational wars, especially the First Peloponnesian War (431–421 BC) and the Second Peloponnesian War (413–404 BC).

The Persian tyrant Xerxes, as we have seen, bridged the Hellespont by making what seemed to be two islands into a single mainland, and channeled through Mount Athos by turning an apparent peninsula into an island. He defeated the Greeks at Thermopylae and forced the Greek navy to withdraw to the island of Salamis. Themistocles (524–459 BC) now evacuated Athens. This was a turning point in Greek nesology and its political comprehension of dominion itself.

Thucydides reports that Themistocles had always said that "they [the Athenians] must make the sea their domain."[47] "The Piraeus appeared [to Themistocles] to be of more real consequence than the upper city. He was fond of telling the Athenians that if ever they were hard pressed on land they should go down to the Piraeus and fight the world at sea."[48] After evacuation of the *asty* to the port, there came, in 480 BC, the evacuation of the power for the seas and islands. At this time, the Corinthian military commander Adeimantus, during a debate about how to conduct a battle at Salamis (480 BC), claimed superciliously about the self-exiled Athenians that Themistocles had no homeland:

The Corinthian Adeimantos inveighed against [Themistocles] . . . because he had no native land. . . . This objection [Adeimantus] made against [Themistocles] because Athens had been taken and was held by the enemy. Then Themistocles . . . declared . . . that he himself and his countrymen had in truth a city and a land larger than that of the Corinthians, so long as they had two hundred ships fully manned; for none of the Hellenes would be able to repel the Athenians if they came to fight against them.[49]

For Themistocles, however, "both polis and land were the men in the ships."[50] This appeal to the ship of state was decisive. For Herodotus, too, as one historian says, "there [had] exist[ed] an Attica which [was] not mainland, that is, an Attica on board the Athenians' triremes."[51] If the Athenians could withdraw from *chōra* to *asty*, they could now also withdraw from *polis* to ships. In fact, in some ways, the *polis* became the ship.[52] The shipboard existence of the state is not only the sign of the Greek navy but, as we will see, also the hallmark of Britain understood as an insular ship of state. Wrote Gilbert Murray in "The Wanderings of Odysseus" (1905): "At present England is the thalassocrat."[53] It is not so much the history of the sea that matters. It is rather the natural and political that, taken together, are the geographic history of the island.

Thus, with the defeat of the Persians in 480 BC, the Athenians could turn their attention to rebuilding the long walls that connected the city to the Mediterranean, making the city more of a nesological entity. With the new walls, the Piraeus would be part of Athens, as it were, and maybe more important. Athens, a mainland power, was now able to turn itself ever more and again into an island naval power. The engineering work was well under way during the time of Cimon. He drained the wet ground west of Athens:[54] some of it around the delta of the Ilissus River and some around the Cephissus River.[55] Not everyone wanted the walls. The Athenian aristocracy feared a democratic navy, for example, because it would threaten the power of the landed aristocracy.[56] And, of course, the plans for new walls threatened the hegemony for which the Spartans hoped. Completed around 440 BC, the walls connected the urban center of Athens to its new port, the Piraeus, almost as if it were an isthmus—"Once completed, the walls would make Athenian access to the sea as secure as if the city had become an island: the realization of Themistocles' Dream."[57] At the same time, they disconnected Athens from other outlying regions, its *chōra*,[58] in much the way suggested by Herodotus's tale about Knidos. Thucydides here wisely comments: "As a result of [the battle that took place at the pass of] Decelea (around 415 BC) near the source of the Cephissus River, Athens, instead of a polis, became a fortress." That abandonment felt wrong, though. The Old Oligarch, in his *Constitution of the Athenians* (fifth century BC), notes implicitly the effect on the *chōra* of its *asty* having become islandic:

Since from the beginning, the Athenians happen not to have lived on an island, they now do [as essential islanders] the following: they place their property on islands while trusting in the naval empire and they allow their land to be ravaged, for they realize that if they concern themselves with this, they will be deprived of other greater goods.[59]

During the Archidamian War, the Athenians could abandon the *chōra* partly because they were storing cattle on Euboea.

Pericles, ruler of Athens, is often credited with transforming the old Delian and Calaurian leagues into the Athenian Empire. According to John Hale in *Lords of the Sea* (2009), Pericles said something like this:

Sea power is of enormous importance. Look at it this way. Suppose Athens were an island, would we not be absolutely secure from attack? As it is, we must try to think of ourselves as islanders; we must abandon our land and our houses, and safeguard the sea and the city.[60]

Thucydides has Pericles say, "Reflect, if we were islanders, who would be more invulnerable? Let us imagine that we are, and acting in that spirit let us give up land and houses, but keep a watch over the city and the sea."[61] The Old Oligarch speculates likewise about the nesological hypothesis and Athenian politics. First: "If they were thalassocrats living on an island, it would be possible for them to inflict harm, if they wished, but as long as they ruled the sea, to suffer none."[62] Second: "If they lived on an island, they would have been relieved of another fear: the city would never be betrayed by oligarchs nor would the gates be thrown open nor enemies invade. (For how would these things happen to islanders?)" Third: "If they lived on an island, no one would rebel against the democracy."[63]

Francis Hartog suggests that Pericles' strategy, to turn Athens into an island, partly derives from Herodotus's description of nomadism in Scythia (north and east of the Black and Caspian seas).[64]

For the city of Sidon in Phoenicia, built on a promontory facing an island, the ship was the numismatic type of the state, as shown on its half shekel, with a pentekonter ship sailing over the waves (435–425 BC).[65] So, too, for the coins of the Phoenician king Adramelek of Byblos, which, like Sidon, was subject to Persia (360–340 BC),[66] and the coins of the Indian king Vashishtiputra Shri Pulumavi (first century AD). The same obtains for Roman currency depicting warships: the brass coins of Hadrian (132–134 AD),[67] for example, and coins illustrating the battle of Actium (31 BC) against the Egyptian navy.[68] For the Athenians, however, the ship was not merely a numismatic *type* of the state; it was also the ship of state per se.

ARISTOTLE'S DROWNING

Herodotus, in his nesological considerations, is much struck by how, when the Nile overflows, the *chōra* (the outlying part of the *polis*) in the various island-cities of the Nile Delta, is converted into a sea and nothing appears but the cities, which look like islands in the Aegean.[69] Generally speaking, Herodotus's acquaintance with the rheology of water was limited to the more moderate conditions of small river islands (for example, the Plataea's River Asopus)[70] and of small tides (characteristic of most of the Mediterranean). Other thinkers considered the matter in abstract terms. The Ephesian philosopher Heraclitus, for example, holds that *Ta panta rhei* (everything flows or, more accurately, everything is a stream). Against this Heraclitean rheology (logic of flow), according to which everything flows, Platonic philosophy was born. For Plato, the Euripus Strait, or the pre-Socratic notion of it, was the geological and hydrological counterpart of philosophy as well as its foundation.

One does not need to argue that geology and hydrology play a foundational role in Socratic morality,[71] or even need to link rheology with mechanics,[72] to understand that Plato attacks the intellectually and morally unreliable rhetoricians and sophists as merely "Euripian." In the dialogue *Phaedo*, he states outright:

When a simple man who has no skill in *dialectics* believes an argument to be true which he afterwards imagines to be false, whether really false or not, and then another and another, he has no longer any faith left, and great disputers, as you know, come to think at last that they have grown to be the wisest of mankind; for they alone perceive the utter unsoundness and instability of all arguments, or indeed, of all things, which, like the currents in the Euripus, are going up and down in never-ceasing ebb and flow.

The waters that "go up and down" (physically) at Euripus resemble the people who hold that nothing (physical) is stable.[73] Plato takes the same position against Heraclitus in his dialogue *Cratylus*, where he quotes the latter's statement about universal flux.[74]

The Euripian flux, understood as a physical phenomenon, also bothered Aristotle the Stagirite, who was from the Chalcidice Peninsula—one leg of the peninsular "island" Mount Athos that Herodotus's Xerxes transformed into an island. The first Greek settlers at Stagira had come from Chalcis, the native land of Aristotle's mother, Phaestis, on the island of Euboea near the Euripus Strait. The flow of water around his mother's island apparently bothered Aristotle, enough so that the problem partly informs works as varied as *Meteorologica*[75] and *Nicomachean Ethics*[76]—which are about the physical (geological) and the moral (political) universe.

The Euripus bothered Aristotle so much that in 322 BC he killed himself over it. One story[77] has it that, when the Athenians charged him with impiety, Aristotle fled to his mother's estate at the Euripus. There the flow of water through the strait at the island of Euboea again stirred him up. Thomas Browne well describes (1646) how it all ended:

That Aristotle drowned himself in Euripus, as despairing to resolve the cause of its reciprocation, or ebb and flow seven times a day, with this determination, *Si quidem ego non capio te, tu capies me*, was the assertion of Procopius, Nazianzen, Justin Martyr, and is generally believed amongst us.[78]

The Latin text translates as "Since I can't take you [Euripus], you take me."

Biographers from the Hellenistic period onward agree that Aristotle died in this way.[79] Strabo the Geographer writes of the Euripus in such a way as to make the story palpable:

Concerning the refluent currents of straits, which also involve a discussion that goes deeper into natural science than comports with the purpose of the present work, it is sufficient to say that neither does one principle account for the straits' having currents, the principle by which they are classified as straits,[80] nor, if one principle should account for the currents, would the cause be what Eratosthenes alleges it to be, namely, that the two seas on the sides of a strait have different levels.[81]

In the next part of *Islandology*, we will see that the problem of tidal ups and downs at such places as the Euripus Strait was more or less resolved by the Danish scientist Tycho Brahe while he was on the island of Hven in the Øresund,[82] working on the position of Earth in the heavens. His assistant, Johannes Kepler, who really disliked islands, wrote the following in his *Astronomia Nova* (1609):

If the earth ceased to attract the waters of the sea, the seas would rise and flow into the moon.[83]

Such work helped set the stage for Shakespeare's *Hamlet*, the setting of which is located just a few kilometers from Hven, and which provides a new understanding of islandology.

Part Three

Hamlet's Globe

Flieget den hellen   Fly to the brightness,
Inseln entgegen,   Towards the islands,
Die sich auf Wellen   Out of the waves
Gauklend bewegen.   Magically raised.

     —Goethe, *Faust Part One* (1808)

Weiß ich nun, wo ich bin!   Now I know where I stand!
Mitten der Insel drin,   Amidst this semi-island,
Mitten in Pelops' Land,   Amidst Pelops' country, Earth—
Erde—wie seeverwandt.   Kindred to the sea.

     —Goethe, *Faust Part Two* (1832)*

*Trans. A. S. Kline, lines 9823–9826.

COLOR PLATE 1 Map of the World as Archipelago. From Muhammad al-Idrisi of Ceuta, *Nuzhat al-mushtaq fi'khtiraq al-afaq* (Tabula Rogeriana; 1325). Commissioned by King Roger II of Sicily, 1154. ("South" is at the top.) Source: Bibliothèque Nationale de France, Paris, MS Arabe 2221.

COLOR PLATE 2 Floating Village near Nasiriyah, Iraq. "The whole country is under water, the villages, which are mainly not sedentary, but nomadic, are built on floating piles of reed mats, anchored to palm trees, and locomotion is entirely by boat." Gertrude Bell, 1916; Gertrude Bell Archive, Newcastle University. Photograph by Nik Wheeler, 1974. Reproduced by permission.

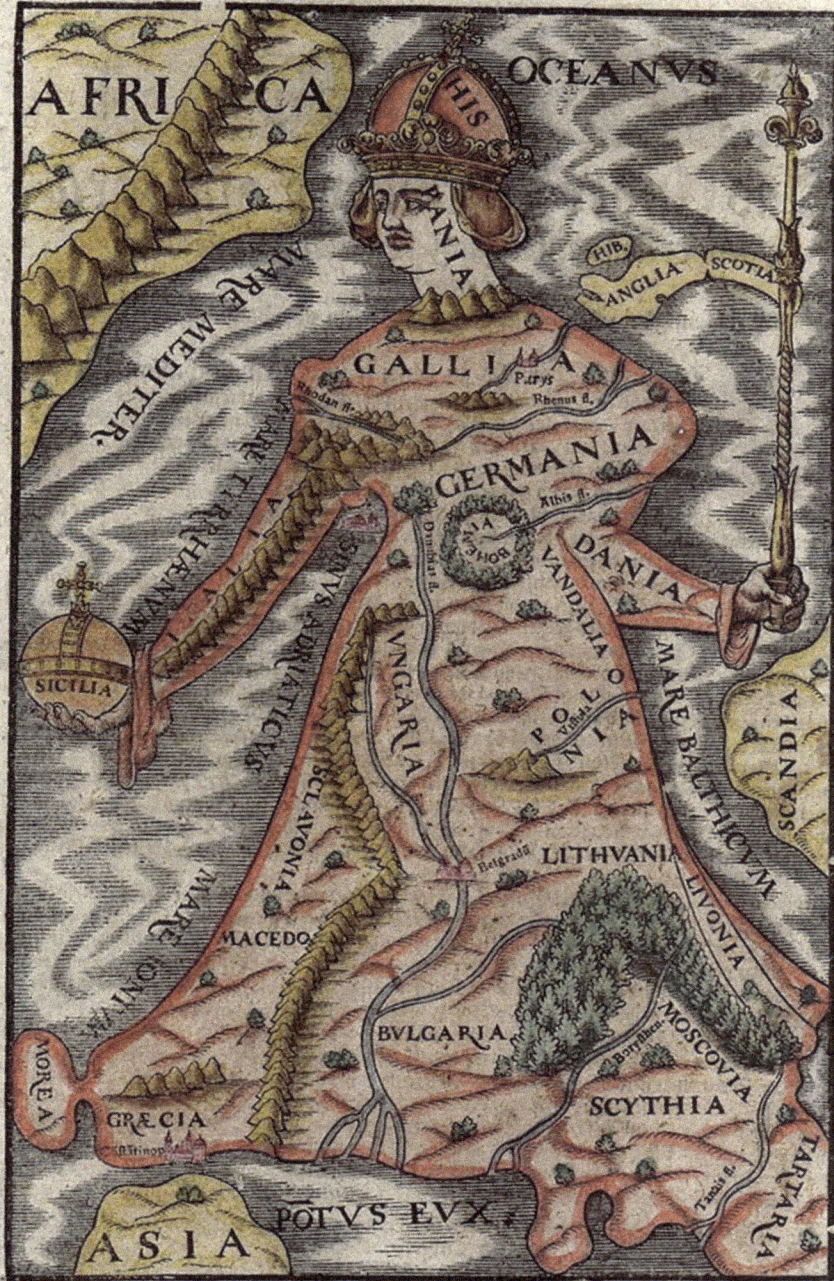

AFRICA

OCEANVS

HIS.

DANIA

HIB. ANGLIA SCOTIA

MARE MEDITER.

GALLIA

Parys
Rhodan fl. Rhenus fl.

GERMANIA

Albis fl.
BOH..IA
Danuu fl.

MARE TYRRHENVM

MARE ADRIATICVS

SALIS

VANDALIA

DANIA

VNGARIA

SCLAVONIA

POLONIA

MARE BALTHICVM

SICILIA

MARE IONIO

MACEDO.

Belgrad.

LITHVANIA

LIVONIA

SCANDIA

MOREA

BVLGARIA

Boristhen..

MOSCOVIA

GRÆCIA
C.Stinop.

SCYTHIA

Tanais fl.

TARTARIA

ASIA   POTVS EVX.

zweyen General Tafeln/vnd in der newen Tafel die allein Europam begreifft. Doch wann man
ansehen will vnd darzu rechnen die grossen Landschafften die gegen Mitnacht gehn/solt wol die
Europa wie breite Europe vbertreffen die länge. Wie aber Ptolemæus Europam beschrieben hat/ist sein länge
fruchtbar es größer dann die breite. Das ist ein mal gewiß/daß Europa ist ein trefflich fruchtbar vnd wol erba-
seye.   wen

COLOR PLATE 3 *Europa Regina*. The map shows Europe as a personified island, or *presqu'île*, in the form of its sovereign royal person. Included in Sebastian Münster's best-selling *Cosmographia* (1544) beginning in the 1570s. The cartographic type of "Europa Regina," first drawn by Putsch (Johannes Bucius Aenicola) in 1537, pretends to unify Europe as an "island-continent" whose political borders match its natural shoreline. (Rotate the map 90° counterclockwise to see "north" at the top.) Heinrich Bunting's map *Asia Secunda Pars Terrae in Forma Pegasir*, in *Itinerarium Sacra Scripturae* (Travels According to the Scriptures; 1581), similarly personifies Asia as an island in the form of the horse Pegasus; Bunting also includes a personified map of Europe. Harvard Widener Library, *Cosmographia*, v.1=HW0BA6, p. 55. Reproduced by permission.

COLOR PLATE 4 Crossing the Red Sea. From the Catalonian illuminated manuscript *Sephardic Haggadah* (fourteenth century). Courtesy of the University Librarian and Director, John Rylands Library, University of Manchester.

COLOR PLATE 5 Øresund. The map shows the island of Hven as well as Elsinore and its Kronborg Castle. Georg Braun and Franz Hogenberg, *Civitates Orbis Terrarum* (1588), vol. 4, map 26. Reproduced by permission of Sanderus Antiquariaat.

COLOR PLATE 6 Britain as Ship: The Heneage Jewel or the Armada Jewel. On the front of the jewel is a miniature of Elizabeth I (1558–1603). On the reverse (shown here) a ship holds steady on a stormy sea. The inscription reads *Saevas tranquilla per undas* (tranquil through fierce waves). Painting by Nicholas Hilliard, ca. 1595. © Victoria and Albert Museum, London. Reproduced by permission.

COLOR PLATE 7 *Demosthenes Practicing Oratory*. From Jean Jules Antoine Lecomte du Nouy, *Démosthène s'exerçant à la parole* (1870). Private Collection / © Look and Learn / The Bridgeman Art Library.

The Distracted Globe

It is in Shakespeare that one finds the best and most beautiful expression of that insular feeling.

—Carl Schmitt, *Land and Sea*

THE GLOBE

Thus, Sciences by the Diverse Motions of this Globe of the Braine of Man, are become Opinions, nay Errores, and leave the Imagination in a thousand Labyrinthes.

—William Drummond, *A Cypresse Grove*

Shakespeare wrote about more places on the globe (Earth) than most other writers. His role as "global author," however, ought properly to include the conceptual and dramaturgical implications of the architecture and setting of his theater. Shakespeare's Globe Theatre, with its peninsular (thrust) stage, was constructed on the south bank of the River Thames. It was just upstream from "Thorney Island," or Westminster, where Parliament met. From there Britannia ruled the waves from the time of the Spanish Armada in 1588, when Shakespeare began to write, until around World War I in 1914. The theater's flag depicted Hercules carrying a globe—the planet Earth—and in the name of the theater, many historians see a reference to the modern "atlas" by Abraham Ortelius, *Theatrum Orbis Terrarum* (1570). Hercules it was who relieved the Titan Atlas of his labor of carrying the globe. Shakespeare's reference in *Hamlet* to "Hercules and his load" also brings to mind the fact that the theater's pillars were called "the pillars of Hercules." They held up the "heavens"—the underside of the roof over the thrust stage, which was painted with stars and other heavenly bodies. (See Illustration 33.)

Shakespeare's Globe Theatre opened around 1600, with Hamlet's "distracted globe" as its premier subject matter. On the one hand, *Hamlet* focuses only on the latest period of the prince of Denmark's life and brainish apprehensions, and only on a few historical periods (classical, medieval, and contemporary) and geographic locations (Rome, Denmark, and Britain). At the same time, though, the play

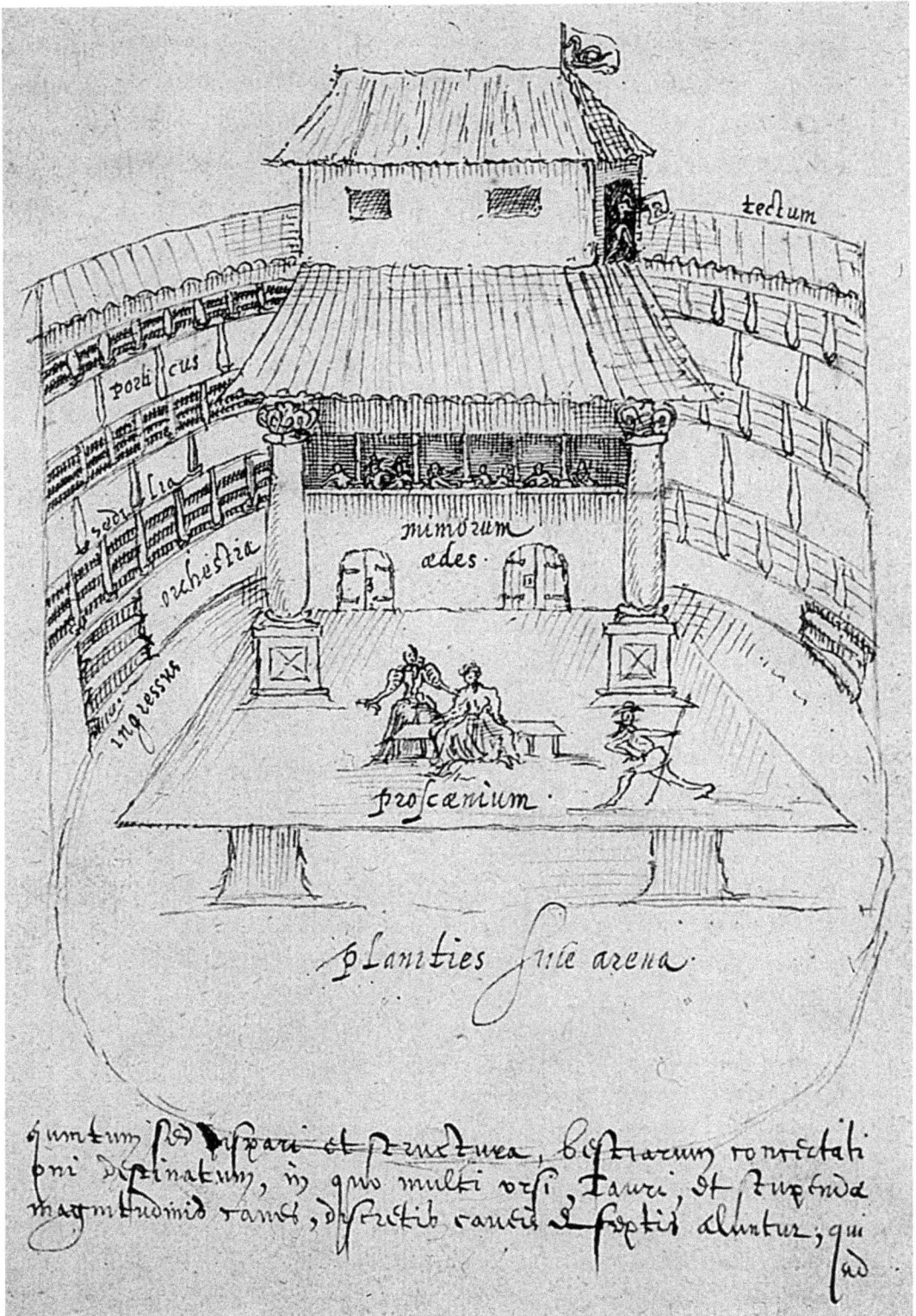

Labels within the sketch: tectum, porticus, sedilia, orchestra, ingressus, mimorum aedes, proscænium, planties sive arena

ILLUSTRATION 33 Swan Theatre. Sketch by Johannes de Witt, 1596, showing the thrust stage and pillars of the roof. The Swan is where Ben Jonson and Thomas Nashe presented their play *The Isle of Dogs* (1597). That performance meant the closing of all theaters in London because of "sedition." In the play, the Isle of Britain as a whole is satirized under the rubric of the Isle of Dogs (originally known as Stepney Marsh), a former island in London bounded on three sides partly by a large bend in the Thames River. Utrecht University Library, Ms. 842, 132r. Reproduced by permission.

has a general or global aspect, exploring, in ways both universal and historical, the linkage between two realms. First is the "physical geography" of the globe (the planet Earth) and hence of the heavens (which delimit, or define, Earth, in islandic terms, as within a shell-like sphere or orbit). Second is the "spiritual topography" of the human brain—the "mind's eye" (according to the play), as defined by its shell-like skull. (See Illustration 34.)

If the Sicilian islander Archimedes, who said,

Give me a place to stand and I will move the Earth[1]

had had, for his engineering purposes, such a spot in the real world as Shakespeare had in his mind's eye for his ideal one, on the south bank of the Thames, the Greek philosopher would have been able to move the Earth (see Illustration 35).

Even so, Archimedes fabricated the first moving hollow brass globe. The Roman poet Claudian writes in "Archimedes' Sphere" (ca. 400):

When Jove looked down and saw the heavens, figured in a sphere of glass he laughed and said to the other gods: "Has the power of mortal effort gone so far? Is my handiwork now mimicked in a fragile globe? An old man of Syracuse has imitated on earth the laws of the heavens, the order of nature, and the ordinances of the gods."[2]

If Shakespeare counts as a "world author," it is partly because he aimed, globally, at this sort of global imitation. "Not marble, nor the gilded monuments / Of princes, shall outlive this powerful rhyme"[3] precisely because, for him, the fulcrum was rhyme and reason.

The adjective *distracted*, which Hamlet uses to describe his own globe (brain), means partly "to have been torn 'asunder.'"[4] The significance is not only neurological (brainish) but geological (planetary). The biblical God of Genesis, after all, *sunders* water from water so that there develop two distinct *districts*, water and earth. The *Coverdale Bible* (1535) reads, "And God sayde: let there be a firmament betwene the waters, and let it deuyde ye waters a sunder."[5] More locally for Shakespeare's setting of *Hamlet* on the Danish island of Zealand, the Norse goddess Gefjun makes a *sundering*, or *sound*, between insular Zealand and peninsular Scania. Her sons pull a huge plough through the earth (*distrahere*) precisely at Elsinore. The *sound*, or *strait*, called Øresund, is literally the *distract*.[6] Danish

ILLUSTRATION 34
Farnese Atlas. This is a Roman copy of a Hellenistic sculpture of Atlas kneeling with a globe weighing on his shoulders, the globe seeming to replace the head. The globe depicts Ptolemaic constellations, including Hercules. Scala / Art Resource, New York.

ILLUSTRATION 35
Archimedes' Lever.
Archimedes (ca. 287–
12 BC) saying, "Give
me a support with a
lever and I will raise
the earth." Stanza
della Mattematica,
1587–1609 (fresco),
Parigi, Giulio
(1571–1635) (attr. to) /
Galleria degli Uffizi,
Florence, Italy /
The Bridgeman Art
Library.

water ↔ land, thus understood in terms of "distractedness,"[7] has hydrological and geological aspects. They are, as we will see, vital to understanding the national historiography and geography of the old Anglo-Scandinavian North Sea Empire.

NORTH SEA EMPIRE
The Place of British Literature

Where is the universal historical home of English national literature? In his presidential address to the Modern Language Association in 1902, the English- and German-language philologist James Wilson Bright reports on two answers to this question: the island of Britain and the continent of Europe.

There is a unity in [the "modern" notion of] English [national] history, whether it is held to begin with the Angles and Saxons on the island [of Britain] (according to Dr. [Thomas] Arnold [father of Matthew Arnold]) or on the continent (according to the English historian [Edward Augustus Freeman]).[8]

Continent here probably means "the mainland of Europe, as distinguished from the British Isles."[9] However, it also means "the main land, as distinguished from islands."[10] Thus one might well ask whether a *European* beginning of English national literature might not yet be insular. Suppose, for example, that the origin and the history of English literature can be linked not only with the European mainland, as readers of *Beowulf* sometimes seem to suggest, but more particularly with a European island, which *Hamlet* sometimes seems to suggest.

Ever since the first publication of *Beowulf* by the Icelandic-Danish scholar Grímur Jónsson Thorkelin (1815),[11] scholars have noted that this originary Old English epic is set elsewhere than on island Britain. Many people believed Thorkelin's claim that the original author was Danish.[12] An Old Norse manuscript of *Beowulf*, he argued, found its way into early medieval England, and there, during the reign of Alfred the Great (871–899), someone translated it into an "Anglo-Saxon dialect of Old Norse." ("That the English Anglo-Saxons spoke a dialect" was part of "the Danish national consciousness" for a century.)[13] English Renaissance scholars knew about the Danish *Beowulf*. Laurence Nowell, tutor to Edward de Vere, probably acquired the manuscript in 1563, a year before Shakespeare's birth, thanks to the offices of William Cecil, a likely model for Polonius in *Hamlet*.

Shakespeare's main source for *Hamlet* was probably Saxo Grammaticus, who set the action of the story in Denmark on mainland Jutland. Shakespeare, however, shifts the action of his play away from the mainland to island Zealand, a setting closely affiliated with the Old English epic. Shakespeare also includes incidents that recall *Beowulf*: for example, the battle on the ice between the Poles and the Danes.[14] Furthermore, the narratological turning point of the plot—its deus ex machina—takes place during his voyage from island Zealand to island Britain, in the sound separating Jutland from Geatland. (Geatland was part of Danish Scania in Shakespeare's time.) The piratic shipboard scuffle in the sound matches the rowing or swimming contest between Beowulf and his friend Breca, in the Kattegat Sund (Sound) off Brännö Island, which is part of the Gothenburg Archipelago.

Eart þú sé Béowulf sé þe wið Brecan wunne' / on sídne saé ymb *sund* flite? [Are you the Beowulf who contested against Breca on the broad sea, contended around the ocean sound?][15]

Né léof né láð beléan mihte sorhfullne síð þa git on *sund* réön [Neither friend nor foe could dissuade from that sorrowful jaunt, when you rowed into the strait (sound)].

Finally, the First Gravedigger in *Hamlet* says that the Danish prince would pass for an Englishman partly because Danes and Englishmen are alike mad. He hints at a millennium of Danish-British history—including immigration, conquest, royal marriage, and rivalrous empires—that helps provide *Hamlet* with its powerful historical position in English literary, political, and geographic historiography.

Zealand and Britain

If one were living in the sixteenth century and wanted to learn about the geographic "distraits"[16] and "straits" of Denmark, one would do well to consult two

Danish scholars. The first would be Elsinore-born, Paris-educated Christiern Pedersen (1480–1554), whose work includes the first complete Danish-language translation of the English-language Bible (Copenhagen, 1550), a fine Latin-language
lexicon (1510), an edition in Latin of the Danish historian Saxo Grammaticus's
Deeds of the Danes (*Gesta Danorum*; 1514), and a Danish-language translation of
this work.[17] The first Danish work published in English-language translation was
Pedersen's Latin-language *Richt Vay to the Kingdome of Heuine* (1533).[18]

The second Dane would be Anders Sørensen Vedel (1542–1616), who completed Pedersen's Danish-language translation of the Latin *Deeds of the Danes* and
oversaw its publication as *Den danske Kronicke* (1575). Vedel was also an active
participant in other important Danish scientific work, having been the tutor to
young Tycho Brahe (1562), who was soon to change the world's understanding
of the heavenly spheres and tidal waters.[19] When Wittenberg graduate Jørgen
Rosenkrantz (1523–1596) supported Vedel's appointment as royal historiographer
(1584),[20] it was hoped he would complete *Deeds of the Danes* so that it would
include the time between Saxo's lifetime and the middle sixteenth century. Vedel
had planned a twenty-four-volume work comprising the history and topography
of Denmark, but he never managed the task. In fact, a national history of Denmark in prose did not appear until the 1630s.[21]

Shakespeare, in *Hamlet* (1599), accomplishes, at the typifying and condensed
level of poetic and dramatic art, the historiographical task that Vedel could not: "It
is I, Hamlet *the* Dane." The play is a shorthand national history of Denmark or,
rather, the British-Scandinavian North Sea Empire, done up in the genre of historical and imperial tragedy. One reason the play is so difficult to contain on stage, as
many directors have noted,[22] is the largeness of the historical scene it conjures up.

The Danish-speaking Saxo's Latin-language *Life of Hamlet*,[23] together with
François de Belleforest's French-language revision of it in *Tragic Histories* (1570),[24]
provided Shakespeare with the main elements of the story. Readers who want to
understand how Shakespeare worked from particular sources to create a universal
artwork of geographic dimension would do well, though, to read more than just
the *Life of Hamlet*. In fact, *Deeds of the Danes* as a whole—from which the *Life of
Hamlet* is an extract—concerns both the Norse national idea of islandness, where
water is identified *with* land, and the Latin notion of water defined *against* land.
The world here is already amphibious—the full and proper home only of creatures
that are able to walk, swim, *and* fly. Such creatures are loons, ducks, cormorants,
and gulls. When Hamlet asks, "Goes it against the main . . . ?" the term *main*
relevantly means both "continent" and "sea."

Saxo is primarily a historian with geographic and topographic dimension. He
seeks to explain not only Danish but also English history and culture in terms of
the relationship between natural place and human politics. For him, the Danish

locus—its water or earth—is the architectonic grounding of all. "I have resolved to begin," Saxo says at the beginning of his *Deeds of the Danes*,

with the position and configuration of our own country [Denmark]; for I shall relate all things as they come more vividly if the course of this history first traverses the places to which the events belong, and take their situation as the starting-point for its narrative.[25]

He then provides a description of the islandological and hydrological parameters of Danish geography. Denmark, he says, is essentially a place of islands. It is all gulf (*sinus*) and strait (*fretum*), a "water"-land (*is*-land):

The extremes . . . of this country are partly bounded by a frontier of another land, and partly enclosed by the waters of the adjacent sea. The interior is washed and encompassed by the ocean; and this, through the circuitous winds of the interstices, now straitens into the narrows of a firth, now advances into ampler bays, forming a number of islands. Hence Denmark is cut in pieces by the intervening waves of ocean, and has but few portions of firm and continuous territory; these being divided by the mass of waters that break them up, in ways varying with the different angle of the bend of the sea.[26]

Islands like Hven and frets (sounds) like the Øresund crisscross Denmark. They "sunder" the landscape and bogs, hindering reliable long-term burial.[27] The same islandness here marks even the Thulean horizon, or "rim of the world."[28]

WHILE SAXO'S Latin-language version of the Hamlet story emphasizes the identity of water and land, Belleforest's French-language revision emphasizes the distinction of the two.[29] Shakespeare mediates between these writers. He reinforces or creates a characteristically bifocal English-language and eventually modern European way of thinking about islands. We will consider this issue in terms of "the sublime" later in this chapter.

For Saxo, Denmark's natural geographic fluctuations parallel human demographic movement across generations. Thus physical and human geography go together: the interrelationship between earth and water parallels the "flow" of kinship between the Danes and the English. The descendants of the brothers Dan (father of the Danes) and Angul (father of the English) are, says Saxo, the same. The first "book" of *Deeds of the Danes* opens:

Now Dan and Angul, with whom the stock of the Danes begins, were begotten of Humble, their father, and were the governors and not only the founders of our race. . . . Of these two, Angul, the fountain . . . of the beginnings of the Anglian race, caused his name to be applied to the district which he ruled. This was an easy kind of memorial wherewith to immortalise his fame: for his successors a little later, when they gained possession of Britain, changed the original name of the island for a fresh title, that of their own land [*Eng-land*].

From Dan, however . . . the pedigrees of [Danish] kings have flowed in glorious series, like channels from some parent spring [*splendido successionis ordine profluxerunt*].[30]

The tribal name *Angles*, Saxo says, derives from the adjective *eng*, which means "narrow," or "strait." Therefore, *Englanders* are "the Danish people who live at the Strait."[31] The Venerable Bede says the same: the English are Danish.

That the English and Danish are one and the same *Volk* is both the jest of the First Gravedigger in *Hamlet* and the gist of *Hamlet*. Consider the exchange between Hamlet and the First Gravedigger. The Danish prince, apparently unrecognized, asks the First Gravedigger why Hamlet was "sent into England"; the First Gravedigger explains that Hamlet "was mad" so that, if he does not recover his wits in England, at least it does not matter much, since "there [all] the men are as mad as he" (act 5, scene 1).

Historically speaking, there is something to the notion that the English are partly Danish. The large waves of Danish conquest and immigration into the British Isles began around 800. Major marriage alliances between the British and the Danes included those of Edith, daughter of King Athelstan of England, with Sihtric, the Danish king of York, around 926. By 1000, the "Danelaw area" in England included even London and Westminster. Danish rule in the British Isles reached its apogee with King Canute the Great, grandson of Mieszko (the first Polish king) and son of Sweyn Forkbeard (king of Denmark and England). Canute, who ruled from London, was in 1030 the most imposing ruler in Latin Christendom (next to the Holy Roman emperor himself). One historian writes that he "held in his hands the destinies of two great regions: the British Isles and the Scandinavian peninsulas. His fleet all but controlled two important seas, the North and the Baltic. He had built an Empire."[32] Canute it was who levied the "Danegeld," referenced in *Hamlet*.

A STORY ABOUT KING CANUTE, reported by the twelfth-century chronicler Henry of Huntingdon, is relevant here. Canute was once at the *eyot* (island) of Thorney[33] (other reporters say he was at Hayling Island, near Bosham, and still others say he was at the Isle of Wight, near Southampton), where he set up his royal throne at the edge of the mudflats at low tide. He then commanded the tide not to rise:

Imperio igitur tibi ne in terram meam ascendas.
[I command you therefore not to rise onto my land.][34]

The tide did not cease. Canute then announced to all those present:

Let all men know how empty and worthless is the power of kings, for there is none worthy of the name, but He whom heaven, earth, and sea obey by eternal laws.[35]

Shakespeare recalls the moment in his urban island play *The Merchant of Venice*: "You may as well go stand upon the beach / And bid the main flood bate his usual height" (act 4, scene 1).

As king of Britain, the Danish Canute set up his "palace" on the *eyot* of Thorney,[36] which lay at the confluence of the Tyburn and Thames rivers. Since that time, Thorney Island has become part of the British mainland. Today it is known as "Westminster"—the seat and metonym of British-style monarchal parliamentary democracy in the world. The island origin is all but forgotten. "It is strange," write Cyril Ionides and John Black Atkins in *A Floating Home* (1918), "that Londoners should know so little, below bridges, of the river that made them."[37]

Some Christian saints believed that they could do what Canute the Great could not and what Xerxes—the would-be Persian *isotheos* (equal to a god) who tried to bridge the Hellespont and turn Mount Athos into an island—believed he could: manipulate the tides. In the early sixth century, one of these, Saint Illtud, residing on tiny Caldey Island in the Bristol Channel, drew a line at low tide and commanded the tidewaters never again to cross it. The waters heeded Illtud's prayer, presumably because God intervened. The result was a coup for the Church. Tiny Caldey Island was now united with the mainland, and the Church got a huge tract of real estate.[38]

Scandinavian Unions

Is thy union here?

—Shakespeare, *Hamlet*

The plot of *Hamlet*, written by an Englishman, seems almost to cover the bases of Danish Canute's North Sea Empire. Old Denmark (Hamlet's father) kills his royal brother Old Norway (Fortinbras's father) and takes his land. Old Denmark's consanguineous brother (Claudius) then kills Old Denmark, taking the queen (Gertrude) and the island (Zealand) with which she is conjunctive. Then Young Norway schemes to take back Zealand (as Claudius fears), which he may have the right to do. ("I have some rights of memory in this kingdom.") It seems that the dispute between the rivalrous brother Scandinavians, Denmark and Norway, will be endless unless there can be some sort of "union" between factions.

In history, at least, there was such a union: the Kalmar Union (1397; see Illustration 36), which the royal houses of Denmark and Norway signed at the Kalmar Strait (then part of Danish Scania), opposite Öland Island.[39]

How this Scandinavian "union" happened matches certain aspects of *Hamlet*'s plot as Shakespeare adapts it from Saxo Grammaticus. In the first instance, the groundwork for the union involved the political maneuvers of the extremely powerful and influential Queen Margaret.

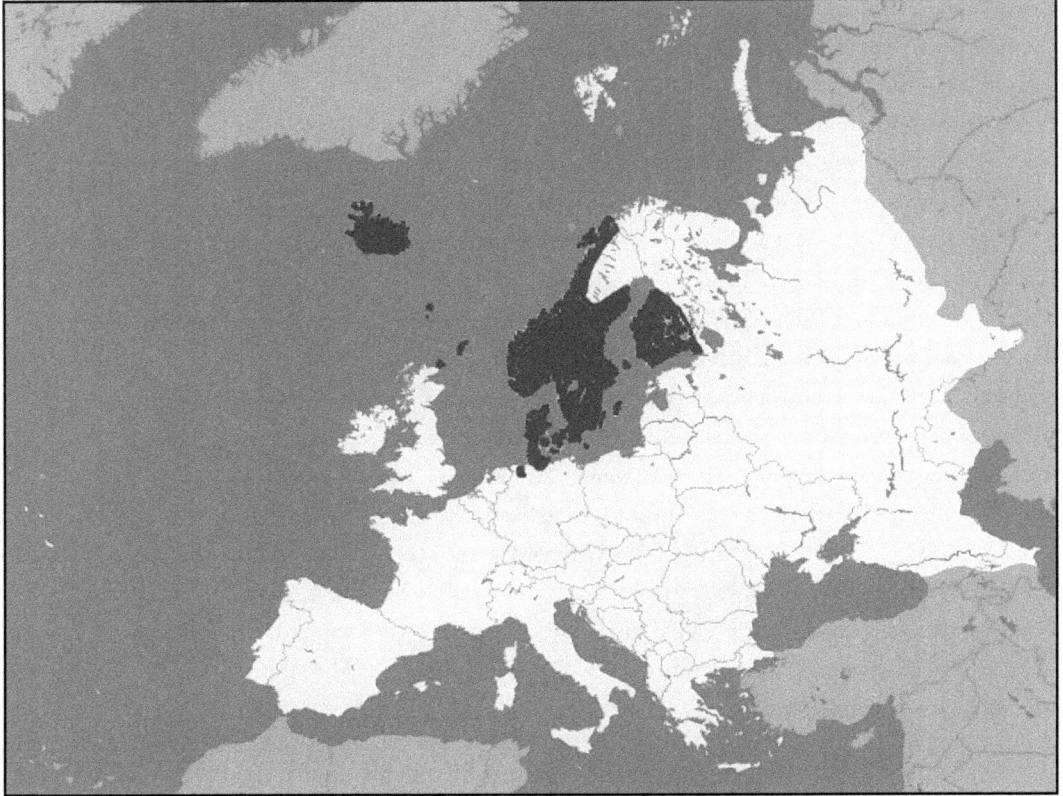

ILLUSTRATION 36
Kalmar Union.
The map shows
the extent of the
union established
by the Treaty of
Kalmar (signed in
1397) in the years
1460–1464, during
the reign of King
Christian I, whose
House of Oldenburg
ruled Denmark
from the time of
his accession until
1814. © Howard M.
Wiseman, Griffith
University,
Queensland,
Australia. Reproduced
by permission.

A Polish-Pomeranian chronicle writer begins his version of the story with a description of the marriage in 1363 of the last independent Norwegian king, Haakon VI, to Margaret, the daughter of the Danish king Valdemar IV.[40] Their son, Olaf, was ten years old when his father died in 1379. Olaf's widowed mother, Margaret, arranged for Olaf, who was now the legitimate ruler of both Norway and Denmark, to be elected to the Swedish throne, but at age seventeen he suddenly died. A few conspiracy theorists believed that the ambitious Margaret had poisoned Olaf in order that she might become "All-powerful Lady and Mistress (Regent) of the Kingdom of Denmark." (By comparison, Shakespeare's Gertrude more or less knowingly offers poison to Hamlet.) In any case, Margaret produced a body and had it buried on Zealand. (This is where Shakespeare's *Hamlet* takes place but not the real story, according to its main source, Saxo Grammaticus.)

What happened next sounds like a trickster's ghost tale. Olaf, or someone calling himself "Olaf," showed up in Poland[41] (where Old Hamlet fought on the ice). How could that be? Was not Olaf dead? One faction said that the body that Margaret buried was not really Olaf's. The real king had escaped and taken refuge in Poland. Another faction said that he had actually died but now was arisen, spirit and body, from the dead to take back what was rightfully his. (By comparison, Hamlet, thought dead, returns to Denmark.)[42] According to the story, Olaf returned to Denmark with letters patent and a seal ring, expecting or hoping to become king. (Just so, Shakespeare's Hamlet returns to Denmark and becomes a kind of very short term king.) Margaret had him killed—or killed again.[43]

Margaret of Denmark now became what Shakespeare's usurping King Claudius calls Gertrude: an "imperial jointress to this warlike state." Saxo's Danish queen Geruthe had been a jointress in the fourfold sense that she was the daughter, sister, mother, and wife of kings.[44] Margaret, however, took a more apparently active role in structuring the kingdoms. With Olaf out of the way, she adopted as her son the Danish-born Eric of Pomerania[45] and then appointed him king of Norway in 1389 and king of Denmark and Sweden in 1396. Next, she arranged for the Kalmar Union (1397) of Norway, Denmark, and Sweden (including Iceland, Greenland, and the Faroe Islands) as well as the Shetland Islands and the Orkney Islands.

Of particular interest to Danish-English history is what followed next. Margaret tried to unite two unions—the Kalmar Union with the united British kingdoms—and, in this fashion, to re-create Canute's "Empire of the North." In 1402, she contacted King Henry IV of England about a *double* sibling marriage that would cement this union of unions. The first marriage would be that of her son by adoption, King Eric of Denmark, with Henry's daughter Philippa. The second would be that of Henry's son, the Prince of Wales (also known as "Prince Hal" in Shakespeare's *Henriad*), with Margaret's daughter by adoption, King Eric's sister Catherine. The former marriage alliance took place. The latter did not.[46]

The history of Margaret's regency tells of events that took place long before the period when Shakespeare was writing his Scandinavian play. But those old times were again on people's minds because of a current dispute about island sovereignty.

In 1471, Scotland had annexed the Orkneys and Shetlands from the Danes. Now, in 1589, when Shakespeare was turning twenty-four, King James VI of Scotland (later King James I of England) *twice* married a Danish princess, Anne (the second time in Elsinore, where Shakespeare sets his story).[47] Princess Anne was the daughter of the Danish king Frederick II, who, like Shakespeare's Claudius, was well known for drunken escapades and adulterous affairs. In England, Frederick was feared for ceaselessly building up the Danish navy, making "the night joint-labourer with the day" (*Hamlet*). The Britons well knew that, although they no longer had to fear the Spanish Armada, defeated in 1588, they still had to contend with Danish naval expansion. The Danish king Christian IV, who had inherited the throne in 1588 from his father, Frederick, was already sending ships to Greenland to find and reestablish the old Danish Eastern Settlement, Hudson

Bay in Canada, and Tranquebar in India—thus challenging British sea supremacy every which way.

The Danish-English marriage between James and Anne reestablished old alliances. At the same time, it reinvigorated old enmities. Who now would get the Orkney Islands? The Shetlands? These archipelagoes had been part of the Kalmar Union of Scandinavia, but in the fifteenth century, the Danish monarch granted them to Scotland as part of a *provisional* dowry for the marriage of the Danish princess Margaret (James VI's great-great-grandmother) to the Scottish king James III. When James married Anne of Denmark in 1589, the Danes claimed that the Orkneys should revert to them; the quarrel potentially pitted the Kalmar Union (Norway and Denmark) against the united kingdoms (England, Scotland, and Ireland).[48]

Few writers of Shakespeare's period would not know these things. Anne of Denmark—queen consort of Scotland, England, and Ireland—was an active patron of the arts.[49] On more than one occasion, she attended plays written by the author of *The Tragicall Historie of Hamlet, Prince of Denmarke.*[50]

HOLMGANG

> The question at issue was decided with sword and battle-axe by a
> holmgang.
>
> —Paul Henri Mallet, *Ancient Scandinavians*

Holm is an English-language word meaning "island."[51] The term is originally from the Norse languages. *Holmgang,* too, is an English-language word, also originally Norse, generally connoting "a duel to the death"[52] and denoting "a duel that takes place specifically on an island [*holm*]"[53] and/or "a duel whose prize is an island." (An ancient Greek counterpart would be *nēsomachia.*) "'Going to the *holm*' (or islet) on which a duel was to be fought"[54] was an old tradition of the North Sea Region, including Britain as well as Scandinavia. John Selden—whose celebrated *Mare Clausum* (1618) we will consider later in this book—wrote about such *holmganga* in his *Duello, or Single Combat* (1610).[55] For him, the term *holmgang* implies a legal or ritual hand-to-hand combat or duomachy that takes place on an island—"two champions in the field, for decision of some controversy."[56]

The practice of *holmgang* informs many other cultures as well. For example, there is the celebrated Japanese *holmgang* of 1612, which took place on Ganryu Island in the Kanmon Straits (see Illustration 37).[57] The straits run between the large islands Honshu and Kyushu, thus conjoining the Sea of Japan with the Inland Sea.[58] Nowhere, however, is the *holmgang* more important than in the Norse regions. Norse and English sagas are replete with these island battles, a famous one of which involves Northumbrian Britain and Norway (Egil and Eric Bloodaxe)

and takes place on Bergonund's Island.[59] The Icelandic *Saga of Egil* reports that Egil, while in Angle-land (Eng-land), struggles on an island with Eric Bloodaxe, who is king of both Norway and Northumbrian England.

Saxo Grammaticus's *Deeds of the Danes* recounts dozens of *holmganga*. Among these Danish-English island duels are the following:

Sciold v[ersus] Attila; Sciold versus Scate, for the hand of Alfhild; Gram versus Swarin and eight more, for the crown of the Swedes; Hadding versus Toste, by challenge; Frode versus Hunding, on challenge; Frode versus Hacon, on challenge; Helge versus Hunding, by challenge at Stad; Agnar versus Bearce, by challenge; Wizard versus Danish champions, for truage [tribute] of the Slavs; Wizard versus Ubbe, for truage of the Slavs; Athisl versus Frowine, meeting in battle; Athisl versus Ket and Wig, on challenge; Uffe versus Prince of Saxony and Champion, by challenge; Frode versus Froger, on challenge; Eric versus Grep's brethren, on challenge, twelve a side; Eric versus Alrec, by challenge; Hedin versus Hogni, the mythic everlasting battle; Arngrim versus Scalc, by challenge; Arngrim versus Egtheow, for truage of Permland; Arrow-Odd and Hialmar versus twelve sons of Arngrim Samsey, fight; Ane Bow-swayer versus Beorn, by challenge; Starkad versus Wisin, by challenge; Starkad versus Tanlie, by challenge; Starkad versus Wasce-Wilzce, by challenge; Starkad versus Hame, by challenge; Starkad versus Angantheow and eight of his brethren, on challenge; Halfdan versus Hardbone and six champions, on challenge; Halfdan versus Egtheow, by challenge; Halfdan versus Grim, on challenge; Halfdan versus Ebbe, on challenge, by moonlight; Halfdan versus Twelve champions, on challenge; Halfdan versus Hildeger, on challenge; Ole versus Skate and Hiale, on challenge; Homod and Thole versus Beorn and Thore, by challenge; and Ragnar and three sons versus Starcad of Sweden and seven sons, on challenge.[60]

A *holmgang* gone awry, challenged by the son of one of the participants, is what motivates the plot of *Hamlet* from the get-go. The story line, summarily defined in Hegelian terms as a tragic *Aufhebung* (sublation), thus begins with a fight (*gang*) between the leading members of rival yet familiar ("brother[ly]") "gangs" of sea-roving pirates that takes place on an island (*holm*), or for an island, or both on and for an island. (The island might be no larger than a "patch of ground," or it might be no smaller than Zealand itself, where Shakespeare relocates his version of the story, or than Britain, where a good part of Saxo's version takes place.) In *Hamlet*, the island struggle—a *holmgang* in the British and Danish sense—repeats itself, with variations, within and between generations and nations, throughout the play.

The piratic tribal structure of the Scandinavians is significant here, as Belleforest depicts it in *Tragic Histories* and as Saxo analyzes it in *Deeds of the Danes*. Belleforest writes:

Now the greatest honour that men of noble birth could at that time win and obtain was in the art of piracy upon the seas . . . and just how much more they used to rob, pill, and spoil other provinces and islands far adjacent, so much more their honours and recognition increased.[61]

Saxo makes the piratic Horwendil (Old Hamlet) a consanguineous brother of Feng (Claudius). Horwendil leaves peninsular Jutland (where Saxo sets the story) to go roving about the high seas: "Horwendil held the tyranny [in Jutland] for three years, and then, to will the height of glory, devoted himself to piracy."[62] Then follows the famous man-on-man combat:

Horwendil soon meets up with his rival Koller [Old Norway]: Then Koller, King of Norway, in rivalry of [Horwendil's] great deeds and renown, deemed it would be a handsome deed if by his greater "strength in arms" he could bedim the far-famed glory of the pirate [Horwendil].[63]

Horatio describes this *holmgang* battle at the beginning of *Hamlet*. Shakespeare makes the scene of contest an island, Zealand. Saxo makes the combat a *holmgang*:

Insula erat medio sita	There was an island lying in the middle
pelago, quam *piratae*	of the sea, which each of the pirates
collatis utrimquesecus	[Koller and Horwendil], bringing his
navigiis obtinebant.	ships up on either side, was holding.[64]

According to the story, Old Hamlet wins the battle, and Denmark gets the territory, whose extent and type are unspecified. Things are settled.

Or so one hopes. *Holmganga* are supposed to solve problems. After all, a magician had once thus challenged King Rorick, Gertrude's father and Hamlet's maternal grandfather:

Suffer a private combat to forestall a public slaughter, so that the danger of many may be bought off at the cost of a few. . . . If I conquer, let freedom be granted us from taxes; if I am conquered, let the tribute be paid you as of old: For to-day I will either free my country from the yoke of slavery by my victory or bind her under it by my defeat. Accept me as the surety and the pledge for either issue.[65]

The duel should have been the end of it, but Shakespeare is concerned with what happens when the participants in a *holmgang* or their descendants break the *holmgang* contract.

In Shakespeare's *Hamlet*, the originary island duel never disappears from the background action. Young Fortinbras wants (back) the territory that Old Norway lost to Old Hamlet.[66] The same trope informs and repeats in other struggles in the play: Old Hamlet versus (his brother) Claudius; Young Hamlet versus (his uncle-father) Claudius; Young Fortinbras versus (his uncle) Old Norway; Young Hamlet versus (his [collactaneous] "brother") Laertes.[67]

When it comes to the matter of the particular islands fought for in *Hamlet*, there is also the "little patch of ground" where and for which Fortinbras is apparently willing to go to war. The word *patch* is essentially islandic. It refers to "part of a surface of recognizably different appearance or character from the rest"—a fair description of an island understood as "land defined against the water that surrounds it."

PLOTS AND PATCHES
Continent Enough

. . . twenty thousand men
That . . .
. . . fight for a plot
Whereon the numbers cannot try the cause,
Which is not tomb enough and continent
To hide the slain? . . .

—Shakespeare, *Hamlet*

There is the broadly panoramic moment in *Hamlet* when the Norwegian captain mentions to Hamlet the "little patch" or plot of earth for which his soldiers will fight—a patch that Hamlet later calls "a *plot* / Whereon the numbers cannot try the cause" (act 4, scene 4). The moment bears comparison with two others in the play about "plots" and brings to mind an overlooked meaning of the English word *ham*. The first moment is Hamlet's meditation on the palling of his plotting, or scheming, to discover and remove Claudius: "Our indiscretion sometimes serves us well, / When our deep *plots* do pall" (act 5, scene 2). The second moment is Horatio's request to Fortinbras that Horatio be allowed to speak after Hamlet's death, "lest more mischance / On *plots* and errors, happen" (act 5, scene 2). The overlooked meaning of *ham*, whose diminutive is *hamlet*, means "little 'plot of land hemmed in by water'":[68] a "little patch." The hamlet in *Hamlet* is the captain's "little patch." How can we bury the soldier?

If the "plot" for which Fortinbras is fighting is large or continent enough and if it is not made of the sort of material that intermittently melts or thaws, then it can provide a sufficient number of burial plots for his soldiers. Otherwise, those soldiers . . .

> Go to their graves like beds, fight for a *plot*
> Whereon the numbers cannot try the cause,
> Which is not tomb enough and continent
> To hide the slain? . . .
>
>   (act 4, scene 4)

By analogy, the plot might also contain, once and for all, the body of Hamlet old or new.

The "patch" Fortinbras seeks is an island of sorts. Does it matter where? Is it part of "Denmark"? Fortinbras informs Claudius that he wants merely "quiet pass / Through your dominions" (act 2, scene 2). That means that he seeks either overland passage (across Zealand) or sea passage past Elsinore at the *els* (neck) of the Øresund, which runs between Scania and Zealand past the guard station where *Hamlet* opens.

Is it Zealand, the main Danish island? Claudius fears at first that Fortinbras intends to take Zealand, where Elsinore is located. Hamlet figures the same. Fortinbras's uncle reports to the Danish court that seizing Denmark had been Fortinbras's original intention. Finally, Fortinbras, who believes he has "some rights of memory in this kingdom," shows up after Hamlet's death to receive the Danish kingdom.

And yet Claudius is "a king of . . . patches." Which *holm* (island), other than Zealand, might Fortinbras be seeking out? Here is a list of potential candidates, some relevant to the interpretations of *Hamlet* we take up later in Part 4 and all relevant to the place where presumably the *holmgang* between Old Hamlet and Old Norway may be "thought of" as having taken place.

Bornholm: Bornholm Island was medieval northern Europe's largest frontier fort. Lübeck returned it to Danish ownership when Shakespeare was young (1575).[69] (See Illustration 38.) Bornholm was also the original home of the Burgundians, whose lineages became the *Nibelungen*, whose mythic story, as we will see, Richard Wagner takes up.

Wiek and Oesel: These are two Danish island fortresses in the northern part of the Baltic.[70] They guard the entrance to the Irbe Strait and hence the Gulf of Riga. Denmark conquered Oesel (or Saaremaa) in 1219, then lost it in 1346.[71] In 1572, when Shakespeare was eight years old, Oesel again became Danish.[72]

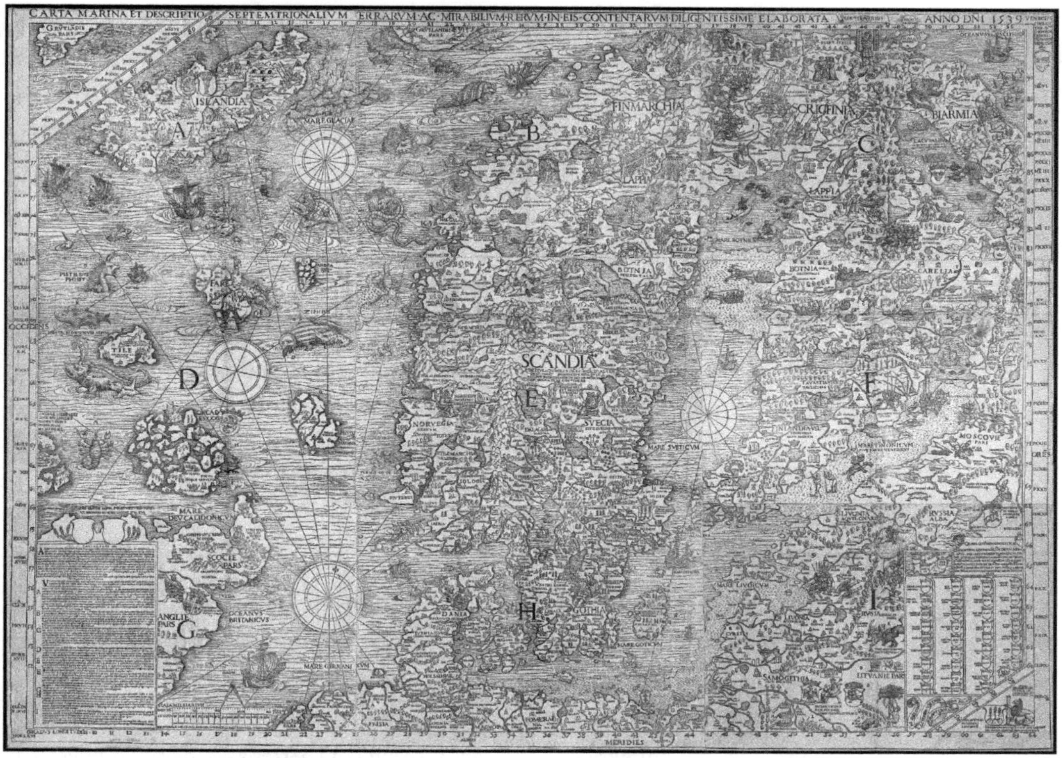

ILLUSTRATION 38
Map of the North
Sea Region. The map
shows Bornholm
Island as almost
contiguous with
Polish Pomerania.
From Olaus Magnus,
Carta Marina (1539).
Reproduced by
permission of Uppsala
University Library.

Pomeranian Islands: These are the island offshore fortresses of German and Polish Pomerania, Usedom and Wolin,[73] as well as Jomsborg and Vineta. Some scholars call these latter fictive;[74] others claim they are real[75] but submerged, like the lost island of Atlantis.[76] This is the island region where nineteenth-century German thinkers rediscovered Germany's national purpose even as they decoded the German national character in the personage of Hamlet.

Graves and Mounds

> Whilest I remained in this Island I caused one of those round hills
> (which in the Plains of Wiltshire are by the Inhabitants termed
> Barrowes, like as in the Midland parts of England they call them Lowes,
> commonly and truly held to be the Sepulchres of the Danes or
> Norwegians and others of that Northern tract invading and possessing
> Brittain) to be opened. . . .
> —James Chaloner, *A Short Treatise of the Isle of Man*

The rhetoric surrounding non-Christian English burial practices pervades *Hamlet*. There is, for example, talk of throwing shards into the grave together with the body. "I thought thy bride-bed to have deck'd, sweet maid," says Gertrude to the body or spirit of Ophelia, "[a]nd not have strew'd thy grave" (act 5, scene 1). This practice is ancient in Britain. In his essay "The Ancient Britons and the Lake Dwelling at Ulrome in Holderness" (1884), T. M. Evans writes that the "custom of throwing shards, and finds, and pebbles into the grave is common both to Romano-British and to Anglo-Saxon interments in England. That it was pagan, and even of very early origin, seems probable, and that it persisted into Christian periods is pretty certain."[77] Evans focuses on a part of Britain (Holderness in Yorkshire) that had been marshland until it was drained in the Middle Ages and whose Danish past is suggested even by the name *Holderness*.[78] He speculates that the "pagan origin" of this practice "had somehow or other so strongly impressed itself upon the public mind that it was no longer practiced in Christian burials."[79] More difficult are the debates about how to bury a suicide like Ophelia. For "Christians," a suicide should be buried only in unsanctified soil.

Relevant here, though, is the manner of Ophelia's death. Here was a "muddy death," with parts of water and earth. This too affects the question of the substance and state of the material in which she will be buried. Dead souls want to be kept down. Put another way, the living want to keep them down. But dead souls sometimes rise up. (The First Gravedigger asks, "What is he that builds stronger than . . . the . . . shipwright . . . ?" [act 5, scene 1].) Laertes says, demandingly, "Lay her i' the earth" (act 5, scene 1), but of what sort?

Should Ophelia be buried in soggy "flat" land subject to flooding and the sort of decomposition of which the Gravediggers speak? The word *flat*, of Norse origin, indicates "the low ground through which a river flows,"[80] "a nearly level tract, over which the tide flows, or which is covered by shallow water,"[81] or, as in *The Tempest*, "a tract of low-lying marshy land; a swamp."[82] The First Gravedigger reminds us of what is at stake here. When Hamlet asks, "How long will a man lie i' the earth ere he rot?" the First Gravedigger responds, "Your water / is a sore decayer of your whoreson dead body" (act 5, scene 1).

One can understand why a brother like Laertes or a lover like Hamlet would want for Ophelia a burial mound—higher ground, a sort of island—to raise her memorial above the flat. From *inside* the grave, therefore, Laertes now demands a mountain to top the skies:

> Now pile your dust upon the quick and dead,
> Till of this flat a mountain you have made,
> To overtop old Pelion, or the skyish head
> Of blue Olympus.
>
> (act 5, scene 1)

Hamlet, referring to the Aloadae brothers' attempt to storm Mount Olympus, wants even more height:

> If thou prate of mountains, let them throw
> Millions of acres on us, till our ground,
> Singeing his pate against the burning zone,
> Make Ossa like a wart!

(act 5, scene 1)

Both speakers recall a common pagan practice typical of the Danes: the construction of a low (Danish *langdøs*), a sort of barrow, burial cairn, or tumulus. Traces of pagan barrows were relatively common in the Renaissance. (They were systematically pillaged and destroyed in the nineteenth century.) The Bartlow Burial Mounds (Danish *Hubbaslow*) in Britain and the Tinghøjen in Denmark are examples (see Illustrations 39 and 40).

ILLUSTRATION 39
Group of the Principal Forms of Barrows (engraving), English School (19th century) / Private Collection / © Look and Learn / The Bridgeman Art Library.

GROUP OF THE PRINCIPAL FORMS OF BARROWS.
a, Long Barrow, *b,c.* Druid Barrows. *d.* Bell-shaped Barrow, *e.* Conical Barrow. *f.* Twin Barrow.

Is Ophelia buried in a "special" place in or near the churchyard? That would likely be a raised moat or barrow. Writes Walter MacFarlane, "Hard by this Church . . . stands a remarkable artificial Mote or little hill rising up like a Piramide."[83] What sort of burial place that *Hamlet's* dramatic plot prepares for Hamlet—by sea or by land, flat or mound, temporary or permanent—is the chief concern of the tragedy.

A TELLING HISTORICAL CHANGE in the English-language meaning of the words *moat* and *dike* is significant insofar as it mirrors the change in meaning of the word *island*. The English *moat* and *dike* once meant only "raised mound of land,"[84] as they did in many other European languages.[85] Uniquely in English, *moat* and *dike* came to mean not (or not only) "the mound created when one digs up the earth" but "the ditch created when one digs up the earth."[86] The specifically English-language islandological transference of meaning here affects how we read one of the most influential statements about what England "is":

> This . . . scepter'd isle,
>
> This earth of majesty, this seat of Mars,
>
> This other Eden, demi-paradise,
>
> This fortress built by Nature . . .
>
> . . . this little world,
>
> This precious stone set in the silver sea,
>
> Which serves it in the office of a wall,
>
> Or as a *moat* defensive to a house,
>
> Against the envy of less happier lands,
>
> This blessed plot, this earth . . . this England, . . .[87]

Moat in this passage indicates not only the English Channel, which is how it is usually read, but also the White Cliffs of Dover. It refers at once to the "land" (as circumscribed by water) and to the "water" (which does the circumscribing). England is the moat created by the natural "digging" of the moat that is the English Channel. Compare how, for the Danes, Zealand was the place created by the "digging" or *udgravning* (Danish) of the Øresund. The English-language verb *to grave* also means "to dig"[88] or "to form by digging."[89] The Coverdale Bible (1535) thus reads, "Stronge diches are grauen on euery syde off it."[90]

On the island of Britain, "big digs"—dikes, ditches, and moats—were nation-forming geoglyphs marking the landscape.[91] King Offa of Mercia (to whom Hamlet alludes when he mentions "Ossa")[92] defined the English by means of Offa's Dyke. This huge building project made a permanent change to the landscape as a boundary marker that went "from sea to sea" (the Irish Sea in the north and the Bristol Channel in the south).[93] Offa's contemporary, the Holy Roman emperor Charlemagne, did the same with his Karlsgraben (literally Charles's "dig," or "grave"), which connected the Rhine and Danube river basins. At the same time, King Gudfred of Denmark built the Dannevirke across the Cimbrian Peninsula,[94] effectively turning "mainland" Jutland into an "island."

One may not want to make mountains of molehills. Nation building in the early period seems variably linked with public works; however, the graves, ditches, and mounds of celebrated royalty became, like ancient Pharaonic projects, almost the definition of early modern nationhood.

THIN ICE

One pities that poor human race that slits its throat on our continent
about a few arpents of ice in Canada.[95]

—Voltaire, Letter to François-Augustin Paradis de Moncrif

Old Hamlet defeated the Poles on solid water, or ice.
. . . in an angry parle,
[Old Hamlet] smote the sledded Polacks on the ice.[96]

—Shakespeare, *Hamlet*

The Danes and Poles fought on shell ice[97] or cat ice (a "thin ice of a milky white appearance in shallow places, from under which the water has receded").[98] Or maybe they fought on patch ice or ice-patch reef (where the water isolates the ice above, which becomes an eggshell-like membrane). In any case, any place where the ice regularly "melts, thaws, and resolves itself into a dew" (to paraphrase Shakespeare) is "not [for long] tomb enough and continent. / To hide the slain? . . ." (*Hamlet*, act 4, scene 4).

Such icy battlegrounds are common in the Baltic region. The Scandinavian Olaus Magnus's map *Carta Marina* (1539; see Illustration 41) depicts

ILLUSTRATION 41
Polacks on the Ice.
Detail from Olaus
Magnus, *Carta
Marina* (1539).
Reproduced by
permission of Uppsala
University Library.

warriors fighting both on shelf ice attached to the mainland and on freshwater ice lands.

By the Renaissance, Londoners knew firsthand about thin salt- and freshwater offshore ice. The Tideway—the hundred-mile-long section of the Thames River running from the mouth of the estuary to Ham in the western suburbs—often froze over during this period, which was known as the "Little Ice Age."[99] The year that Shakespeare was born (1564), for example, Queen Elizabeth attended a celebration on the ice. In 1598, workers building the Globe Theatre used sleighs to move the timbers from Burbage's Theatre (in Shoreditch) across the frozen Tideway.[100] "Behold the wonder of this present age / A frozen river now becomes a stage" reads a seventeenth-century poem. Its author goes on to warn against slipping through the ice: "Folks do tipple without fear to sink / More liquor than the fish beneath do drink."[101] The "icemen" of the Royal Humane Society[102] kept watch over the places where ice met water and the times when it became water.

During the winter that marked the end of the Little Ice Age, at London's last Frost Fair (1814), printers distributed a handbill reading, "Behold the Liquid Thames frozen o're."[103]

The same delicate opposition of "land that is also water" to "water that is also land," together with that of solid to liquid water, greatly informed the geography of Saxo's *Deeds of the Danes*, a work that made its way into the culture of England and France in the sixteenth century. There, "ice ↔ land" or "water-land" always unreliably threatens to melt and thaw away into a dew. "O, that this too too solid flesh would melt / Thaw and resolve itself into [adieu]!" says Hamlet (act 1, scene 2). *Zealand*, the name of Denmark's great island, itself means "sea-land." *Pomerania*, from Old Slavic, contains *po*, meaning "by/along," and *more*, meaning "sea."[104] *Åland* (which names the islands at the entrance to the Gulf of Bothnia) comes from *ahvaland*, which means "land of water." *Öland* (which names the island at the Kalmar Strait) comes from the words *ö*, meaning "island," and *land*, meaning "land."

Because most places in the Baltic Sea freeze solid for a good part of each year, any "patch" of *aqua solida* masquerading as *terra firma* turns out to allow only for a "watery grave"—not continent enough.[105]

In setting Old Hamlet's foundational battle with the Poles on ice (see Illustration 42), Shakespeare references more than the experience of river life east of the previously mentioned Ham. (This regional name is included in the eleventh-century *Domesday Book*.) He might also seem to reference three long-standing formative historical myths about war and ice.

First, there is the ice battle between Asians and Europeans that informs the oldest surviving theater text in Western drama, *The Persians*, by Aeschylus (472 BC),

ILLUSTRATION 42
Fortinbras's Soldiers on the Ice. Still from Kenneth Branagh's movie *Hamlet* (1996). Branagh conveys on the visual plane the panoramic sense of Hamlet's soliloquy about the patch. Hamlet speaks the lines from a mountain overlooking an icy valley, where the soldiers seem to march off to a watery grave. Source: Harvard College Libraries.

in which the military outcome—and hence the fate of "world civilization"—depends on whether ice remains solid or melts into adieu. The Persian soldiers retreat after their defeat at the Battle of Salamis (480 BC) and ford the unusually frozen River Strymon. The ice melts under their feet, and the Herald reports thus:

> That night, ere yet the season, breathing frore,
>
> Rush'd winter, and with ice incrusted o'er
>
> The flood of sacred Strymon: such as own'd
>
> No god till now, awe-struck, with many a prayer
>
> Adored the earth and sky. When now the troops
>
> Had ceased their invocations to the gods,
>
> O'er the stream's solid crystal they began
>
> Their march; and we, who took our early way,
>
> Ere the sun darted his warm beams, pass'd safe:
>
> But when this burning orb with fiery rays
>
> Unbound the middle current, down they sunk
>
> Each over other. . . .[106]

ILLUSTRATION 43
Battle of the Ice. From the Russian illuminated manuscript *Life of Alexander Nevsky* (ca. 1560–1570).

The second—the Battle of the Ice—was the result of the clash between the Catholic Danish king Valdemar the Conqueror (from mainland Jutland), together with other Catholic Scandinavian and Teutonic invaders, and Russian troops led by Alexander Nevsky (see Illustration 43). It took place in the spring of 1242 on the mostly frozen strait connecting the two parts of Lake Peipus. Sergei Eisenstein's film *Alexander Nevsky* (1938) shows the Danish-led troops falling through the ice (see Illustration 44).[107]

The third battle involves the biblical crossing of the Red Sea, when, according to some interpretations of the biblical text, God arranges for "melting" to destroy the Egyptians. (See Illustration 45.) The book of Exodus describes how the Children of Israel passed through the Red Sea.

	And the LORD overthrew the	יהוה את מצרים
	Egyptians in the midst of the	וינער
Exodus 14:27	sea . . .	בתוך הים
	. . . but the children of Israel	הלכו ביבשה
	walked upon dry land in the	ובני ישראל
Exodus 14:29	midst of the sea . . .	בתוך הים

How did the biblical God "turn the sea into dry land" (Psalm 66:6) or "ma[k]e the sea dry land" (Exodus 14:21)? The sense is not so much that He parted the waters in such a way that the Israelites could walk on the sea floor. The sense is rather more that God transformed the water itself into *terra firma* in such a way that the Israelites could "walk on water." When the Hebrew tribes leave Egypt, God congeals ("freezes") the waters. ("And with the blast of thy nostrils the waters were gathered together, the floods stood upright as an heap, and the depths were congealed in the heart of the sea" [Exodus 15:8].) In this way, the biblical God provides an islandic "wall" (חומה; Exodus 14:22) around the Israelites like the protective wall that is the waterway protecting the coastline city of Egyptian Thebes. In her essay "Edges and Otherworlds," Catherine Clarke writes that "water—normally fluid and insubstantial—becomes a solid robust material."

ILLUSTRATION 44
The Battle of the Lake. Sketch by Sergei Eisenstein for his film *Alexander Nevsky* (1938). © DIOMEDIA / Mary Evans / Ronald Grant. Reproduced by permission.

"The language," she adds, "seems to revel in the paradox of water as an architectural material."[108]

Thanks to the freeze, the Hebrews cross safely to the other side. When the Egyptian troops chase them, the frozen waters melt. The Egyptian troops sink "as lead in the mighty waters" (Exodus 15:10). The prophet Nahum, writing soon after the Assyrian king Ashurbanipal conquered Thebes (663 BC), asks the Jewish people, "Art thou better than [Thebes], that was situate among the rivers, that had the waters round about it, whose rampart was the sea, and her wall [חומה] was from the sea?" (Nahum 3:8). The Old English poem *Exodus* (ca. 1000),[109] which retells the story of the crossing of the Red Sea partly as the history of the Saxons crossing the English Channel, likewise melts the solid liquid.

וַיָּעַר ה' אֶת מִצְרַ֫יִם בְּתוֹךְ הַיָּם: וּבְנֵי יִשְׂרָאֵל הָלְכוּ בְּיַבָּשָׁה בְּתוֹךְ הַיָּם

וַיָּעַר ה' אֶת מִצְרַ֫יִם בְּתוֹךְ הַיָּם וּבְנֵי יִשְׂרָאֵל הָלְכוּ בְּיַבָּשָׁה בְּתוֹךְ הַיָּם

SUBLIME COAST

. . . that beetles o'er his base . . .

—Shakespeare, *Hamlet*

In *Hamlet*, the geography of the scene where the ghost of "Old Hamlet" appears to "Young Hamlet" conflates earth with water on the vertical plane: the cliff above juts out over the water below. It is a vertiginous landscape, a dizzying waterscape, where the ground "beetles o'er his base" (act 1, scene 4). Not just the cliff is beetling o'er; the entire "firmament" (act 2, scene 2) is o'erhanging its base. The world overhangs in the same way a "promontory" (for Hamlet, the Earth) juts out into the sea or a promontory stage juts into the audience of such a theater as the Globe. Robert Greene writes in his play *The Honourable History of Friar Bacon and Friar*

Bungay (1592), "Welcome . . . To Englands shore, whose promontorie cleeues, / Shewes Albion is another little world."[110] That world is the Globe. Consider how Shakespeare allies his stage promontory as a global peninsula in the epilogue of his last play, *The Tempest*. The magical dramatist Prospero, his powers now "all o'erthrown," speaks to the audience from this promontory: he is "here [on the island] confined by you [the audience]" and, like a ship waiting for the sea wind, he begs for the "breath" of a clapping audience to drive him back home to Italy:

> Let me not,
>
> . . .
>
> . . . dwell
> In this bare island by your spell;
> But release me from my bands
> With the help of your good hands:
> Gentle breath of yours my sails
> Must fill, . . .
>
> (act 5, scene 1)

Edgar Allan Poe well captures *Hamlet*'s hydrological spirit of the coastal scenes in the short story "A Descent into the Maelström" (1841), set "close upon the Norwegian coast," at the Moskenstraumen (a marine region of powerful eddies and whirlpools at the Lofoten Archipelago). There the overhanging cliffs vertiginously beetle out over their base:

I looked dizzily, and beheld a wide expanse of ocean, whose waters wore so inky a hue as to bring at once to my mind the Nubian geographer's account of the Mare Tenebrarum [sea of darkness]. A panorama more deplorably desolate no human imagination can conceive. To the right and left, as far as the eye could reach, there lay outstretched, like ramparts of the world, lines of horridly black and beetling cliff, whose character of gloom was but the more forcibly illustrated by the surf which reared high up against it its white and ghastly crest, howling and shrieking for ever.[111]

Being caught up in the maelstrom is the substance of Poe's tale. This trap is the fate that Hamlet's friends fear for him.[112]

Israel Gollancz in *Hamlet in Iceland* (1898)[113] and Giorgio de Santillana and Hertha von Dechend in *Hamlet's Mill* (1969)[114] signal this aspect of the Shakespearean vortex. It was a German idealist philosopher and geographer, however, who best understood this aspect of Shakespeare's coastal imagination. Immanuel Kant, who defined *geography* in terms of the relationships between political theory and natural history in *Critique of the Power of Judgment* (1790), lays the groundwork for understanding how we experience such spectacles of the sublime. He had in mind places that Germans of his time, who found themselves in *Hamlet*, generally found, on the aforementioned islands of the Baltic Sea. Writes Kant:

Bold, overhanging, and, as it were, threatening cliffs, thunderclouds piled up the vault of heaven, borne along with flashes and peals, volcanoes in all their violence of destruction, hurricanes leaving desolation in their track, the boundless ocean rising with rebellious force, the high waterfall of some mighty river, and the like, make our power of resistance of trifling moment in comparison with their might.[115]

The first German translators of *Hamlet*, August Wilhelm von Schlegel and Christoph Martin Wieland, present this sublime aspect of the Zealand coast. Most German adaptations of *Hamlet* tend to emphasize, more than the English text, the play's vertiginous Baltic Sea setting.[116]

Wie, wenn es hin zur Flut Euch lockt, mein Prinz,	Und wie dann, Gnädiger Herr, wenn es euch an die Spize des Felsens führte, der	What if [the ghost] tempt you toward the flood, my lord,
Vielleicht zum grausen Gipfel jenes Felsen,	sich dort *über die See hinaus bükt*, und	Or to the dreadful summit of the cliff
Der in die *See nickt über seinen Fuß*?	dann eine noch fürchterlichere Gestalt annähme, welche euern Verstand verwirren	That *beetles o'er his base* into the sea,
Und dort in andre Schreckgestalt sich kleidet,	und in sinnloser Betäubung euch in die	And there assume some other horrible form,
Die der Vernunft die Herrschaft rauben könnte	Tiefe hinunter stürzen könnte? Denket an diß! Der Ort allein, ohne daß noch andere	Which might deprive your sovereignty of reason
Und Euch zum Wahnsinn treiben? O bedenkt!	Ursachen dazu kommen dürfen, könnte einem, der so viele Faden tief in die See	And draw you into madness? think of it:
Der Ort an sich bringt Grillen der Verzweiflung	hinab schaute, und sie von unten herauf so gräßlich heulen hörte, einen Anstoß von	The very place puts toys of desperation,
Auch ohne weitern Grund in jedes Hirn,	Schwindel geben.	Without more motive, into every brain
Der so viel Klafter niederschaut zur See		That looks so many fathoms to the sea
Und hört sie unten brüllen.		And hears it roar beneath.
—Schlegel (1800)	—Wieland (1760)	—Shakespeare

One English-language production that captures the sublime landscape in the play is Hay Plumb's silent film *Hamlet* (1913). Shot partly at the headlands of Lulworth Cove, "the Ghost was let loose on the stones and boulders of the Dorset shore."[117] The ghost appears, in the cinematographic process, as a double exposure (see Illustration 46). Kant's near contemporaries, Henry Fuseli and William Blake, capture the spirit otherwise (see Illustrations 47 and 48).

ILLUSTRATION 46 Hamlet and the Ghost at the Shore 1. Still from Hay Plumb's movie *Hamlet* (1913). British Film Institute. Source: Harvard University Libraries.

ILLUSTRATION 47 Hamlet and the Ghost at the Shore 2. Henry Fuseli, *Hamlet, Horatio, Marcellus, and the Ghost of Old Hamlet*, 1780s. Engraved for John Boydell's Shakespeare Gallery. © The Metropolitan Museum of Art. Art Resource, New York.

ILLUSTRATION 48 Hamlet and the Ghost at the Shore 3. William Blake, *Hamlet and His Father*, 1806. © Trustees of the British Museum. Reproduced by permission.

In Germany, Hamlet's distractedly brainish globe sometimes needs curbing. Friedrich Ludwig Schröder thus imposes on *Hamlet*'s plot French classical notions about the unity of time and place. Because everything must happen in a single day, Schröder's Laertes leaves the docks at Elsinore, heading for France, and then, immediately, is blown back to Denmark by a pseudo-sublime tempest, as if the Norse physics of land and sea, air and fire, were believably subject to neoclassical norms.[118]

Island Words

HAM

-*holm* and -*ham* often tend to interchange.
> —James B. Johnston, *The Place-Names of England and Wales*

Hamlet [is a diminutive] of *Ham*.
> —John Rastell, *Les Termes de la Ley: Or, Certaine Difficult and Obscure*
> *Words and Termes of the Common Lawes and Statutes of this Realme*

Amleth is the name of the character in Saxo Grammaticus's story.[1] *Hamblet* is the name in the French-language version by François de Belleforest. Why does Shakespeare change the name specifically to *Hamlet*? Put otherwise, what might *Hamlet* have signified before Shakespeare made the name globally recognizable?

For one thing, Saxo and Shakespeare both ally the meaning of the proper name *Hamlet* (foolish) with the meaning of the Roman proper name *Brutus* (dullard). (Lucius Junius Brutus killed his tyrant uncle Lucius Tarquinius Superbus, and Marcus Junius Brutus killed his adoptive father Julius Caesar.)[2] The name *Hamlet* recalls also Shakespeare's son Hamnet, who died a few years before Shakespeare produced the play. (Elizabethans often pronounced "Hamnet" as "Hamlet"; the two spellings were almost interchangeable.)[3] Furthermore, Shakespeare jokes about how Hamlet's weak legs—or "hams"—match his name, meaning "little leg."[4] Beyond that, one son (Helgi) of the murdered king in the *Saga of Hrolf Kraki* takes on the counterfeit proper name *Ham* in order to protect himself.[5] Are these statements not reasonable responses to our question?

There is, however, another sort of answer to the question, What role does the proper name *Hamlet* play in the play *Hamlet*? This answer looks at the English-language word *ham* in a way that students of this play have not. It connects the islandic "patch of ground"—or "ham"—for which so many people in *Hamlet* struggle, with Hamlet's name, *ham-let*, which in English means "little patch of ground," and also associates Hamlet's "ham" with the play's overall islandology.

According to John Minsheu's *Guide into Tongues* (1617), a *ham* is a "little plot of ground."[6] According to the *OED*, it is "an enclosed plot of . . . ground."[7] On the Frisian Islands, a *ham* is "a pasture or meadow enclosed with a ditch," like an island, and *Hämel* means "little ham." "*Hamlet* [is a diminutive] of *Ham* [meaning "plot of land"]," writes the printer, playwright, and legal theorist John Rastell in the widely distributed work quoted in the epigraph to this chapter.

Shakespeare rarely uses the word *ham*, but he does use its cognate, *hem*.[8] A *hem* is a circumferentially insular textile patch bordered by a *hem*line, so a "real estate patch," or *ham*, is bordered by a ditch, sewer, or sea. The burial place, or "ham," of Timon in *Timon of Athens* is thus twice "hemmed" in at the coast:

> Timon is dead;
> Entomb'd upon the very hem o' the sea;
>
> (act 5, scene 4)

A moat it is, to trouble the island of the mind that speaks itself as "the Dane." Timon is buried, like Beowulf, in a tumulus, or burial mound, by the sea. (See Illustration 49.)

ILLUSTRATION 49
Sir Walter Scott
Visiting Smailholm
Tower. Turner, Joseph
Mallord William
(1775–1851) / Private
Collection / Photo
© Agnew's, London,
UK / The Bridgeman
Art Library.

ELSINORE

... I do think, *or else* ...

—Shakespeare, *Hamlet*

Shakespeare sets his play on a Danish island (Zealand) instead of the Danish mainland (Jutland) or an English island (Britain), as Saxo does in *Deeds of the Danes* and as Belleforest does in *Tragic Histories*.[9] What explains these departures from the sources? Why does Shakespeare want a circumferential "hem," or "ham," for his play?[10]

Why does he set the play at Elsinore, for that matter? Saxo had not done that. What difference does it make? It is not as though Shakespeare keeps the change quiet. Hamlet uses the word *Elsinore* four times—twice to welcome visitors to the island[11] and twice to ask them why they came to it.[12] "What make you at Elsinore?"

No reader of *Hamlet* that I know has attempted to answer this question. Some critics have tried to defuse it, asserting that Shakespeare erroneously confused "Elsinore" with "Jutland." Such an error seems unlikely to me. Saxo distinctly sets his story in Jutland; he also ends it with the memorable statement that "the field in Jutland where [Saxo's] Ammel [Hamlet] was buried is still called after him 'Ammel's Heath.'"[13] Elsinore was enough in the news during Shakespeare's day to ensure that he would make no such error.

Moreover, some "news of the day" would seem to have encouraged the relocation of the story to Elsinore. First, King James's marriage to Anne of Denmark took place there[14] at a time when Denmark was threatening to wrest back from Scotland the once Danish Orkney Archipelago[15] (where ships of the Spanish Armada had landed in 1588).[16] Second, the famous Tycho Brahe set up his astronomical observatory on the island of Hven just a few kilometers from Elsinore. Third, English theater companies traveled to Elsinore during the mid-1580s. Fourth, the Danish navy at Elsinore had designs on the British navy.

More important than these factors, though, is the physical and spiritual geography of Elsinore on Zealand. Elsinore is located where the Øresund, only two miles wide, "sunders" island Zealand from mainland Scania (nowadays part of Sweden but then part of Denmark).[17] (See Illustration 50.) It guards the way through the narrow neck of water into and out of the interior sea (*mare internum*) that is the Baltic (which means "the 'belted,' or sundered, sea").[18] The word *els* in Elsinore means "neck," even in the English language, where the term is usually spelled *hals*.[19] The town Elsinore joined at the neck the two great maritime regions of the North Sea Empire: the "interior" Baltic Sea (with its various mainlands and islands) and the "open" North Sea (with its island of Britain, the Orkney and Shetland archipelagoes, and the Isle of Man).

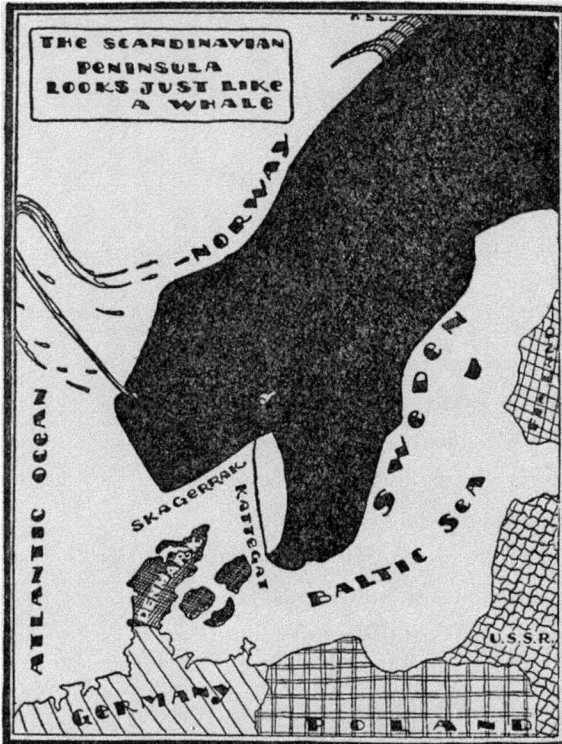

ILLUSTRATION 50 *The Scandinavian Peninsula Looks Just Like a Whale*. Illustration by Mary Sherwood Wright Jones. From V. M. Hillyer, *A Child's Geography of the World*, rev. Edward G. Huey (1929; New York: Appleton-Century Crofts, 1951). Writes Hillyer of Skagerrak and Kattegat: "When anything annoyed my uncle he used to cry out, 'Skagerrack and Kattegat!' It sounded terrible, for I didn't know then that Skagerrack and Kattegat were merely the names of the narrow waterways around Denmark from the North Sea into the Baltic and that Kattegat simply meant 'the cat's throat' and Skagerrack meant 'Skager throat'" (252). Permission granted by Collection Selechonek.

This physical geography of *Hamlet* matches up, as we will see, to its philosophical aspect and informs its spiritual and corporeal motivation. In many other works of literature, the "match" between spiritual and physical realms is "merely" allegorical or metaphorical, as for Mark Twain's description of the island of Krakatoa[20] in the Sunda Strait and Cicero's description of political commotion in terms of the water moving both ways through the Euripus Strait.[21] In *Hamlet*, though, the similitudes all verge, at the limit (or *orizon*), on identity.

ORISONS

The faire Ophelia, Nimph in thy Orizons?

—Shakespeare, *Hamlet*, Fifth Quarto

Hamlet speaks the words "Lady in thy *orizons*, be all my sins remembered" (First Quarto, 1603; emphasis mine) immediately after he ends his soliloquy "To be or not to be" (act 3, scene 1)—or after Ophelia interrupts it.

Hamlet's line to Ophelia comes down to us in several versions. One is "Lady in thy *orizons*, be all my sins remembered" (First Quarto, 1603). Another has the term as *orisons* (First Folio, 1623). In yet another version, the spelling is *horizons* (Second Folio, 1632; Third Folio, 1664; Fourth Folio, 1685). The version quoted as the epigraph, from the Fifth Quarto (1637), ends the line about *orizons* with a question mark.

No matter which version we choose, the word *orizons* has several sounds and soundings. The same sound, which I represent here by means of the sign ☉, suggests several otherwise apparently disparate meanings at the same time, overlapping soundings that variously bear on the geographic and spiritual location of the play. The use of the sign ☉ sheds light on Shakespearean punning and islandology. Learning from both is the principal purpose of this chapter. (See Illustration 51.)

PRAYER: At first hearing, the sound ☉ seems to be only that of *orizon*, or *orison*, in the sense of "prayer" or "exhortation." This makes sense enough: Hamlet

For Shakespeare's Prince, and the Princess of Wales, To England dear. Her royal spirit quails; From skating faint, she rests upon the snow; Shrinking from unclean beasts that grin below.

ILLUSTRATION 51
Anthropocentric Map of Denmark Showing the "Sound." From Aleph (William Harvey), *Geographical Fun: Humorous Outlines of Various Countries* (London: Hodder & Stoughton, 1869). In 1863, Princess Alexandra of Denmark married the Prince of Wales (later King Edward VII of the United Kingdom). Alexandra enjoyed ice-skating, and here she is shown as a skater resting on the snow. David Duff, *Alexandra: Princess and Queen* (London: Collins, 1980), 143. The epigraph reads: "For Shakespeare's Prince, and the Princess of Wales, To England dear. Her royal spirit quails; / From skating faint, she rests upon the snow; / Shrinking from unclean beasts that grin below." Reproduced by permission of the Yale Map Collection.

wants Ophelia to remember his sins in her prayers. (Claudius, too, might have done well with some help along those lines.)[22]

SHORELINE: A term homophonically like *orison* (prayer) is *ore sund*, meaning "shoreline" or "edge of a brook."[23] An "ore sund" is where Ophelia becomes something like a mermaid, or animate floating island—"Her clothes spread wide; / And, mermaid-like, awhile they bore her up" (act 4, scene 7). Ophelia is "like a creature native and endued / Unto th[e] element [water]" (act 4, scene 7).

The English-language *sound* means the "action or power of swimming,"[24] but Ophelia, though she floats along for a while, does not have the nesological requirement. Her clothing becomes waterlogged. "Too much of water hast thou, poor Ophelia" (act 4, scene 7).[25] Hers is a "muddy death," neither on land nor on water. Antoine-Augustin Préault's bronze bas-relief *Ophelia* (1842) is relevant here

ILLUSTRATION 52
Ophelia 1. Antoine-
Augustin Préault,
Ophelia, 1842.
© RMN-Grand
Palais / Art Resource,
New York.

(see Illustration 52). It depicts the horizon between land and water creature as one between animate and inanimate creation. Ophelia's body, almost mermaid-like, appears to flow in and out of solid and molten substance. Préault's sculpture recalls Hector Berlioz's *La mort d'Ophélie* (1840),[26] whose libretto compares Ophelia with a *naïade* (from Greek *naein*, "to flow"). The mermaid Ophelia, as imagined by Romantic artists, fits well with a compelling textual variant of what Hamlet calls Ophelia—namely, "Lady" (First Quarto, 1603). In some other versions of the play, he calls her "nymph" (Second Quarto, 1604; First Folio, 1623; Fifth Quarto, 1637). John Everett Millais, whose family came from the island of Jersey—thus painted his *Ophelia* (1851–1852) as if the subject were a water nymph on the edge of life between water and earth (see Illustration 53).[27]

HORATIO: "Horatio" sounds like *oratio* (oration) and hence like *orison* (prayer). What is more, the two are often spelled the same.[28] "More an antique Roman [like Cicero and Livy] than a Dane," Horatio knows best how to manage sound at the Sound: "[B]les'd are those / Whose blood and judgment are so well commingled / That they are not a pipe for Fortune's finger / To sound what stop she please" (*Hamlet*, act 3, scene 2).

ØRESUND: The sound ⊙ is much like that of *Øresund*, which names the Øresund, "the 'sound' or 'neck'"—"the *els*"—where the guard town of Elsinore is located and guards the narrow entrance and exit of the Baltic Sea. The Øresund is where—ever since the time of Eric of Pomerania, king of the Scandinavian Kalmar Union—passers-through were compelled to pay Sound Dues. John Boroughs, in *Sovereignty of the British Seas* (1633), well recalls "the King of Denmark at his ward house in the Sound."[29] That guardhouse is very like the platform where the action of *Hamlet* begins and ends.

THE SOUND OF ⊙: The English-language toponym *Sound*, the voiced sound of which I here indicate by ⊙, names one particular place, the Øresund, and no sound [*sund*] else. (See Illustration 51.) At the same time, the Danish name of that

ILLUSTRATION 53
Ophelia 2. John
Everett Millais,
Ophelia, 1851–1852.
© Tate, London,
2013. Reproduced by
permission.

place, *Øresund*, and the full English-language name, *Øresund Sound*, of the same place, both contain the sound ☉.[30] Sounds come when one breathes through a narrow aperture, like a reed or like the human throat, which joins head with body. Hamlet warns Rosencrantz and Guildenstern that they cannot sound him. "You would *sound* me from my lowest note to / the top of my compass . . . / . . . yet cannot / you make it speak" (act 3, scene 2).

Likewise, Hamlet warns them about how they can fret him: "Though you can fret me, yet you / cannot play upon me" (act 3, scene 2). Soon enough, Rosencrantz and Guildenstern, called back to Elsinore, will be caught up in the *fretum* (strait) that is the Øresund itself. In fact, some experts in geographic toponymy say that the Øresund got its name, which sounds like "aurum-sound," from the figural resemblance of the cartographic outline of the Øresund's shorelines with some anatomical outline of a human *aurum* (ear).[31] The ear is the narrowing orifice that receives poisonous words ("poison in jest," for example) and actual poison (the "hebenon" that Claudius pours into the ear of Old Hamlet).[32] In *Hamlet*, one can die if one's ear canal is not well guarded against dangerous words ("And let us once again assail your ears, / That are so fortified against our story" [act 1, scene 1]) as a country can be conquered if a Danish sovereign allows a Fortinbras to pass through the Øresund.

Rosencrantz and Guildenstern, visitors to Elsinore, do not really "hear" what Hamlet is saying. He is warning each of them that, as he is in Elsinore at the Øresund, he is in danger of losing his neck (*els*). The Øresund is *the* mythic place of Danish sundering.[33] Polonius's head will be sundered from the rest of his body if he is mistaken about Hamlet's madness; Yorick's skull has been sundered from the rest of his skeleton; and Hamlet's "head should be struck off" if Claudius has his way. Just so, island Zealand was sundered from Scania at the neck (*els*) that is Elsinore.[34] (See Illustration 54.)

HORIZON: The soliloquy "To be or not to be" partly concerns the "horizon" where life and death meet. It is as if one were "as near to heaven by sea as by land," as Humphrey Gilbert, facing death at sea near Newfoundland, is reported, in *Voyages* (1583), prayerfully to have said.[35] A few weeks before speaking these words, at Sable Island,[36] the outer limit of sea and land, Gilbert had founded the British Empire by claiming Newfoundland for Queen Elizabeth (1583). (That was one year before his half-brother, Walter Raleigh, claimed Roanoke on what we now call the coast of North Carolina.)

ILLUSTRATION 54
Map of Denmark and Sweden. The map, which emphasizes the island aspect of Denmark, represents the fortifications at Elsinore on both sides of the Øresund: Danish Zealand and Danish Scania. Sebastian Münster, *Cosmographia* (Basel, 1567). Harvard Widener Library, *Cosmographia*, v.4=HW0BA9, p. 1329. Reproduced by permission.

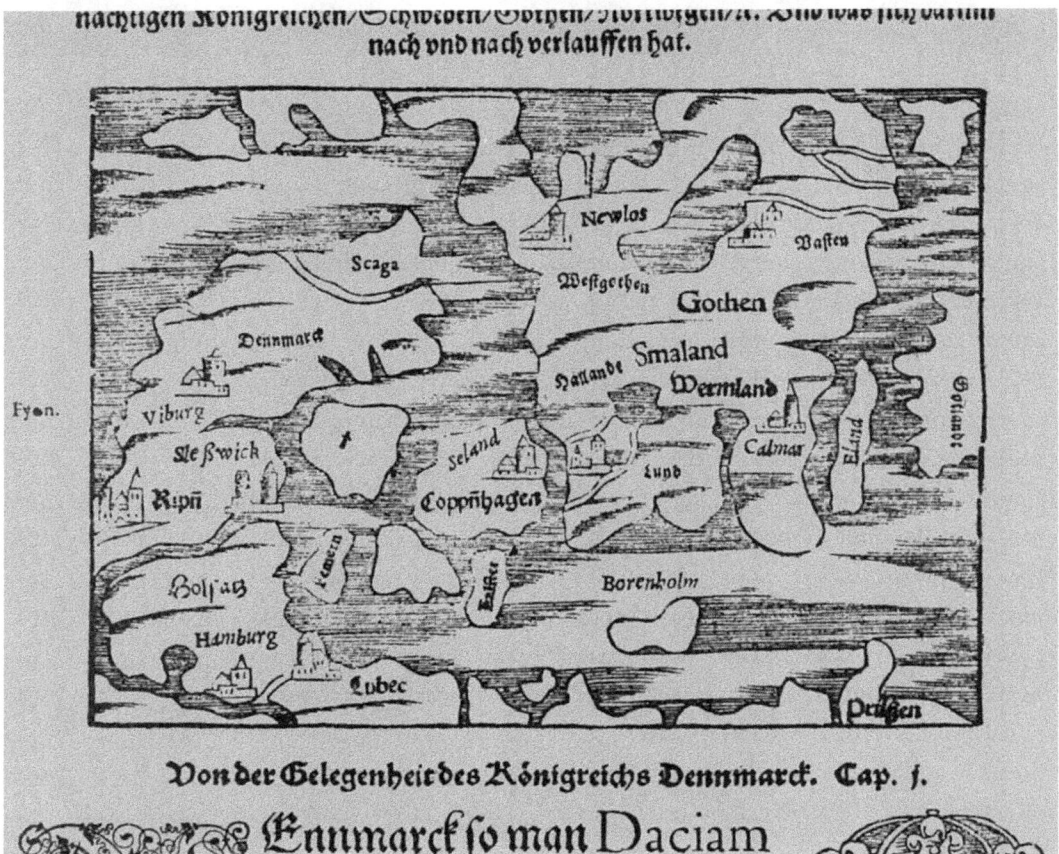

The sense in which Hamlet's exhortation to Ophelia indicates both "horizon" and "prayer" at the same time is suggested by the frequent homographic identity between *orison* (prayer) and *horizon* (limit)[37] and by the similarity in pronunciation of these two otherwise apparently disparate words. Hamlet's words sunder (i.e., cut off) his oration on being and not-being, marking its horizon, or "bourn," as Hamlet calls it,[38] even as they define the living in terms of the dead or vice versa.[39] The tragedy of Hamlet, as we will see, requires prayer for boundary, orizon for horizon. The German philosopher Friedrich Nietzsche, writing during the latter part of a century in which many German thinkers had seen their own fate as a nation defined in terms of Hamlet's dilemma,[40] writes thus in *The Gay Science* (1882):

At hearing the news that "the old god is dead," we philosophers and "free spirits" feel illuminated by a new dawn . . . finally the horizon seems clear again . . . finally our ships may set out again . . . the sea, *our* sea, lies open again . . . maybe there has never been such an "open sea."[41]

The only passage to the open sea is death.

Dire Straits

PASSAGES, GEOLOGICAL AND ZOOLOGICAL

The Animal Figure of Mankind, I will similize to an Island, the Blood as the Sea that runs about, the Mouth as the Haven which receive the Ships of Provision.

—Margaret Cavendish, *The World's Olio*

The analogy between the human body and the Earth, common in human thinking, takes the form in *Hamlet* of correlating various human canals, including those aural and vaginal, with geological passageways, including the sound between Zealand and Scania.

An understandably neurotic Hamlet, for example, wants full *egress* out through Gertrude's birth canal even as he also fears it. He imagines his uncle-father Claudius, *Els*inore's king, "paddling in [his aunt-mother Gertrude's] neck [*hals* or *els*]"—in effect, "necking" with her.[1] Hamlet wants nothing of that "ha[l]sing and kissing," as Richard Bernard calls necking in *The Self-Tormentor* (1598).[2]

By the same token, Hamlet wants *ingress* into Ophelia's vaginal canal. He asks, "Lady, shall I lie in your lap?" (act 3, scene 2), which refers to "country matters."[3] (Perhaps he also wants ingress into Gertrude's, even as he likewise fears that.) Yet "To a nunnery go," he says to Ophelia. He wants both that Ophelia become a nun and that she become a prostitute. In Saxo's *Deeds of the Danes*, Ophelia's counterpart cannot have sexual intercourse with Hamlet without transgression because she is his collactaneous sister. One etymology for *Ophelia*—a name that Shakespeare adds to the story—derives it from *o-philia* (sibling love).[4]

"Let not ever / The soul of Nero enter this firm bosom" (act 3, scene 2), Hamlet says to himself as he enters his mother's chamber. He is thinking about the tyrannical Roman emperor (54–68 AD), nephew to Emperor Claudius (41–54 AD)—Nero, who had sexual intercourse with his mother Agrippina and killed her.

Tyrants like Nero regularly commit incest—often on islands.[5] Shakespeare's Antiochus in *Pericles, Prince of Tyre* is one. Hamlet of Zealand, however, is not.

Like Shakespeare's Pericles, his Hamlet seeks another way (albeit that it requires his mother's death for him to ascend, if only for a moment, to the throne).

Hamlet's two main exhortations to Ophelia—"To a nunnery go" and "*Lady in thy orizons, be all my sins remembered*" (First Quarto, 1603; emphasis mine)— match each other both because Hamlet is thinking of Ophelia as a nun at her prayers (orisons) and because he is thinking about passage through a corporeal or geographic strait (Øresund). In much the same way, we remark that Hamlet calls "prison" the place (Denmark), whence his mother would not have him go ("Let not thy mother lose her prayers, Hamlet. / I pray thee, stay with us, go not to Wittenberg" [act 1, scene 2]). For him, Denmark is like a corporeal amniotic womb or a geographic inland sea from which he "lack[s] advancement" through its birth canal or geographic strait (Øresund).

In that sense, Hamlet is like a human being not fully born or not fully born into adulthood. He is thus like the child players who "exclaim against their own succession," resembling the premature "eyas," a baby hawk whose training is incomplete—that is, it is still in the "eyrie," or nest (act 2, scene 2). He recalls Ophelia, insofar as she is a mere "nymph." He is even like Osric, who is a "lapwing"—the kind of bird that carries its own eggshell on its head even after hatching.[6] ("This lapwing runs away with the shell on his head," Horatio says to Hamlet about Osric, half forgetting that, before his return from adventure at sea, Hamlet too was something of a foppish courtier, or partly pretended to be one.) "Some [human beings] yet are Embrio's, yet hatching, and in the shell," writes William Chillingworth (1638).[7] *Hamlet*, as the *Bildungsroman* (coming-of-age novel) that Goethe made it out to be in his pathbreaking novel *Wilhelm Meister's Apprenticeship* (1795–1796), presents just such a one.[8]

Before his piratic release at sea, Hamlet is stuck, or "pall[ed]," in interior seas and straits or in wombs and birth/vaginal canals. The rhetoric linking "marine channel" with "birth canal" pervades the rhetoric of anatomical and navigational textbooks of the period.[9] Straits like the Øresund are associated with bodily orifices—with vaginas, throats,[10] mouths,[11] and cloacas.[12] John Florio describes the womb in his *Worlde of Wordes* (1598) as a narrow neck, and Thomas Gibson does the same in his *Anatomy of Humane Bodies Epitomized* (1682).[13]

The analogy here of vaginal canal and marine channel with geological ravine, gorge, or defile helps to explain *Hamlet*'s linkage with the story of Oedipus the Tyrant, where the son's attempt to enter, and/or the father's attempt to exit, Delphi, which means "womb" (Greek *delphos* [δελφός]), gives rise to the action in the first place. Returning from the oracle at Delphi, Oedipus finds himself at the narrow passage that leads into and out of the interior. Laius is coming the other way, and the passage is too narrow for father and son to pass each other. Neither

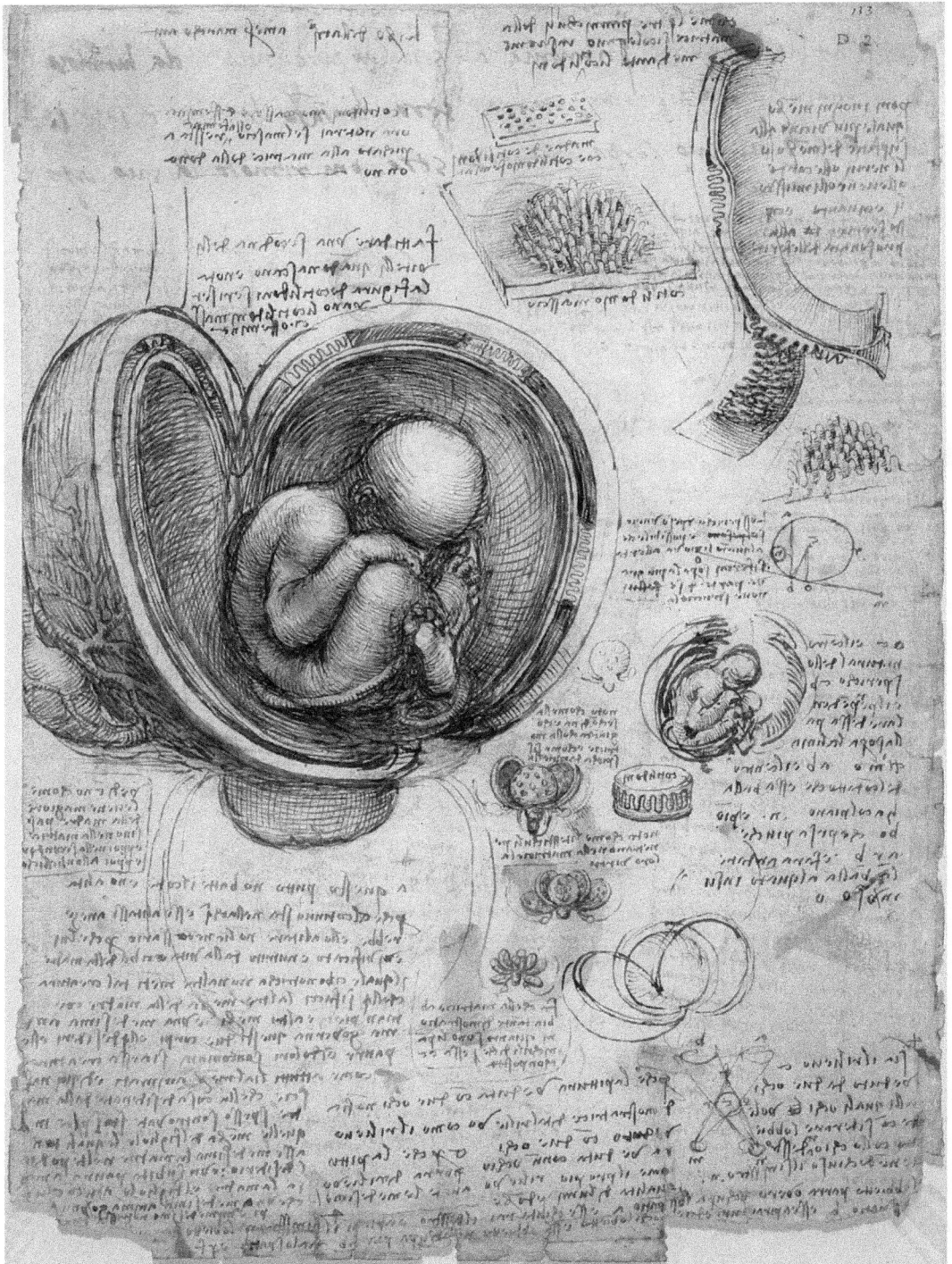

ILLUSTRATION 55 *Foetus in the Womb*. Leonardo da Vinci (ca. 1510–1513). The drawing depicts the womb as a sort of nutshell or egg—a sort of fetus prison—and compares it with various sorts of (other) botanical specimens. The Royal Collection © 2011 Her Majesty Queen Elizabeth II / The Bridgeman Art Library.

wants to give way. (In *Hamlet*'s terms, one is popping between "the election and . . . choice" of the other.) The son, Oedipus, kills the father, Laius, then gets on with his life, or thinks that he does. Soon afterward, however, he meets a *sphinx*. "She" regularly kills her victims by having sexual intercourse with them and then *constricting* them in the *sphinct*er muscles of her vaginal passage. She catches men in the way the ravine leading into Delphi does. Again, Oedipus gets away, or thinks that he does. Then he meets and marries a queen, Jocasta, who turns out to be his mother. So the young man Oedipus has now returned—by way of passages through the ravine at Delphi, past the fearful constriction of the Sphinx, and in sexual intercourse with his mother—to the same womb from which he had exited in the first place, when he was born.[14]

The ancient Greek traveler and geographer Pausanias reports that the constricting ravine where Oedipus killed his father was called the "cleft way."[15] *Cleft* here means both the "parting of the thighs" and the "pudendal cleft"—the passage through which both father and son must pass one way or another.[16]

The tragically essential "passage" in *Hamlet*, simultaneously from being to nonbeing (perhaps death) and from nonbeing (birth)—"To be or not to be"— recalls Sophocles' Tiresias, who says to Oedipus, "On this day you will be born and die."[17] Not until Hamlet returns from his sea voyage, again via the Øresund, is he finally "fit and season'd for [this essential] passage." (See Illustration 55.)

HVEN IN ØRESUND

The place-name *Hven* (*Heven, Ven, Hveen*), often pronounced like "heaven," names the islet in the Øresund just offshore from Elsinore.[18] Hven is now part of Sweden but during most of its history was Danish or autonomous. The spelling *Ven*, which modern-day tourists encounter, is the relatively recent Swedish spelling, deriving from the Great Swedish Spelling Reform (1906) that replaced *hv* with *v*. Until 1958 the population of the island kept the older spelling, which the best-known etymology, from *Hwaethaen* (surrounded by [sea]), supports. Books about Hven[19] and maps of the island (including the two shown in this book) support "Hven," as do many modern academic works.[20] As pointed out in *Hamlet*, Elsinore is to the "north-north-west" of Hven.

Hven is where the "modern" understanding of the heavens began to take hold in an intellectually sustained fashion, and Shakespeare's contemporaries knew it. The association of Hven with heaven involves even the first Danish work published in the English language: Christiern Pedersen's Latin-language *Richt Vay to the Kingdome of Heuine* (1533).[21]

The story of the role that Hven came to play in bringing down the heavens actually begins in 1572, when Shakespeare was only six years old. A "new star" appeared in the Milky Way. It shone brightly and was visible day and night for

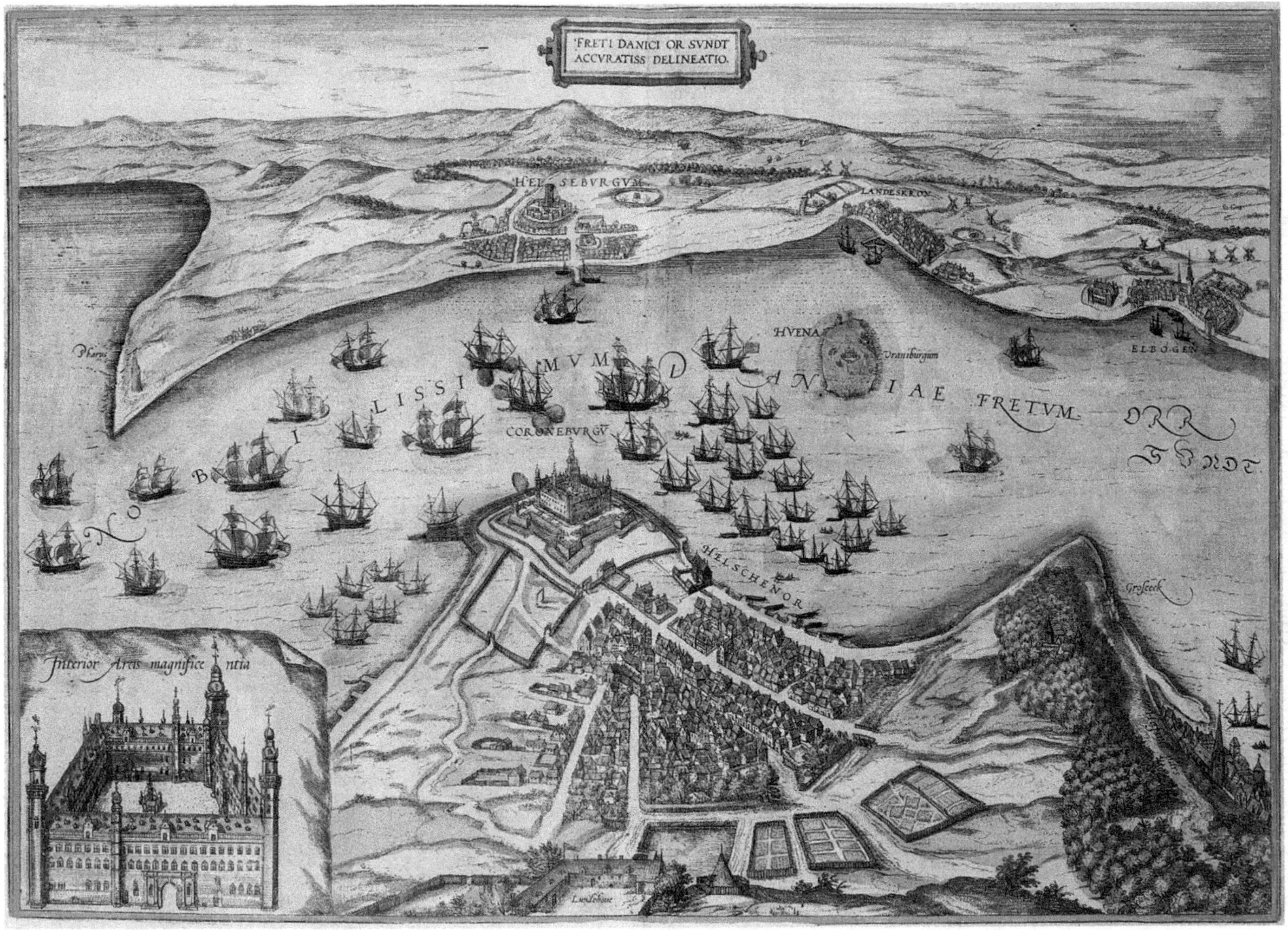

ILLUSTRATION 56 Øresund. The map shows the island of Hven as well as Elsinore and its Kronborg Castle. Georg Braun and Franz Hogenberg, *Civitates Orbis Terrarum* (1588), vol. 4, map 26. Reproduced by permission of Sanderus Antiquariaat.

two years. As an adult, Shakespeare set his *Hamlet* against the backdrop of this supernova, the "star that's westward from the pole." The supernova was the subject of much debate throughout the period during which Shakespeare grew up. In England, astronomers Thomas Allen and Thomas Digges argued against the then-prevailing, dominantly Christian, view of the heavenly "firmament": that only its sublunary region was changeable. They hypothesized that the nova had to be beyond any sublunary region.

In Denmark, where the star also shone, the astronomer Tycho Brahe published his epoch-making *On the New and Never Previously Seen Star* (1573).[22] He then established (1576) a research institute, Uraniborg (Castle of the Heavens), on the island of Hven. This "castle" was one of the most expensive think tank[23] and national research projects in world history.[24] The map titled *Freti Danici Or-Sundt* (Danish Strait Øresund) by Georg Braun and Frans Hogenberg (1590s) gives as much prominence to Hven's Uraniborg as to Elsinore's Kronborg, which was rebuilt in Grand Renaissance style around the same time as Uraniborg's construction. (See Illustration 56, Color Plate 5.)

Tycho's work on Hven "revolutionized" the older understanding of the heavens. Before Shakespeare's time, many people supposed that the heavens were composed of shell-like orbs, "concentric hollow spheres . . . believed to surround [or isolate] the earth and [to] carry the planets and stars with them in their revolution."[25] The heavenly spheres were supposed to be "concentric, transparent, hollow globes imagined by the older astronomers as revolving round the earth and respectively carrying with them the several heavenly bodies [moon, sun, planets, and fixed stars]."[26] That is what Geoffrey Chaucer meant when he wrote about "the cope of heven."[27] By 1590, though, many people were compelled to reconceive their notions about the *horizon* (limit) of heaven, and, by implication, the nature of *orison* (prayer).

Thanks to Tycho's work on Hven—and to that of later counterparts[28]—the view of the universe changed. The stars were no longer considered part of a fixed "eggshell." The English poet John Donne thus wrote in his *Anatomy of the World* (1611) that "new philosophy calls all in doubt, / . . . / The sun is lost, and th'earth, and no man's wit / Can well direct him where to look for it."[29] The old celestial cover, or "canopy," was in *Hamlet* transformed from "fretted roof" into "pestilent congregation." One can hear the old guard fretting about the new frets of heaven:

> . . . this most
> excellent canopy, the air, looks you, this brave o'erhanging firmament,
> this majestical roof fretted
> with golden fire, why, it appears no other thing to
> me than a foul and pestilent congregation of vapours.

(act 2, scene 2)

ILLUSTRATION 57
Archimedes' Lever.
Archimedes (ca.
287–212 BC), ancient
Greek mathematician
and inventor,
reputed to have said,
"Give me a lever
and I will move the
Earth." Woodcut of
Archimedes putting
saying into action
from title page of *The
Mechanic's Magazine*,
London, 1824 /
Universal History
Archive /UIG /
The Bridgeman Art
Library.

This *fret* is not only "ornamental interlaced work."[30] It is also "marine narrows"—called "fretum Öresundicum" on the regular maps of this region (see Illustration 56, Color Plate 5).

The changing view of Earth and heaven was figured even into the architecture of Shakespeare's Globe Theatre: "Over the stage," Frank Kermode reminds us, "was a cover known as 'the heavens.'"[31] The shell was broken at the Globe. Thomas Elyot, in *Image of Governance* (1541), writes that the Globe "was a place made in the form of a bow that hath a great bent."[32] In *Hamlet*, the sailor's tack between the island of Hven (Tycho's "Uraniborg") and the island of Zealand (Shakespeare's "Elsinore") looks mad, but it is the only way to get between one place and the other. "I am but mad north-north-west," says Hamlet. "When the wind is southerly I know a hawk from a handsaw" (act 2, scene 2). The approach to the island of Britain might look straight enough, but it is crooked: "Everything is bent / For England" (act 4, scene 3). One may think that "the bark is ready, and the wind at help" (act 4, scene 3), as Claudius does, but these are finally indirections—windlasses and assays of bias—that, through a process of rebirth, find out the direction that makes Zealand into England or England into Zealand. (See Illustration 57.)

HAMLET'S EGG

Íeg-land or *Éigland*: "Land . . . belonging to water."
—Julius Zupitza and George Edwin MacLean, *An Old and Middle
English Reader: On the Basis of Professor Julius Zupitza's
"Alt- und mittelenglisches Übungsbuch"*

Not only do the English-language words *egg* and *edge* sound alike; they are often linked etymologically by the same etymon: the Old Norse *eddja'*. *Egg* means an "embryo surrounded [or "hedged," as *Hamlet*'s Claudius might put it] by a 'hard shell.'" *Edge*, spelled *ecg* in Old English, is "a sharp cutting edge" (compare the French *côte* [coast]) or outer limit (horizon).

The English-language verb *to egg*, linked directly with *edge* and hearkening back to Old Norse *eggja*, means "to goad on, incite." Thus Thomas Drant, writing in the mid-sixteenth century, says, "Ile egge them on to speake some thyng, whiche spoken may repent them."[33] In this sense of *to egg*, the Ghost of Old Hamlet *eggs* Young Hamlet on to the edge of the cliff. Rosencrantz and Guildenstern "give [Hamlet] a further edge / And drive his purpose on to these delights" (act 3, scene 1). And Hamlet constructs "The Mousetrap" (the play within the play) to egg Claudius on to reveal his inner self. Finally, Hamlet ponders greatness in terms of Fortinbras being egged on, or "pricked [on]," even for the sake of "an egg-shell":

> . . . a delicate and tender prince
>
> . . .
>
> Exposing what is mortal and unsure
> To all that fortune, death and danger dare
> Even for an egg-shell.
>
> (act 4, scene 4)

When Hamlet says, "I could be bounded in a nut shell" (act 2, scene 2), he means, in part, that he is shelled in. Just so, a coastline surrounds a prison island, wooden ribs enclose a prison ship (which holds Hamlet on the passage to England),[34] and the celestial spheres surround the Earth.[35] Yet Hamlet's brain is enclosed within its skull, and Hamlet himself is, like Osric, a chick—half-born, at once both attached to and detached from the egg or womb, as if he were a "lapwing."

Thomas Nashe catches the gist of being on the edge of the egg when, in *Christ's Tears over Jerusalem* (1593), he writes about "My young novice . . . not yet crept out of the shell." It is as if Nashe were referencing the dilemma of novice Isabella, in *Measure for Measure* (1603–1604), who fears both to exit from and to enter into the islandic Clares' nunnery. Isabella is as unsteady a "novice" as Hamlet's nymph Ophelia would be were she to become a "nun."

Hamlet's multivalent words to Ophelia—"Lady [or Nymph] in thy *orizons*, be all my sins remembered" (First Quarto, 1603; emphasis mine)—mark the beginning of a series of exhortations to her: "Get thee to a Nunnerie."[36] The series expresses the same link between physical and spiritual geography that marks the liminal or edgy meanings of the English-language term *orizons*. A "nunnery" is not only a part of the world but also apart from it, like an island, and, in fact, the location of nunneries on islands has always played a role in British thinking, especially in relationship to Denmark,[37] and in Danish thinking.[38] For England, there is the case of Anglo-Saxon Saint Cuthbert, who chose a spe-

cifically tidal island, Lindisfarne, on which to found his monastery. Writes the
Venerable Bede:

Owing to the flow of the ocean tide, called in Greek *rheuma*, twice a day become an island
[*insula*], when the tide ebbs from the uncovered shores, becomes again contiguous with
the land.[39]

In the North Sea Region, there were many nunneries on islands, and many of
them were actually called "insula."[40] Unmarried Danish women in dangerous or
delicate situations, like Ophelia, often went to live in such islandic convents.[41]
Queen Margaret I, who established the Kalmar Union, gave St. Agnes Priory
on Gavnø Island to the Dominican nuns as a residence for just such unmarried
women of high position.

For Shakespeare, however, the nunnery is no simple refuge: actual deliverance
or liberation from the island prison that is womb-like or nunnery-like informs
the plot of such plays as *Measure for Measure* and *The Winter's Tale*. In *Measure
for Measure* (1603–1604), which is the dramatist's most extensive treatment of
nuns and monks, a fetus (the offspring of Juliet and Claudio) awaits deliverance
throughout the play. Only at the end does it issue forth.[42] The stage action of *The
Winter's Tale* (1610–1611) takes place on the island of Sicily and a "Seacoast in Bohemia,"[43] and the offstage action takes place at "Delphi Island." That Shakespeare
makes mainland Delphi into an island has puzzled critics.[44] However, when we recall that *delphos* means "womb" and that *The Winter's Tale* concerns *imprisonment*
in the womb, the puzzle reveals itself. The pregnant Queen Hermione is in prison
when she delivers her child Perdita. The nursemaid remarks that Perdita, although
freed from the "prison" that is the womb, remains a prisoner on the island:

> This child was prisoner to the womb and is
> By law and process of great nature thence
> Freed and enfranchised, not a party to
> The *anger* of the king [Leontes] nor guilty of,
> If any be, the *trespass* of the queen [Hermione].
>
> (act 2, scene 2)

The parallel between Perdita and Hamlet, for whom Denmark is a prison, is distinct. Hamlet finds himself in the same "dire straits" as Perdita. His father (Old
Hamlet), like Perdita's father (Leontes), was angry. (See Old Hamlet's "*angry* Parle";
compare the emphasis in the quotation from *The Winter's Tale*.) Hamlet's mother
(Gertrude), like Perdita's mother (Hermione), is presumed guilty of trespass. (See
Hamlet's statement to his mother, "Your *trespass* . . . speaks.") Zealander Hamlet,
like Sicilian Perdita, is an island prisoner. From this perspective, *The Winter's Tale*,

at the level of "romance," works out a series of problems—dramatic, personal, and historical—that Shakespeare had represented earlier, in tragic format, in *Hamlet*.[45] The role of the "resurrection" at the stormy seacoast in *The Winter's Tale* merely incorporates and transcends (*aufhebt*), at a comedic level, the deus ex machina role of the pirate episode in the North Sea in *Hamlet*. In that offstage episode, Hamlet is reborn from the sea. Henceforth he is on the way to liberty—and to death.

THE GOODWIN SANDS

> The Goodwins I think they call the place—a very dangerous flat, and
> fatal, where the carcasses of many a tall ship lie buried. . . .
>
> —Shakespeare, *The Merchant of Venice*

In *Fratricide Punished*, a German-language version of *Hamlet* that circulated at the beginning of the seventeenth century,[46] the turning point of the story is not, as in Shakespeare's plot, a *chassé-croisé* between two floating islands, one a prison ship and the other a *piratica*. The turnaround takes place instead on a definitively grounded island, where Hamlet and his captors find themselves stranded after a tempest. The brigands plan to kill Hamlet there:

Hamlet: This is a pleasant spot, here on this island [*Insel*]. Let us stay here a while . . .
First Bandit: . . . from this island you will never depart.

. . .

First Bandit: It is our orders from the King; as soon as we get your highness on this island
  we are to kill you.[47]

Thanks to Hamlet's trickery, the brigands shoot each other instead. When Hamlet in *Fratricide Punished* later tells "Horatio" about the incident, he specifies the whereabouts of the island:

Now it happened one day we [on the way from Denmark to England] had contrary winds and we anchored at an island not far from Dover.[48]

Dover marks the main, specifically English, entrance to or exit from the island of Britain. Englishmen and Danes who settled in this area knew well its bicultural and geographic aspect.[49] There are the firm-land guard islands, like Thanet and Sheppey. Each in its own way guards the mouth of the Thames in something like the way that Hormuz Island watches over the opening of the Persian Gulf and that Elsinore on Zealand watches over the Øresund narrows into the Baltic Sea. Both British islands have been the subject of literary, political, and geological investigation for centuries. (Sheppey was of special interest to Charles Darwin's teacher, Charles Lyell.)[50]

There are also near Dover the dangerous infirm islands of the region. These include the shoals of the Nore, partly exposed at low tide, and the Goodwin Sands.

The Goodwins lie at the entrance to the Strait of Dover from the North Sea, whence Hamlet would have had to sail. They had already provided Shakespeare with one happy ending, that in *The Merchant of Venice*. The nice toponym *Goodwins* names the place where the ships belonging to the merchant-prince Antonio presumably went down, lost at sea. It is also where Antonio's ships eventually turn up, with their "good win."

Another place in the region of the Goodwin Sands, now presumably lost under the sea,[51] is Lomea, which the Romans called *Infera Insula* (low island).[52] This was a beautiful stopping-off place for ships crossing the English Channel. Shakespeare would have known this actual location thanks to his association with the Twyne family. Laurence Twyne wrote the *Pattern of Painful Adventures* (1576), partly about Apollonius of Tyre. (It was republished in 1607 just before *Pericles, Prince of Tyre* came out in 1607–1608.) Thomas Twyne produced the edition of his father John's *De Rebus Albionicis, Britannicis, atque Anglis Commentariorum* (1590), in which the Goodwin Sands' Lomea, that lost Atlantis, has a signal position for mysterious rebirth.

BREAKING OUT

"How can someone be born when they are old?" Nicodemus asked.
"Surely they cannot enter a second time into their mother's womb to be born!"

—John 3:4

How *can* a human being be reborn without becoming his or her own father or mother and hence committing a sort of incest of the kind that Hamlet seeks to avoid? This is an old question. In the New Testament, the Pharisee Nicodemus asks Jesus how, without actually entering his mother's womb a second time and then returning from it, a man might be reborn. Jesus provides the one answer that avoids the way of Neronian or Oedipal incest: "Very truly I tell you, no one can enter the kingdom of God unless they are born of water."[53] What the wise counselor Gonzalo in *The Tempest* believes rightly about the boatswain—"I'll warrant him for drowning, though the ship were / no stronger than a nutshell" (act 1, scene 1)—one might say about shelled-in Hamlet at the moment before his sea change.

In the story of Hamlet as Shakespeare refashions it, Hamlet's encounter with the pirates[54] might well have made for his own "watery grave." Instead, it allows him to break out of his bondage. Reborn from the salty waters, Hamlet returns—a new man—to the kingdom he left. Back on island Zealand, he warns Claudius: "I am set naked on / your kingdom" (act 4, scene 7).[55] "What a king [at the last] is this!" (act 5, scene 2).

Hamlet, reborn, now awaits a fitter death than the meager one that Claudius had planned for him in England. This time he will not be shipped out to sea. Instead, he will be shipped properly *into* the *land*. The First Gravedigger draws out for us the parallel here between burial at sea and burial in earth that informs the play as a whole:

> However, age, with his stealing steps,
> Hath claw'd me in his clutch,
> And hath shipped me intil the land,
> As if I had never been such.
>
> (act 5, scene 1)

The Norse-inspired pun on *intil* (into or until) and *till in* (plough in) combines land with water. It matches burial at sea with shipment into earth even as it allows, finally, for a fitting sacrifice as a prelude to burial.

Relevant here is the tradition, common in Britain and Denmark, of burying people in the ground together with ships, as at Snape and at Sutton Hoo (Suffolk), where in the seventh century King Raedwald of East Anglia maintained both a Christian court and a pagan temple. (See Illustration 58.)

The islandic turning point of *Hamlet*, somewhere in the North Sea, is no accident, but goes to the life-and-death heart of the plot.

ILLUSTRATION 58
Excavation of Sutton Hoo Burial Ship, 1939. Photograph by William Phillips, flickr.

Liberty 12

His liberty is full of threats to all.

—Shakespeare, *Hamlet*

EXILE, THEATRICAL AND CORPOREAL

The Globe Theatre's motto was *Totus mundus agit histrionem* (the whole world plays the actor). The reference is partly to the lines spoken by the melancholic Jaques in Shakespeare's *As You Like It* (ca. 1599). In that pastoral play, actors are described in terms of their goings out and comings in:

> All the world's a stage,
> And all the men and women merely players:
> They have their exits and their entrances.
>
> (act 2, scene 7)

In *As You Like It*, Duke Senior, victimized by his usurping brother Duke Frederick, has exited the court and exiled himself in the islandic Forest of Arden, whose name, *Arden*, is derived from Celtic *ardu* (high [land])[1] or "island." (The family of Shakespeare's mother, Mary Arden, came from there.)[2] In much the same way, Duke Prospero of Milan is victimized by his usurping brother Antonio and is exiled on the nameless island where the action of *The Tempest* takes place.

For such exilic representations, the Globe's stage architecture was excellent. Whereas the architecture for earlier English dramatists had often been an "island stage"—that is, surrounded by the audience on all sides like a modern theater-in the-round[3]—Shakespeare's stage was "peninsular," or thrust—that is, surrounded by the audience on three sides and usefully connected to the backstage.

The more architectonic theme of exile, which informs almost all Shakespearean plays,[4] suggests influence from the new translations of Plutarch that were defining exile anew in insular and philosophical terms for the British of the Renaissance. One thematic question of the time was whether isle exile was to be avoided or to be welcomed. On the one hand, island exile was to be avoided. In *Precepts of State-*

craft (ca. 100), Plutarch gives precepts to a colleague, Menemachus, about entering public life, emphasizing the dangers by warning him that a single misstep in the political realm could lead to island banishment.[5] On the other hand, island exile was welcome. In *On Exile* some years later, Plutarch consoles this same Menemachus, who has now been "ostracized" and banished.[6] He begins by pointing out that Sardis (Menemachus's home) is anyway a prison-like shell (*ostrakon*).

I do not think that there are many natives of Sardis who would not choose your fortune even with exile, and be content to live as you do in a strange land, rather than, like snails who have no other home than their shells [ὀστράκοις], enjoy no other blessing but staying at home in ease.[7]

What is more, says Plutarch, life itself is, for the soul, an exilic imprisonment in the body. Socrates thus chose death instead of geographic exile: he wanted finally to exit from the shell that is the body:

But the truth is the soul is an exile and wanderer, being driven about by the divine decrees and laws, and then, as in some sea-girt island [*nēsoi*], gets joined to the body like an oyster to its shell, as Plato says, because it cannot call to mind or remember from what honour and greatness of happiness it migrated . . . exchanging heaven and the moon for earth and life upon earth. . . .[8]

According to Plutarch, then, the wise teaching of the Sicilian island philosopher Empedocles—that "human life on earth is exile from heaven"[9]—means both that "no native land is such by nature"[10] and that the entire universe is native to us:

For nature makes us free and unrestrained, but we bind and confine immure and force ourselves into small and scanty space.[11]

Insofar as Zealand Island is a prison to Hamlet, it is because Hamlet's own thinking makes it so. The mind erects its own calciferous shell. The mind's eye becomes its own island prison. In *On the Nature of Things*, Lucretius describes the wonders of "the three cornered island" of Sicily only to say that "it seems to have produced on its shores nothing more famous, more sacred, more wonderful, or more dear than this man [Empedocles]."[12]

ÞING AND KING

Every hamlet used to have its *t[h]ing*-place, where disputes and conflicts were settled.

—Geir Helgesen and Uichol Kim, *Good Government:*
Nordic and East Asian Perspectives

A thing, my Lord?

—Shakespeare, *Hamlet*

Hamlet, as we have seen, is motivated by a series of *holmganga*—struggles on and/or for an island. What is needed to arrest these struggles is a *tingholm*[13] (*þingholm*)—a relatively democratic assembly held on and for an island. If, in *Hamlet*, "the play's the thing / Wherein I'll catch the conscience of the king" (act 2, scene 2), then the *thing*, as the play-within-the-play ("The Mousetrap") that catches Claudius out, points not only in the direction of Hamlet's individual liberation from avuncular-paternal tyranny but also in the direction of political, even elective, liberty: the *ting*, or *thing* (*þing*) of a Danish sort.

The *tings*, or island-based people's assemblies, of the North Sea Region were a marked historical development.[14] They variably granted kings their "legitimacy,"[15] and, to that extent, they limited their authority and marked the regal horizon. Recently, political scientists[16] have been calling *tings* "the cradle of democracy."[17] The older view, which is more accurate, was that they were "predemocratic."[18] At the *ting*, generally speaking, "any free man was entitled to speak." (Consider Hamlet's "peasant [who] comes so near . . . the courtier.") Its main power, though, resided in "the right to choose and depose a King."[19]

In England at the height of its empire, John Ruskin's friend W. G. Collingwood helped to reintroduce the study of the *ting*. Collingwood's watercolor of Iceland's Althingi is included in his *Sagasteads of Iceland* (1899; see Illustration 59). Around the same time, Israel Gollancz was publishing *Hamlet in Iceland* (1898),[20]

a work that focused partly on the Scandinavian influence on *Hamlet* and thus broadened English understanding of its own political history.

Hamlet, concerned with political succession, provides for succession by *ting*. "The Mousetrap," so called, is "the thing / Wherein I'll catch the conscience of the king" (act 2, scene 2). Hamlet hopes there to demonstrate to the people that the king (Claudius) is nothing, as it were, no *thing*.

Hamlet: . . . The king is a thing—
Guildenstern: A thing, my lord?
Hamlet: Of nothing.[21]

Histories and legends of the Norse *ting* have it, in fact, that "the earliest *þings* were . . . started by people trying to escape the monarchic system in Scandinavia."[22] (The *Faereyinga Saga* says that the Faroe Islands were settled by emigrants who left Norway in the ninth century wanting to escape the tyranny of the Norwegian king Harald I and establish a freer society: they established a principled *ting* on the peninsula of Tinganes in Tórshavn.)[23] Stories about Thorgny the Lagman (lawspeaker) accord with this. In a public speech delivered at the Uppsala *ting* in the tenth century, Thorgny warned the Swedish king Olaf (Olof Skötkonung) to watch his step. "It is not the *King* that holds power, it is the *ting*."[24] Unless the king did the *ting*'s "thing," the *ting* would drown the king in the water-land morass.[25] King Olaf agreed that, from then on, he would follow the rule of the *ting*. This was long before Magna Carta and much more democratic.

In Saxo Grammaticus's *Deeds of the Danes*, the election of kings follows this pattern. The people "acclaim" Amleth (Hamlet) to be their king after he kills Feng (Claudius): "rex alacri cunctorum acclamatione censetur" ("he was appointed king by prompt and general acclaim").[26] In the pre-Shakespearean stories, the hero, once liberated, becomes king by acclamation of the *ting*. One modern Danish poem commemorating Hamlet—inscribed on a stone in Jutland (Hamlet's native land, according to Saxo)—puts it thus:

Amled ypperste	Amled the greatest
Oldtids-snille	Craftiest of Old
Teed sig tåbe	Acted a fool
Til H'vnens time	Until the hour of revenge
Kaaret paa ting	Elected at the Ting
Af jyder til konge	By Jutes to be king
H jsat han hviler	Raised high he rests
Paa Ammel Hede	At Ammel Heath[27]

Shakespeare's Hamlet of island Zealand had no such fate as that of Saxo's Am-leth of mainland Jutland, partly because the story of Hamlet in islandic Elsinore proclaims, in unlike tragic terms, the movement from tyranny, or monarchy based on kinship or marriage, toward relative democracy.

In *Hamlet*, "The toe of the / peasant comes so near the heel of the courtier, [that the peasant] / gaffs [the courtier's] kibe" (act 5, scene 1). Hamlet's friend-ship with Horatio suggests the same accelerating class homogenization as do these gaffs. The crowd's cry as they rush into the palace, "Choose we: Laertes shall be king" (act 4, scene 5), suggests popular rule. The problem of the play, as Hamlet first sees it, is partly that Claudius the tyrant "popp'd in between the election and [Hamlet's] hopes," which still depended on kinship. The ending, which Hamlet assumes, points toward the right of election like that of a *ting* man. Hamlet's dy-ing voice "in the [upcoming] election of a new ruler [to replace Claudius and, to the limiting extent that he is king when he speaks these words, also to replace Hamlet] lights / On Fortinbras" (act 5, scene 2). The Danish Hamlet stands with, and eventually for, the *thing* (*ting*, *þing*).

No matter who "the king / that was and is the question of these wars" (act 1, scene 1),[28] the new dispensation that is Fortinbras may yet differ from them both. Long live the king in Denmark and Britain. *Tingholm* rules out *holmgang*.

The new "Scandinavian dispensation" is already hinted at in the thirteenth-century Icelandic *Saga of Gunnlaugr Snake-Tongue*. Concerning the island duel between the poets Gunnlaugr and Hrafn for the love of Helga, we read:

> But this was the last duel [*hólmganga*] that took place in Iceland [Island], for the very next day a statute was passed [by the *Althingi*] making such hostile [island] meetings illegal.[29]

The law council, having accomplished its purpose, now abolishes duels. (The next day, Gunnlaugr and Hrafn agree to continue their holmgang elsewhere—in Scan-dinavia—but the idea is set.)

What of the British after the time of Canute's *ting*? They went on to establish Parliament on Thorney Island. Although they eventually buried the rivulet tribu-taries of the Tyburn that had made this place an island—as if to hide the Danish raison d'être for the Parliament's starting there—the Tyburn even now continues to flow, underground. It is called the King's Scholars' Pond Sewer.[30] In her post-humously published novel *Thrones, Dominations*, the mystery writer Dorothy L. Sayers says about the subterranean rivers of London:

> You can bury them deep under, sir; you can bind them in tunnels . . . but in the end where a river has been, a river will always be.[31]

Traces of the old *ting* politics are likewise "subterranean" in British and North Sea culture, but the ghosts of the old *tings* yet stalk the lands of the old North Sea

Empire. Consider briefly constitutional issues, toponymical suggestions, and modern Scandinavian institutions.

The constitutional issues, for Britain, involved the role of post–Danish Anglo-Saxon institutions like the *witenagemot* (meeting of wise men) in the development of Britons' understanding of their "indigenous" political assembly and eventual parliament.[32] The *witan* changed with the introduction of the Norse *ting*. Aelfric of Eynsham (Oxfordshire), an islandic patch of land (*ey*) around twenty feet higher than the surrounding Thames floodplain, writes, around 1000, that

no man can make himself king, but the people has the choice to choose as king whom they please; but after he is consecrated as king, he then has dominion over the people, and they cannot shake his yoke off their necks.[33]

Whether or not the Anglo-Saxon *witan* ever had full legitimate authority to choose the king, or even to ratify a chosen one or to depose him,[34] the Danish *ting*'s limitations on the power of the king were usually decisive.

Toponymical traces of the institution of the *tingholm*, assembly places, in British life are clear. The British Isles still have many of these "islet[s] on which the . . . *ting* met."[35] One example is Ting Holm Island in the freshwater Loch of Tingwall on Mainland Island in the Shetland Archipelago (see Illustration 60).[36] (Denmark gave Mainland Island to Scotland in the fifteenth century, but its sovereignty became a matter of dispute between Britain and Denmark at the time that Shakespeare was writing *Hamlet*.)[37] Shetlanders believe that the *ting* on their Mainland Island was the first democratic assembly in the world.[38] Another example is the Isle of Man's *Tynwald*, which Manxmen claim is the oldest parliament

ILLUSTRATION 60
Ting Holm Island. Ting Holm, on a lake on Mainland Island, is in the Shetland Archipelago. The causeway is a relatively recent addition. Photograph by Kevin Osborn. Reproduced by permission.

in the world with an unbroken record of meetings.[39] Such toponyms exist throughout Britain.[40]

In Shakespeare's day, there were still *ting* elections on Ting Holm and elsewhere in the archipelagoes. An instance is suggested in "Ane letter under the commown seale of Zetland of the electioun of Nichole Ayth . . . to the office of lawman generale of all Zetland, quhilk is of the dait, In the Ting holm of Ting Wale" (1532).[41] Archeologists continue to find old *ting* places, including Thynghowe (*ting* mound), an important site long used for resolving political issues recently unearthed in Sherwood Forest, Nottinghamshire.[42]

The regional and national assemblies of Scandinavia and related regions are still often located on islands and/or named by them. These islands/names include Iceland's *Alþingi*, Åland's *Lagting*, the Faroe Islands' *Løgting*, and Greenland's *Landsting*. The Danish National Parliament, or *Folketing*, in Copenhagen, is on Strandholmen (or Slotsholmen) in the strait between the islands of Zealand and Amager.[43] For Britain, the chief case is Westminster itself. Already in the charter of Canute there is mention of the "hustings weight";[44] the hustings, or courts, were formerly held in such places as the English island of Sheppey in the Thames Estuary.[45]

MAGNA CARTA ISLAND

Among the British isles relevant to the development of democracy in Great Britain, from Ting Holm to Westminster, is the island where the Magna Carta was signed. History has it that it was there that, on June 15, 1215, the English barons documented the limitations of the king's power. They compelled King John to sign the charter at the "water-meadow" called Runnymede. *Water-meadow* means "an area of grassland subject to controlled irrigation"—hence, an "island."[46] *Runnymede* is derived from the Anglo-Saxon words *runieg* (regular meeting) and *mede* (mead or meadow). In his *Short History of the English People* (1874), John Richard Green

recalls the place: "An island in the Thames between Staines and Windsor had been chosen as the place of conference; the King encamped on one bank, while the Barons covered the marshy flat, still known by the name of Runnymede, on the other."[47] Other historians say that the charter was actually signed on the nearby small island of Ankerwyke (which sometimes goes by the name *Magna Carta Island*). However, the history of the geomorphology of the place suggests that it does not really matter which particular "place" it was. The island and water-meadow where the "parle" occurred may have been one place at that time.[48] In any case, *Runnymede* names the places where the Anglo-Saxon *witan*, or parliament, sometimes met as far back as the reign of Alfred the Great,[49] whose place in the mythology of British naval power we will consider in the next chapter.

After King John died, in October 1216, his successor, Henry III, reissued the charter, with variants, first in November 1216 and then in 1217. The truce of 1217 is depicted in Illustration 61.[50] In the margin of the manuscript, Matthew of Paris, a contemporary, drew a diagram emphasizing the fact that the meeting took place on "water-land"—and, given Danish-British history, it almost had to have been.[51]

Part Four

Sea and Land

Hamlet *Is* Germany

RULE, GERMANIA!

What has become of the "natural liberties" of primitive humanity saved
from the deluge in the ark of England?

—Moisey Ostrogorsky, *Democracy and the
Organization of Political Parties*

How do the islandological issues of land and sea, especially as presented in *Hamlet*
and illustrated by ancient Greek history,[1] come to bear on the political and cultural history of Germany? German thinkers themselves have remarked that any
answer to this question would need to bring to bear the question, Whose sea is
it? and, more particularly, the ancient Roman way of answering it. The ancient
Romans called the body of water that we moderns call the Mediterranean (mid-
land, middle-of-the-earth) by the name *Mare Nostrum*—meaning "our sea." But
whose was it?[2]

The Romans often claimed the land surrounding what they understood
as a *mare internum*, as Sallust calls it,[3] partly because it was bordered by their
land. It was as if "*our* sea" were an island of water surrounded by a sea of our
territory—as if the Mediterranean were a "liquid continent."[4]

In that sense, the Romans did not claim the waters in the way they claimed
a tract of land. (The general Roman idea was that the sea was common to all
men—and so, too, were men of its shorelines.)[5] Nor, since ships in those days
closely hugged the shoreline, did most Romans think of their "dominion" in a
strictly maritime way. The "Roman sea" was meant ideally to be open—free to
travel except for straits and bottlenecks—and belonging to no one: it was a "loose
fish." Only with the advent of maritime republics a thousand years later—long
after the decline of ancient Rome—did claims to "pan-maritime dominion" arise
in a serious way. These claims meant more than requiring tolls for the use of water
lanes—the Danish tolls for passage through the Øresund at Hamlet's Elsinore, for
example, and the German tolls for passage through the Rhine River narrows at
the river maidens' Lorelei. With Christopher Columbus's "discovery" of the New

World in 1492, dominion of the sea came to mean even more than claiming the internal sea, many of whose shores the Venetians or others already dominated. Now it meant claiming *all* the world's seas—and hence, by implication, all its islands. Owning the main (i.e., the sea) now meant owning the mainlands.

There were several stages in the development of arguments for and against such claims for world dominion.

First, in 1493, the pope put forth the view that the high seas outside the Strait of Gibraltar—the neck of water guarding the entrance to the internal Mediterranean Sea and the exit from it—were closed, meant for control by the major Catholic nations. That is the gist of various *Papal Responses* (1493–1529).[6] The Portuguese king, for example, was now called "King of Portugal and the Algarves, within and beyond the sea in Africa, Lord of Commerce, Conquest and Shipping of Arabia, Persia and India."

Second, there was the response from Protestant Holland, a "lowlands" of dikes and islands. That "the seas are open" was the argument of Hugo Grotius's *Free Sea* (Mare Liberum; 1609), published two years after Shakespeare first staged *The Tempest*, a play partly about dominion of an ambiguously located island. In *The Free Sea*, Grotius laid the practical framework for international law, arguing that the sea was "international territory" and all groups were free to use it. He also put forth the view that "uninhabited" islands belonged to those who first landed there.[7]

Third, there came a powerful, predictable English restatement of the first thesis, that a single power could own the seas—and a prediction that Britain, given time, would be that power. This restatement came in the form of the British historian John Selden's Latin-language *Mare Clausum* (1618). When this volume was finally published as *The Closed Sea* (1635)—King James had feared an earlier date of publication[8]—Selden's argument finally went public:

That the Dominion of the British Sea, or That Which Incompasseth the Isle of Great Britain, is, and Ever Hath Been, a Part or Appendant of the Empire of that Island.[9]

The sea, Selden argued further, was capable of appropriation in the same way that land was. Just how far and wide, then, should the "British Sea" extend?

Marchamont Needham makes clear the implications of this sort of question in the epistle dedicatory to his English-language translation of *The Closed Sea*. Needham's poetical "Neptune to the Commonwealth of England" (1652) poses a powerful rhetorical question:

> What then should great Britannia pleas,
> But rule as Ladie o're all the seas. . . .[10]

Here was the overt claim to rule the world's main (ocean) and, by implication, to rule also the world's main(lands).

Fourth, a practically acute if intellectually obtuse Dutch response attempted to synthesize the open-sea viewpoint (Grotius) and closed-sea viewpoint (Selden). Cornelius Bynkershoek's synthetic *Dominion of the Sea* (1702) suggests that dominion at sea should extend out from the coast only so far as the nation-state could fire a cannonball from its *shores*.

The British, however, preferred to measure their claim according to how far they could fire cannonballs from onboard English *ships*. The difference involves a political outlook as well as a practical stratagem. England, it turned out, wanted to view itself not only as a fixed island but also as a floating ship. By way of example, consider the famous Armada Jewel (ca. 1580s; see Illustration 62, Color Plate 6), which depicts the ark of the English Church with the inscription *Saevas tranquilla per undas* (tranquil through stormy waves). James I's medals show the same ark.[11]

ILLUSTRATION 62
Britain as Ship:
The Heneage Jewel
or Armada Jewel.
On the front of the
jewel is a miniature
of Elizabeth I
(1558–1603). On the
reverse (shown here)
a ship holds steady
on a stormy sea. The
inscription reads
*Saevas tranquilla
per undas* (tranquil
through fierce waves).
Painting by Nicholas
Hilliard, ca. 1595.
© Victoria and Albert
Museum, London.
Reproduced by
permission.

Such claims gave some credence to the British argument that they held, and should hold, near-universal maritime dominion. In this regard, a musical masque titled *Alfred*, produced in 1740 by Thomas Augustine Arne with words by James Thomson and David Mallet, was typical in that it hearkened back to the old Elizabethan idea of England as floating island.

Alfred celebrated the Anglo-Saxon Alfred the Great's ninth-century defeat of the Danes, a then-presumptive turning point in British history and, in particular, the origin of what Britishers regarded as their world naval power. The action of the masque takes place "after the [roving] Danes had made themselves masters of most of the Kingdom of Wessex" in 878. King Alfred "found himself obliged to retire to the little isle of Athelney [*Æthelinga Ieg*, or Island of Princes] in Somersetshire."[12] On Athelney, still sometimes called "the ark of England," Saint Cuthbert, from the tidal island Lindisfarne, appears to King Alfred; a wise peasant dwelling on the marshy water-land island insists to Alfred that victory over the Danes entails understanding the water table even better than the Danes do:

> Content thee, wife:
> This island is of strength. Nature's own hand
> Hath planted round a deep defence of woods,
> The sounding ash, the mighty oak; each tree
> A sheltering grove: and choak'd up all between
> With wild encumbrance of perplexing thorns,
> And horrid brakes. Beyond this woody verge,
> Two rivers broad and rapid hem us in.
> Along their channel spreads the gulphy pool,
> And trembling quagmire, whose deceitful green
> Betrays the foot it tempts. One path alone
> Winds to this plain, so roughly difficult,
> This single arm, poor shepherd as I am,
> Could well dispute it with twice twenty Danes.[13]

The little "ham" (or raised land) on which the peasant and his wife live "hems" or "hedges" them in, yes, but at the same time, this moving water-land provides the foundation for the *idea* of Britain as "fortress ark," an idea that informs the British tradition during the following centuries.[14]

For a long while, England had been a fortress. Shakespeare's John of Gaunt in *Richard II* says it well: England is "this scepter'd isle" that is blessed as a "fortress built by Nature for herself" (act 2, scene 1). But now it was also a moving island, a ship, from which to take or absorb the world. "Give me but one firm spot on which to stand," said Archimedes, "and I will move the earth."[15] The Sicilian islander did not have in mind only a ship, spaceship, or volcano, it would now seem, but also a global political order.

In *Alfred*, of course, victory falls to the British. The most famous lines of "Rule, Britannia!" are these:

> All thine shall be the subject main,
> And every shore it circles thine.
> "Rule, Britannia! rule the waves:
> Britons never will be slaves."[16]

The English nationalist audience of Arne's *Alfred* adored the notion of a sovereign who holds dominion over both the *main* (seas) and the *mainland* (continents) that the main surrounds.

> Britons, proceed, the subject Deep command,
> Awe with your navies every hostile land.
> In vain their threats, their armies all in vain:
> They rule the balanc'd world, who rule the main.[17]

The masque became an oratorio in 1745 and then an opera in 1753. All three forms of the spectacle ended with the plea for global rule. People worldwide now had the same reason to beware of the British restatement of the papal claim to maritime dominion.

In this regard, the German principalities were especially interested in British thinking and self-representation. For one thing, many Germans identified with the British claim because Britain was, in a manner of speaking, German. The first Hanoverian king, George I (Georg Ludwig), was a German, and Arne and Thomson actually wrote *Alfred* to commemorate his succession in 1714; its first performance took place at the home of George's son, Friedrich (already Marquess of the Isle of Ely), who was first in line to inherit the British throne.

Not surprisingly, therefore, German writers and composers, soon after *Alfred*'s success, began to imitate the new German-British naval jingoism. Variations of Arne and Thomson's "Rule, Britannia!"—the most famous chorus from *Alfred*—included the *Occasional Oratorio* (1746), composed by the German-British George Frideric Handel (from the Duchy of Madgeburg),[18] as well as the *Five Variations on "Rule Britannia"* (1803)[19] and *Wellington's Victory* (1813)[20] by Ludwig van Beethoven (from the Electorate of Cologne). They also included the *Grand Variations on "Rule Britannia"* (1817)[21] by Ferdinand Ries (from the Electorate of Cologne) and the waltz *Homage to Queen Victoria of Great Britain* (1838)[22] by Johann Strauss (from Vienna). The same year of the *Homage*, the young Richard Wagner (from the Brühl in Leipzig) wrote his overture *Rule Britannia* (1837–1838).[23] He believed that "the first eight notes of [Arne's] *Rule Britannia* typified the British character."[24] For all of these composers, there was not yet a Germany, but, thanks partly to them, one was in the making,

No one, including Wagner, likes to imitate forever the national songs of another *Volk*. One salve for such discomfort is to claim that the other *Volk* is one's own. Here, that meant the argument that the English as a whole (not only the royal family) were actually Germans "planted on an island." (Saxo Grammaticus, after all, had claimed likewise that the English were island-planted Angles.) Another solution to the problem for the Germans, a largely continental group of principalities, was for them to somehow found an islandic culture of their own. Wagner, as we will see, in "The Ring in the Reef," saw the solution as meaning that he had to create an ideal musical and poetic tradition of extraordinary achievement and political consequence. This creation would seek, in the wake of Wagner's essay *German Music* and his opera *The Flying Dutchman*, to translate, in a German realm of art, the real power of Britain as a free, floating, islandic empire.

PLANTED ON A FREE ISLAND

By the unerring and unalterable laws of nature, the people who live in
an island are or may be entirely free.

—*Encylopaedia Britannica*

For many centuries, England had thought of itself as ruling the seas. Since the seas surround mainlands and islands, ruling them meant ruling the whole Earth. By the nineteenth century, this notion was codified in some British colonial law:

By the law of England the King of England is Lord of the four seas and his power over the ocean has extended from time to time by the growth of the Navy and the maritime ascendancy of Great Britain.[25]

Such a global mythology or legal fiction was relatively well known from the medieval period.[26] How it played out in the modern period centrally involves German-British history.

For hundreds of years, most Germans had thought of Germany as relatively landlocked and happily so. The German astronomer Johannes Kepler developed his dislike of Hven into a distaste for all islands that was typical of many Germans until the Romantic period. In fact, Kepler turned down an invitation from King James I of England to visit Britain because of this Teutonic dislike. Half-jokingly, he writes to his astronomer friend Matthias Bernegger:

Therefore shall I cross the sea, where Lord Wotan calls me? I, a German? A lover of firm land [*Festlandes*], who dreads the confinement of a narrow island [*der engen Insel*]?[27]

Neither Angle (Englishman) nor Dane, the German Kepler was a landlubber.[28]

However, Kepler's attitude toward Hven did have a neurotic aspect. In fact, Kepler never visited Hven, but met Tycho Brahe in person only after Tycho had left the island in 1597 and taken up work at a "new Hven," the observatory at Benátky on the Jizera River. There Kepler worked as Tycho's assistant until the latter's death in 1601. Whether or not Kepler had a hand in Tycho's death—which seems unlikely to me[29]—he did steal his research papers[30] and used them to derive the law of ellipses that helped ground his own *Astronomia Nova* published in 1609. That same year, Kepler developed something like a guilty conscience. In a strange pseudo-fictional work, *The Dream* (Somnium), his proxy, Duracotus, who hails from the island of Thule, works with Tycho on Hven. "I was delighted beyond measure by the astronomical activities," says Duracotus, "for Brahe and his students watched the moon and the stars all night with marvelous instruments." After working with Tycho, Duracotus travels to the spherical island that is the moon. Kepler kept *The Dream* under wraps. However, he added lengthy notes to it over the course of the next fourteen years. By the time of his death there were 223 such notes.[31] *The Dream* was published posthumously.

Starting with the general Enlightenment in the later eighteenth century, Germans were becoming less happy with living relatively landlocked. They would now try to discover themselves as a "free nation" by projecting a *Festland*—of the sort that Goethe's celebrated *Faust* wrests from the sea by the employment of dikes—and by finding their culture mirrored in the archipelagoes not so much in ancient Greece as in the North Sea Region. That is to say, German thinkers sought to replace an earlier eighteenth-century admiration for the Greeks—which the British scholar Eliza Marian Butler in 1935 called "the tyranny of Greece over Germany"[32]—with a newer admiration for the Norse. Among these thinkers were many for whom the Mediterranean archipelago of ancient Greek times and later of the medieval Venetian period were crucial historical and conceptual moments in the "unfolding" of human culture, but required, in the modern era, radical dialectical incorporation and transcendence.

Johann Gottfried von Herder, in *Another Philosophy of History for the Education of Mankind* (1774), praised the Mediterranean islands: "How famed are you, Mediterranean Archipelago, in the history of the human spirit! The first commercial state, founded entirely on trade."[33] In the modern world, though, what is there? There is England and the English island people. What explains English success? Herder concludes that the English are lucky in their islandic geography. He argues in *Ideas for the Philosophy of History of Humanity* (1784–1791)[34] that the English are not only "hemmed in" by their island but also, on that account, "liberated" by it. In fact, their island habitat is, for Herder, the main difference between the states of German and English culture. In *On the Resemblance of Medieval English and German Poetry* (1777), he says that the English and the Germans comprise one people, and, in *Towards Philosophy and History*, he writes, "We [Germans] feel the English are men of our kind, bone of our bone."[35] The general idea is that "the English are Germans planted on an island."[36] Kant makes the same point in terms of language:

Latin language is dominant in Spain, France, Italy; German in the remaining cultivated countries. Slovonic is still Asiatic. All are inoculated with German blood. In England, it [German] is the foundation. Spirit (Roman) and discipline (German). Way of ruling.[37]

The Germans and English are as closely related as the First Gravedigger in *Hamlet* jokingly says that the Danes and English are,[38] and as Saxo Grammaticus seriously says that they are.

From this point of view, the Germans, in order to enter the modern world (England's), were required to become, in a manner of speaking and thinking,

islandic. This meant partly that Germany would need to cease admiring the inland sea that is the Mediterranean, with its gates at Gibraltar and the strait of that
name. It would need to focus more on the Baltic, with its Øresund area. To put
it otherwise, Germany would need to subsume an older definition of culture in
terms of the Mediterranean within a newer one in terms of the Baltic.

The task would not be easy: the German language lacked even a single word
to refer reliably to the Baltic region,[39] but never before in world history had such
a team for such a task been assembled.

Goethe, for one, went some way toward accomplishing this novel "topographical hermeneutics"[40] by shifting the view from the Mediterranean to the Nordic
regions in his "Helena Act" of *Faust Part Two* (1827). The better title for this act
would be "A Classical-Romantic Phantasmagoria": the "classical" Helen of Troy
is teleported, over both time and space, to modern "romantic" Germany, where
she speaks modern German using ancient Greek syntax. *Faust*'s Mephistophelean
Phorkys—the son of Pontus (Sea) and Gaia (Earth)[41]—himself transports Helen
to Faust's Teutonic fortress.

The northern sea comes to replace the southern sea in Herder's influential
works. In *Outlines of the Philosophy of the History of Man* (1784–1791), Herder
writes that the northern Baltic will replace the southern Mediterranean: "The Baltic [will be] the eye of all Northern Europe."[42] The island of Rügen, which became
German only in 1815, served the purpose when it came to an island for the German
national spirit.

Rügen was an interesting and influential choice. The island had been Danish
from 1168 until 1325 and for centuries had been the seat of an international trading empire that included even the Arabic world. (Herder says that "Vineta, in the
island of Rügen, was the Amsterdam of the [S]lavians.")[43] The larger Baltic area
had been known as the *Mare Rugianorum*,[44] and the nearby Baltic island Usedom, with its Venice-like lagoon, still shows the remains of the great pre-Christian
temple at Arkona, about which Saxo Grammaticus wrote. The geology of Rügen
presented, in fact, a sublime landscape of the sort that Immanuel Kant, a teacher
of Herder in Königsberg (the island-city home of both in Prussia), had described
(see Chapter 9), a landscape that, for the Germans, was exemplified in the coast
of *Hamlet*, as it was for Shakespeare's translator August Wilhelm von Schlegel
(from Hanover). Finally, though the island of Rügen evinced German national
consciousness, it was not yet, strictly speaking, part of "Germany." It devolved to
German ownership only in 1815—and it is one place where people looked, if they
were so inclined, for *Hamlet*'s "little patch."

According to the German nationalist poet, preacher, and scholar Ludwig Gotthard Kosegarten, who hailed from the coast of nearby Swedish Pomerania, Rügen

was also associable with the sunken Atlantis of distant prehistory.[45] Kosegarten's *Uferpredikten* (Shore-Sermons) envisaged homilies and religious lectures taking place at Wittow on the island. His poem "Eclogue" (1796) and his two-volume novel *Ida of Plessen*,[46] along with both *Jucunde: Eine ländliche Dichtung* (1797) and the "Hymn to the Island of Rügen," are set there:

> Germany's splendor! . . .
> Full like the sea is Germany's power
> And which is consolation, like this coastal wall
> For fate and time.[47]

Ernst Moritz Arndt, born and raised on Rügen itself, followed suit and became thereby the national bard for the unity and liberty of the German nation. His *Towards a History of the Bonded Peasantry in Pomerania and Rügen* (1803) helped trigger both a general liberation of German serfs[48] and entrance into a recognizably modern world. Arndt's nationalist work included translations into German of works written in Swedish, Old English, and Old Scottish, as well as monographs on the Orkneys and Shetlands.[49] His dreamy projects for a linguistic alliance with the Scandinavian Northlands entailed a pan-Germanic realm that included the Netherlands and the Frisian Islands. Wilhelm Müller's poem cycle *Shells from Rügen Island* (1825) includes "Der Adler auf Arkona" (The Eagle on Arkona), with a similar call for German national unity and liberty based partly on Saxo's description of the pagan temple at Rügen's Arkona.[50] In "Rügen, the Island Vineta"—one of the poems that Henry Wadsworth Longfellow collected in his monumental *Poems of Places*—Müller, who hailed from Dessau, writes, "Deep beneath the gleaming surface sunken / Ruins of that city still remain."[51] As late as World War II, the German expatriate Hans Henny Jahnn, author of *River without Banks* (1949–1961) and admirer of works by Müller, was discovering on Danish Bornholm the same Romantic tensions between paganism and Christianity. His *Die Insel Bornholm* (1941) shows marked sympathies with the National Socialist (Nazi) sort of flirtation with Nordic landscape and seascape:

The sea shapes the island. It is the second landscape. It lies deep beneath the hills. All the roads that lead to the coast end in a tremendous view whereby a blue, grey, gleaming or dull, overcast or wind-whipped surface spreads before one as in an enormous valley.[52]

For Wilhelm von Humboldt, who was the founder of the Prussian educational system during the first decades of the nineteenth century, the island locus of Rügen was likewise sublime and sacred: the universe writ small.[53] Karl Friedrich von Schlegel (brother of August Wilhelm) expresses a like interest in the island in his *Philosophy of History* (1829),[54] and Friedrich Schleiermacher, Henriette Herz, and

Ehrenfried von Willich all projected onto it a Romantic cult of political friend-
ship. Writes Henriette Herz:

Amidst all the strange changes within and around me, [Rügen] is the only fixed point on
which I have focused for a long time and always with the same happiness. It is the only
piece of life that I see before me, like a small island in a desolate sea.[55]

Rügen's becoming "German" in 1815 was, for all these thinkers and many more, a
crucial point on the road toward German nationhood.

GERMANY IS HAMLET!

Deutschland ist Hamlet! [Germany is Hamlet!]

—Ferdinand Freiligrath, *Ein Glaubensbekenntnis*

During the earlier years of the *Sturm und Drang* literary movement in Germany
(the latter decades of the eighteenth century), Christoph Martin Wieland translated Shakespeare's island play *The Tempest* (1762–1766), and *verdeutscht* (German-
ified) the title as *Die Sturm; oder Zie zauberte Insel* (The Storm; or, the Enchanted
Island; 1762–1766). By the time Wilhelm's translations were completed, mainland
Germany's island trajectory was well set.

Heinrich Laube, following in the footsteps of these writers, in 1837 produced
A Voyage to Pomerania and the Island of Rügen,[56] in which he compares the coast
of Rügen with that of Britain as understood by Shakespeare in such plays as
Richard II.[57] Laube went on to prominence in the Junges Deutschland (Young
Germany) movement, which much influenced the young Richard Wagner.

Appropriation of Rügen went hand in hand also for Goethe's title character
in *Wilhelm Meister's Apprenticeship* (1795–1796), often said to be the first *Bildungsroman*, or coming-of-age novel. Wilhelm, with an eye on the Baltic, "defend[s]
Shakespeare [as an author who was] writing for an island people."[58] What's more,
he singles out Danish Hamlet as a *typical* Germanic figure.[59]

Ferdinand Freiligrath's freedom-loving line, "Germany is Hamlet!," was dangerous for the poet: the authorities expelled him from Germany. And yet, by
1865, a good part of the German intelligentsia already believed that the English-
Danish hero of *Hamlet* not only was a German but also was the *essential* German. Hermann Ulrici, professor of philosophy at Halle (Saxony), sought in 1865
to conceptually *entenglisiren* (to un-English) Shakespeare and, at the same time,
ihn verdeutschen (to make him German).[60] Wilhelm Oechelhäuser, founder of the
German Shakespeare Society in Weimar, felt a similar need to show that Shakespeare was more "deutsch-national" than "Englander."

Oechelhäuser was not only founder of the German Shakespeare Society. He
was also the head of Deutsche Continental Gas-Gesellschaft (now Gas Conti).

A good part of German industry and literary nationalism were one, and in the twentieth century their union in politics was always worth watching. By 1911, a year that marked the zenith of the British Empire, Friedrich Gundolf wrote his epoch-marking *Shakespeare and the German Spirit* (1911).[61] He demonstrated to the satisfaction of most readers how and why the German nation had discovered itself in Shakespeare and his Scandinavian play. Now all that remained was for the Germans to "Germanify" Britain—for it to dispose of the tension between real conditions and ideal ambition. Put otherwise, what needed doing now was the dispossession of Englanders.

The Nazi opportunist Carl Schmitt writes in *Land and Sea* (1942) that the worldwide British Empire was centered in an island nation that resembles an untethered Leviathan:

Therefore, the island of Britain, the metropolis of a world empire raised on a maritime destiny, would be uprooted and lose its territorial character. Like a fish, it was able to swim to another spot of the globe. It was no more and no less than the mobile centre of a world empire, the possessions of which were strewn in no coherent pattern over all continents.[62]

By this time, though, there was a global Leviathan that the Germans needed or wanted to fear. Germany not only had its dream of a territorialized Rügen; it also had its nightmare of floating aimlessly like an iceberg in the North Sea.[63] That is the stuff of such works as Richard Wagner's operas and Arnold Fanck's mountain movies.

The Region of Illusion

14

... the region of illusion [*Sitz des Scheins*], where many a fog-bank [*Nebelbank*], many an iceberg [*bald wegschmelzende Eis*] seems to the mariner, on his voyage of discovery, a new country, and, while constantly deluding him with vain hopes, engages him in dangerous adventures, from which he never can desist, and which yet he never can bring to a termination.

—Kant, *Critique of Pure Reason*

THE RING IN THE REEF

At the close of the Middle Ages a new impulse led the nations forth to voyages of discovery; no longer the land-locked sea of the Hellenic world, but the Ocean that engirdles the earth.

—Richard Wagner, on his *Christopher Columbus Overture*, quoted in *The Musical Times*

At around the time that Richard Wagner was writing *Christopher Columbus*, which concerns the way that all land on Earth had become an island (engirdled by water), and *Rule Britannia*, which concerns the way that Earth was ruled by the island of Britain, he experienced a change of mind that transferred his attention back to Germany. In Heinrich Laube's *Journal for an Elegant Society* (*Zeitung für die Elegante Welt*), he published his first essay, the pathbreaking "German Opera" (1834).[1] Wagner's meeting with Laube in the 1830s was part of the turning point toward Germany, away from Britain's "islescape" and island ideology. At the time, Laube was working on two interrelated projects: reforming German thought in such a way as to encourage the development of a modern German national identity, and presenting to the German public the "islescape" of the German island of Rügen. Within a few years, in his journal *Zeitung für die Elegante Welt*, Laube was publishing Wagner's "German Music" and "Autobiographic Sketch" (1843).

Wagner had once looked to islandic British themes and settings—*Rule Britannia*. He had once intended to set his opera *The Flying Dutchman* (1843), with its story of an islandic specter ship, in Scotland. His source, Heinrich Heine's *Memoirs of Mister von Schnabelewopski* (1833), had a Scottish setting,[2] and Wagner's original plan even had Scottish names. Now, however, he removed the setting to the North Sea off Norway, which much impressed him, as he later recounted.[3]

Heine shows another understanding of linkages between England and German culture,[4] based partly on the time he spent on islands in the North Sea, especially Norderney and Heligoland in the mid-1820s. These days one thinks of these islands as German, but that was not the case in Heine's time. Norderney had belonged to Prussia, but when Heine visited, it was under Dutch sovereignty, and he introduced in his various writings on the island the figure of the Flying Dutchman to prepare the way for his groundbreaking discussion of modernity.[5] (Richard Wagner, who took the legend from Heine, had quite other purposes.) Sometimes these writings are celebrated as "the most intense poetic representations of the sea in all of German literature,"[6] in which one finds references to Greco-Roman, Norse, and Judeo-Christian mythologies. (Heine's works on the "sacred land" of insular Heligoland suggests the Danish context: the island had been Danish during the previous six centuries, but in Heine's day it was under English sovereignty. German rusticators and political refugees from the French rebellions of 1830 and 1848 often went there. In his writings about Ludwig Börne in 1840, Heine clarifies this relationship between British Heligoland and European politics.)[7] Laube's *German Opera*, moreover, helped set Wagner on a political and nationalist trajectory leading to *The Ring of the Nibelung*, a four-part "total work of art," world-historical in its mythic scope, that was variously written between 1848 and 1874. Its ambition was to update, match, and surpass the accomplishments of Shakespeare—both the Globe Theatre and its productions.

Wagner's preliminary sketch for *The Ring of the Nibelung* was "The Wibelungen: World History as Told in Saga" (1848), which was based partly on Norse texts informed by the idea of the *ting* (Norse assembly), including the *Sigrdrifumál*.[8] The sketch starts where Wagner claims to find the beginning of Western history: an Aryan island:

At the epoch which most Sagas call the "Sint-Fluth" or Great Deluge, when our earth's Northern hemisphere was about as much covered by water as now is the Southern, the largest island of this northern world-sea would have been the highest mountain-range of Asia, the so-called Indian Caucasus: *upon this island, i.e. these mountains, we have to seek the cradle of the present Asiatic peoples, as also of those who wandered forth to Europe* [emphasis mine]. Here is the ancestral seat of all religions, of every tongue, of all these nations' Kinghood. . . . Now, when the waters retreated from the northern hemisphere to flood the southern once again, and the earth thus took its present guise, the teeming population of that mountain-isle descended to the new-found valleys, the gradually emerging plains.[9]

Wagner's *Nibelung Myth as the Scenario for a Drama* (written a few weeks later in 1848) sets out the trajectory in terms of the four-part opera that was completed three decades later.

The Rhine Gold, the first opera in the series of four, begins with an ideal "waterscape" location. The scene effectively matches that admired by painters who, in earlier decades of the nineteenth century, had painted Rügen: Philipp Hackert, Carl Gustav Carus, Karl Friedrich Schinkel, Carl Blechen, Friedrich Preller the Elder, and Caspar David Friedrich.

Caspar David Friedrich was a German-speaking Swedish Pomeranian from Greifswald who had studied in Denmark, where he observed the Danish understanding of water as land or land as water. He later demonstrated that observation in such works as *Chalk Cliffs on Rügen* (1818).[10] *The Ring of the Nibelung*, which is about the foundation of the German spirit, begins at precisely such a spot but one translated to the Rhine. The Rhine likewise has the originally Celtic hydronym, *renos*, meaning "that which flows." Since river hydronyms change more slowly than other words,[11] they provide critical topography with a place to start.[12]

The plot of *The Rhine Gold* begins, then, at the bottom of the Rhine: *auf dem Grunde des Rheines*. The music starts from a wavering four-minute-long E-flat drone that provides a backdrop for a watery *Riff* (reef), a Norse-based word that indicates "a ridge or bank of rock, sand, shingle, etc., lying just above or just below the surface of the sea or another body of water."[13] Originally a Scandinavian term, *Riff* appears many times (in stage directions and elsewhere), and it is often echoed in the *Reif* (ring) that is the main motive of *The Ring of the Nibelung*.

Das Licht lösch' ich euch aus,	Your light I'll put out,
entreiße dem Riff das Gold,	wrench the gold from the reef
schmiede den rächenden Ring.	and forge the avenging ring.[14]

Moreover, *Riff* was often used as the toponym of the Danish-Swedish Øresund, through whose Old Norse narrows Wagner himself passed. And in his "Autobiographic Sketch" (1843),[15] he singles out a particular Norwegian region of reefs that, he says, inspired his first opera, *The Flying Dutchman* (1843), which he would claim as "really my own":

The voyage through the Norwegian reefs made a wonderful impression on my imagination; the legend of the Flying Dutchman, which the sailors verified, took on a distinctive, strange colouring that only my sea adventures could have given it.[16]

The setting for *The Rhine Gold* (1869) is this Scandinavian reef transformed to German ends thanks to Wagner's reworking of Heine's use of *Riffe* in *Die Loreley* (1823). Heine's poem is set at the great rock Lorelei that marks the narrowest point—or *hals*, *els*, or "neck"—on the Rhine, which empties into the North Sea. *Die Loreley* includes the famously sonorous, mermaid-like siren-woman, who is both seductive and dangerous:[17]

Den Schiffer im kleinen Schiffe	The seaman in his tiny yacht
Ergreift es mit wildem Weh;	It grasps with wilding woe,
Er schaut nicht die *Felsenriffe*,	He looks not at the rock-reefs as he ought,
Er schaut nur hinauf in die Höh.	He looks only up from below.[18]

It was at the *Felsenriffe* (rocky reefs) of Lorelei on the Rhine that Ferdinand Freiligrath composed the famous poem "Germany Is Hamlet!" (1865).[19]

Relevant to the setting of the opening scene in *The Rhine Gold* is the architecture of the theater in which Wagner wanted to set it. Not only did he make orders for the stage designers and painters for the scene; he also built the theater at Bayreuth with the definite purpose of presenting all *Weltgeschichte* (world history) on stage in a single work. He carefully constructed a specific playhouse with a double proscenium and hidden orchestra pit. (Shakespeare, by comparison, built the Globe Theatre with a peninsular stage for plays that variously touched on many parts of the Earth as well as on the idea of the earthly globe.) The most characteristic architectural aspect of Wagner's theater was the mystic gulf (*mystiche Abgrund*), as he called it, between the audience and the stage that he believed would encourage a mythic or dreamlike atmosphere like that pioneered by landscape painters earlier in the century.

The Rhine Gold, with its ring (*Reif*) at a reef (*Riff*), calls for a landscape at once particular (German) and universal (human). For the first production as part of the entire four-part *Ring of the Nibelung* cycle (1876), Wagner hired a landscape painter, Josef Hoffmann,[20] instead of the regular set designer. Hoffmann's designs were too historically specific for Wagner, however. The composer sent Hoffmann's sketches to other painters and set designers[21] for whom Friedrich had been influential (see Illustration 63).

One painter whom Wagner approached was Arnold Böcklin.[22] Nothing much came of that for the sets, so far as we know. A few years later, however, Böcklin produced six versions of a nonhistorical (symbolical) painting, *Toteninsel* (Isle of the Dead; 1880–1886); see Illustration 64.[23] It was a great success, with copies soon decorating the walls of tens of thousands of German homes and offices. *Toteninsel* was also the inspiration for Fritz Lang's movie *Die Nibelungen* (1924–1925). The one-time painter Adolf Hitler bought one of the six originals of Böcklin's work in 1933. *Isle of the Dead* "adorned" the Berghof (Hitler's residence in the Bavarian Alps) and then the New Reich Chancellery in Berlin.

ILLUSTRATION 63 Opening Scene of *The Rhine Gold*. Sketch by Josef Hoffmann, 1876, for act 1, scene 1 of Richard Wagner's *The Rhine Gold* (1867). No photographic record survives of the actual set, which was further designed and then painted by the brothers Gotthold and Max Bruckner of Coburg and by Karl Brandt of Darmstadt. Permission granted by Collection Selechonek.

ILLUSTRATION 64 *Toteninsel* (Isle of the Dead). Arnold Böcklin, 1880. Erich Lessing / Art Resource, New York.

By the mid-1930s, Hitler had ordered the construction on Rügen of the largest concrete building ever built in Germany. This was the Colossus of Prora, a beach vacation spot devoted to "Kraft durch Freude"—three miles long with 11,000 rooms. Not far off is the island of Usedom with its village seaport Peenemünde, where, under the watchful eye of Wernher von Braun, German forces tested top-secret weapons, including the V-2 rocket. (Some historians say nuclear weapons testing also took place on Rügen.)[24] While a student on the North Sea island of Spiekeroog, von Braun had read Hermann Oberth's 1923 *By Rocket into Planetary Space* (*Planetenräumen*),[25] and now, on the islands of the Baltic, he put into practice what his teacher, a founding father of astronautics, had started.

There is no easy line from ideal or aesthetic works to politics, but people try nevertheless to draw one. Peter Schjeldahl, for example, writes: "Like [the composer] Richard Wagner, [the painter Caspar David] Friedrich has long had something to answer for . . . as Hitler's favorite old master [in painting]" (see Illustration 65).[26] Whether or not Schjeldahl exaggerates here, Rügen and the islands were, for most Germans, magic places, like the one figured on stage in the early German translations of Shakespeare's *Tempest*, which was usually called a *zauberte Insel* and which Wagner helped rework into a powerful originary national myth. Mainline German leaders, however occultist, seemed especially fond of the place.

ILLUSTRATION 65
Rocky Reef on the Sea Shore. Caspar David Friedrich, ca. 1824. Source: Staatliche Kunsthalle, Karlsruhe.

Among them was "the man who gave Hitler his ideas,"[27] Jörg Lanz von Liebenfels, publisher of the proto-Nazi *Ostara* (Newsletter of the Blond Fighters for Male Rights; 1908). On Rügen, Liebenfels even set up a castle for his New Templar movement.[28]

An attempt to break Rügen's magical enchantment of the German national consciousness before it could do great harm came with the German Huguenot novelist Theodor Fontane, whose early childhood was passed in Neuruppin, on the shores of the Ruppiner See, in the March of Brandenburg. His late novel *Effi Briest* (1889–1894)[29] includes this passage:

I do not want to torment you with Rügen, and so we will give it up. Agreed.[30]

Fontane spent time, from age seven to age twelve, in the Pomeranian town of Swinemünde on the Baltic Sea and on the islands of Usedom and Wolin in the Szczecin Lagoon. Since the time of Saxo Grammaticus, Swinemünde, as we have seen, had been a focus of Baltic national aspiration. For Fontane, it was the seat of poetry itself[31] and became the model for *Effi Briest*'s Kessin.[32] Fontane's clarion warning to "give . . . up" the mythic Rügen—his SOS to German culture—failed to impress the National Socialists. The search for insular space and planetary room went on. Oberth, at age ninety-three, published his *Primer for Those Who Would Govern* (1987)[33] half a century too late.

WHAT OF LOVE? Richard Wagner set his *Rhine Gold* beings in an imaginary world reef where Alberich denies love (see Illustration 66). Soon after composing this work, Wagner wrote his greatest tragic love opera, *Tristan and Isolde*, the plot of which includes two offshore islands. (Wagner uses the unusual term *Eiland*.)[34] First, as part of the background story, is the island where Tristan and Morold (Irish Isolde's brother) engage in a typically Norse *holmgang*. Tristan kills Morold. Tristan's servant Kurwenal reminds Isolde's servant Brangäne about this killing:

Herr Morold zog	Sir Morold crossed
zu Meere her,	the sea to us
in Kornwall Zins zu haben;	to exact tributes from Cornwall:
ein Eiland schwimmt	on an island swimming
auf ödem Meer,	in the sea's expanse,
da liegt er nun begraben![35]	there now is he buried!

Second is the island off Brittany where Kareol, Tristan's castle, is located and where the entire last act takes place. For Wagner's operas, of course, it does not matter where on Earth we suppose such ideal islands to be locatable. For understanding Wagner on islandness, the factually geographical particulars are secondary. It does not matter much whether the island where his *holmgang* takes place is Enys Samson in the Isles of Scilly, offshore from mainland Cornwall on the way from Ireland. In the same way, little depends on whether Kareol is the modern Île Tristan at the mouth of the Pouldavid Estuary at Douarnenez, just offshore from mainland Brittany.[36] Spiritual cartography here trumps physical cartography as it does for Sophocles and Shakespeare, except in cases, often Celtic (Wales for Shakespeare, Brittany for Wagner), where spirit and hydrological geography go hand in hand.

ILLUSTRATION 66
Rhine Maidens with Alberich Postage Stamp. Alois Kolb, 1933. The stamp was issued by the Reichspost of the German Empire. Permission granted by Collection Selechonek.

Wagner, then, has in mind an island of the mind or a theatrical globe. In particular, his *Tristan and Isolde* seems to be an allegorical interpretation—or, as Wagner might call it, a *Handlung*[37]—of Kant's *Critique of Pure Reason*, which we considered in Chapter 1, but with a full Wagnerian twist that comes partly from Wagner's dependence on Arthur Schopenhauer's *World as Will and Representation* (1844),[38] which is largely a revision of Kant's epistemology for other purposes. The amatory struggle in *Tristan and Isolde*, seen in this way, is the tension between two poles inflected from earlier German idealism: a real "noumenal" Night, hence unknowable and unreachable, like the mythic island of Avalon, and an unreal "phenomenal" Day, hence to that extent knowable and reachable but also unsatisfying.[39]

Both the tragedy of *Tristan and Isolde* and that of *The Rhine Gold* recall the English *Alfred*, in which an English farmer falls into "the deep Atlantic / Midst Equinoctial gales," driven there by the sirens. The Englishman marries the mermaid and there's an end of it:

> My comrades and my messmates,
> Oh, do not weep for me,
> For I'm married to a mermaid,
> At the bottom of the deep blue sea.
> *Refrain*:
> Singing Rule Britannia,
> Britannia rules the waves
> Britons never, never, never shall be slaves.[40]

The British, it would seem, are part mermen. Certainly the Orkney and Shetland islanders have their selkies: "I am a man, upo' da lan' / An' I am a silkie I' da sea."[41] Although the Germans likewise have their mer-wives,[42] for Wagner there is no such exogamous (trans-species) marital bliss.

S.O.S. EISBERG

First, there is still land, but with puddles that are turning into ponds
and straits; then there is only the dark water far and wide, with islands
that quickly crumble [*zerbröckeln*].

—Bertolt Brecht, "On Looking at My First Plays"

Cartographers have long called the continent, or mainland, *terra firma*, as if there
were on Earth an infirm place, *terra infirma*, meaning "islands" or "bodies of
water,"[43] or as if there had been a time when the Earth had been infirm or inchoate
(the land and the water taken together, as in Genesis's beginning).[44] Relevant here
is a passage in the *Histories* of Polybius where he discusses a lost treatise entitled
On the Ocean by Pytheas. Pytheas, from the island of Pomègues in the Frioul Archipelago near Marseilles, was the first person I know of to write about the polar
ice cap, which he describes thus:

Regions [of the Earth] in which there was no longer any proper land nor sea nor air, but a
sort of mixture of all three of the consistency of a jellyfish [*pleumōn thalattios*] in which one
can neither walk nor sail, holding everything together, so to speak.[45]

Pytheas associates the place of which he speaks, at once earth and water, with the
Scandinavian-Germanic "Neverland," called "Thule Island." Thule traditionally
lies just outside the borders of the known world—on the "margins," or cartographic coastline, of the Earth. As such, it defines the Earth.[46]

Worth mentioning here is the attendant view, both science and fiction,[47] that
the Earth is hollow; humanoid beings live "inside"—like ghosts (the Ghost of Old
Hamlet) and cave people (Alberich). In post-Renaissance centuries, the hollow
Earth theory was pioneered by the French bridge engineer Henri Gautier. One
of his diagrams shows the strait, or opening, between the liquid world above and
the liquid world below (see Illustration 67). Caught in the middle—at Euripus
(*Euripe*) itself—is the islandic land.

Germans of the late nineteenth and early twentieth centuries had their oc-
cultist Thule Society (established in 1912 as the Study Group for Germanic
Antiquity).[48] For them, "Ultima Thule," linked with hollow Earth, was the lost
ancient landmass—an Atlantis. This was the Aryan home near Greenland and
Iceland—precisely the locations of Arnold Fanck's 1933 globe-centered film *S.O.S.
Eisberg*. Among the Thule Society's members were the influential Nazis Rudolf
Hess and Alfred Rosenberg.[49] Its establishment followed that of its British counterpart. The Thule Society partly modeled itself on the British Viking Club, estab-
lished and founded by W. G. Collingwood in 1892.[50]

The hollow Earth theory, along with the general fascination with Atlantis and
Heinrich Himmler's enthrallment with the counterfeit "ancient Frisian" manu-

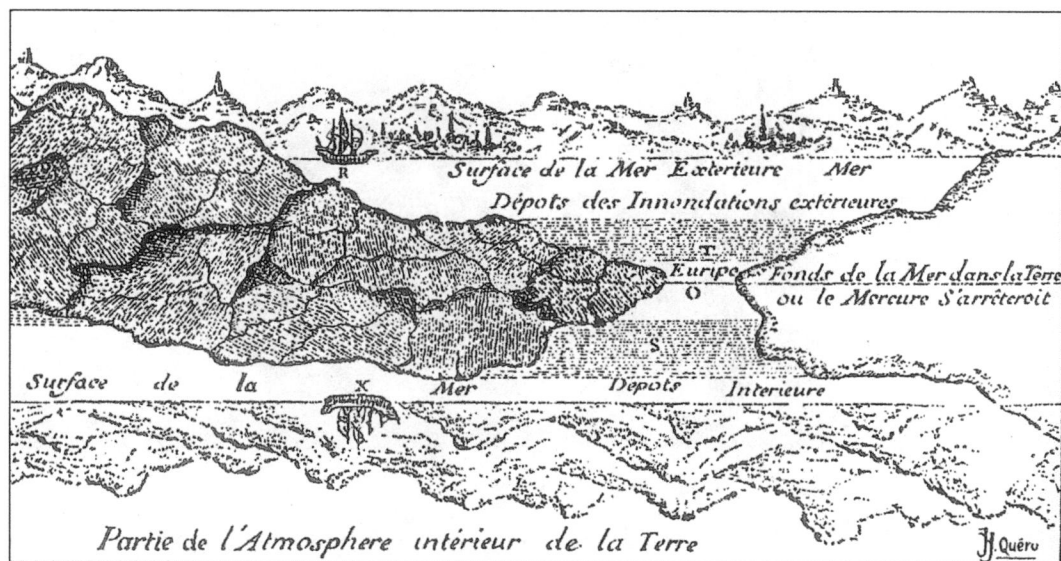

ILLUSTRATION 67
Where Lower and
Upper Worlds Meet.
From Henri Gautier,
*La bibliothèque des
philosophes et des
scavans tant anciens
que modernes, avec
les merveilles de la
nature, où l'on voit
leurs opinions sur
toute sorte de matières
physiques, comme
aussi tous les systèmes
qu'ils ont pu imaginer
jusqu'à présent sur
l'univers et leurs plus
belles sentences sur la
morale, et enfin, les
nouvelles découvertes
que les astronomes
ont faites dans les
cieux* (Paris: André
Cailleau, 1723–1734),
3:203–204.

script *Oera Linda*,[51] helps to explain Hitler's rationale for establishing experiments in long-distance location technology on Rügen. His advisors "considered it helpful to locate the British fleet, because the curvature of the earth would not obstruct observation."[52] If one did not expect to find an opening to Alberich's Nibelheim cave opening to hollow Earth on Rügen, at least one could seek it in the northern climes. Midway through World War II, the theorist of space (*Raum*), Karl Neupert, was still promulgating the hollow Earth theory in his widely read *Geokosmos* (1942).[53]

Seeking out a specifically Germanic Thule is one theme of the film *S.O.S. Eisberg*. On the one hand, Fanck was a doctor of geology[54] who single-handedly invented the "mountain movie"—a German counterpart to the American Western. On the other hand, *S.O.S. Eisberg* (literally, ice mountain)[55] opens with a distinct variant of the usual cinematic logo of Universal Pictures (see Illustration 68), a reminder that we are now in the age of global aviation. The logo shows a plane—not unlike those from World War I—flying around the islandic globe, the whole scene filmed by an Archimedean camera in space. "Give me a place to stand and I will move the Earth."[56]

Fanck's *S.O.S. Eisberg* was the first of two "Universal-Deutsche" co-productions intended to encourage German-American "geopolitical" cooperation in the movies.[57] It is a "documentary adventure movie," in the tradition of Varick Frissell's *Viking* (1931), that Fanck shot partly on Nuljarfik—a small "rock island" off the west coast of Greenland, a Danish protectorate.[58] The story starts with an expedition to Greenland with three purposes. The team wants to learn more about

ILLUSTRATION 68
Airplane Encircling
the Globe. Still from
the opening credits
of the movie *S.O.S.
Eisberg* (1933). Source:
Harvard College
Libraries.

floating ice and glaciation at fast-moving Rink Glacier. They also intend to find a lost explorer, who is modeled on the real-life Alfred Lothar Wegener, the actual developer of continental drift theory,[59] who had led a polar expedition to the interior of Greenland, at a site called *Eismitte* (mid-ice), in 1929. Finally, they want to "find themselves"—a common theme in all of Fanck's mountain movies.

Soon after arriving in Greenland, the expedition team is stranded on a floating ice island—a no-man's *terra infirma* slowly diminishing in size—somewhere between the Old and New Worlds. Luckily, they are able to send off an SOS signal:

DIT DIT DIT DAH DAH DAH DIT DIT DIT   · · · — — — · · ·

a mnemonic device that reads the same upside down and right side up as well as left to right and right to left. The message is picked up by the New World's English-speaking Labradoreans in New-Found-Land. (At the time, Newfoundland had not yet joined the Canadian Confederation and the name *Labrador* still recalled an older name for the Americas as a whole.)[60]

If the explorers in *S.O.S. Eisberg* were prone to misunderstanding the nature of an iceberg, or *Eisberg*, as an "ice mountain," Labradoreans and Newfoundlanders, who called such presumed mountains "islands of ice," were not so inclined.[61] Consider here Kant's warning in the *Critique of Pure Reason* that ice, including a so-called ice mountain, is often not what is appears to be. "Many an iceberg [*bald wegschmelzende Eis*]," writes Kant in the translation by the Victorian Englishman John Miller Dow Meiklejohn (1855), "seems to the mariner, on his voyage of discovery, a new country [*neue Länder*]."[62] Meiklejohn renders Kant's *bald wegschmelzende Eis* as "iceberg." In 1848, Francis Haywood had better rendered it as "masses of ice ready to melt away."[63] Norman Kemp Smith (1929) renders the German phrase as "many a swiftly melting iceberg."[64] Paul Guyer and Allen Wood (1998), who translate the phrase as "many a . . . rapidly melting iceberg," do no better than Haywood at capturing the double ambiguity in the relationship of island to mainland and of solid to liquid.[65]

In *S.O.S. Eisberg*, the island-stranded Germans' SOS call is received in Labrador and then relayed from the New World to the Old World. From that time forth, the plot ceases to be a narrative of being lost and becomes one of being found.

The loyal wife of one the expedition members—Ellen Lawrence, played by Leni Riefenstahl—now sets off for Greenland. (In the English-language version of the movie, her voice is dubbed.) Riefenstahl's part is modeled on the real aviator Amelia Earhart, who had crossed the Atlantic solo from Newfoundland to Londonderry, Ireland, in 1928. Ellen lands on the *Eisberg*, but her airplane is wrecked.[66] Now ace German pilot, Ernst Udet, sets out to the rescue. (In the German version, Udet plays himself: he was a hero of World War I—the Luftwaffe's second-highest-scoring fighter; see Illustration 69.)[67]

The fact that the Universal Studios producers of *S.O.S. Eisberg* chose to employ German and American co-production teams and a presumably mid-Atlantic location makes for telling cinematic geopolitical history. The movie opened in Berlin in August 1933 shortly after Franklin D. Roosevelt delivered his inaugural

ILLUSTRATION 69
Airplane and Iceberg.
Still from the movie
S.O.S. Eisberg (1933).
Source: Harvard
College Libraries.

address as president of the United States on March 4. Hitler was then being elected to presidential office and well on his way to becoming dictator with the help of an enabling act (March 13, 1933). Not surprisingly, differences between the American and German versions of *S.O.S. Eisberg* come down to more than the directorial styles of Arnold Fanck (German version) and Tay Garnett (American version). Garnett, who had been an American fighter pilot during World War I, toned down Ernst Udet's brilliant fighter-pilot antics as they appear in the German version.[68] The American version is likewise less comfortable with icebergs or ice mountains: Garnett's version ends with exploding ice and the consequent destruction of human lives; Fanck's version closes with native peoples kayaking happily to the iceberg in order to save the stranded German explorers.

Universal Pictures advertised the movie mainly in terms of menace. One poster reads: "See the crashing masses of white death—crumbling worlds of ice menacing man and beast alike!" The photography is spectacularly beautiful.[69] The effective "star" of *S.O.S. Eisberg*, said some reviewers, was the floating ice-mountain-island, and Universal Studios president—American-born Carl Laemmle Jr., son of Universal Studios founder, German-born Carl Laemmle— headlined the film thus: "Nature Is the Star."[70]

Human war figures in *S.O.S. Eisberg* as a natural phenomenon. The sights and sounds of ice explosions fill the cinema hall. Solid water, not *terra firma*, turns the world upside down, crashing down into the liquid water, flipping over in vast mountains. This is the space into which Universal Pictures' Archimedean airplane flies, as in the movie's initial branding logo: "Give me a place in space, and I will move the world."

The important composer Paul Dessau, the *Kapellmeister* at the Berlin State Opera, did the sometimes ambiguously Wagnerian music for *S.O.S. Eisberg*.[71] He had worked earlier on island themes and knew Kurt Weill's music for Lion Feuchtwanger's 1928 play *The Petroleum Islands* (which opened at the Berlin State Theater a year after the pacifist Feuchtwanger and the radical Bertolt Brecht collaborated on *Calcutta, 4th May*).[72] After World War II, Dessau gave up the earlier, ambiguously Romantic ways of writing about and writing music: he adopted and adapted Arnold Schoenberg's atonality. When the German Democratic Republic (GDR; East Germany) debated in the 1950s whether to include the works of Wagner in its regular canon, for example, Dessau declared that Wagner did not have relevance for the GDR in political terms. "The work of Wagner," he said, "is filled with poisonous intoxication."[73]

In 1933, the principal creators of the optimistically collaborative "German-American" production of *S.O.S. Eisberg* were already going their separate political ways when the movie opened. Leni Riefenstahl's Nazi-backed *Victory of Faith*, which documents the Nazi Party rally at Nuremberg in 1933 in propagandistic mode, was shown across Germany in 1934. *S.O.S. Eisberg*'s director, Arnold Fanck, went on to make islandic propaganda films for Hitler: his *Ein Robinson* (1940) was largely filmed on Robinson Crusoe Island in the Juan Fernández Archipelago;[74] and his *Atlantikwall* (1944) showed the fortifications that Nazi military units had built in 1942–1944 along European island coasts.[75] Ernst Udet, the flying lead of *S.O.S. Eisberg*, committed suicide in 1941.[76] In 1948, George Frederick "Buzz" Beurling—Canada's best fighter pilot in World War II—called Udet "the greatest flier of all time," adding, "The way pianists can enjoy music by hearing a note in their heads, that's the way I am about angles."[77]

Paul Dessau left Germany in order to collaborate with Bertolt Brecht, who was then residing on the island of Fyn in Denmark. Brecht and Dessau worked together on the anti-Nazi stage play *Fear and Misery of the Third Reich*, also known as *The Private Life of the Master Race* (1938). That work includes a version of the resistors' song of the marshland soldiers (*Moorsoldaten*), which was originally written by inmates of one of the concentration camps—perhaps Börgermoor[78]—in the mossy mire of the fens, the swampy Emsland in Lower Saxony, with its boulder-clay hummocks and dwelling mounds.

There is only the dark water far and wide, with islands that quickly crumble [*zerbröckeln*].[79]

Carl Laemmle Jr., who produced *S.O.S. Eisberg* for Universal Studios, ceased co-operative ventures with Germany, the birthplace of his father, immediately after Hitler came to power. That was that.

AIRSHIPS

England is no longer an island.

—Lord Northcliffe, *Daily Mail*

Of artificial islands that float or fly in the air there are at least two sorts. First, there is the tethered "floating island" that hovers above the land to which it is anchored; an example is Alexander Asadov's massive construction.[80] Second, there is the untethered island that moves through the air in something like the way ships move nautically through the sea; an example is the airplane, first understood as, and called, an "airship" or "flying ship."

"Flying ship" is what the so-called father of aeronautics called his creation. In 1670, the Italian Jesuit mathematician Francesco Lana de Terzi published *Prodromo*,[81] in which he imagines his would-be aerial inventions as steerable floating islands "flying through the air in the way that a boat sails through water."[82] A few years later, the English scientist Robert Hooke in *Philosophical Collections* (1679) was writing about "a Demonstration, how it is practically possible to make a Ship, which shall be sustained by the Air."[83] The French inventor Jean-Pierre Blanchard's air balloon craft of 1785 was called a "flying boat" in the press.[84]

Thus did the English novelist H. G. Wells, in *The War in the Air* (1908), writing of "the ships of the German air-fleet rising one by one,"[85] have centuries of speculation behind him. The Aëronautical Society of Great Britain published a translation of Lana de Terzi's *Prodromo* in 1910, highlighting the notion that the new airships were flying boats.[86] During World War I, pilots also thought of their airplanes as flying ships: Charles Frederick Snowden Gamble, for example, writes about "flying boats" in *Story of a North Sea Air Station* (1928).[87] (Snowden Gamble was not only a squadron leader in the Royal Air Force but also the author of *The Air Weapon* (1931), an account of British military aeronautics, and the popular radio spokesperson for the "enchantment of aviation" programs broadcast on British Imperial Airways radio during the 1930s.)[88]

"Air ships" were unlike "water ships" inasmuch as they had no natural shoreline to hit up against. The actual advent of fleets of military airships—and especially their active participation in World War I—changed the way people, especially the British, thought about sea, land, and island. Lana de Terzi well knew that military forces could use airships to attack cities from the air;[89] Jonathan Swift's fictional flying island Laputa in *Gulliver's Travels* (1736) brought this idea of aerial bombardment to the public at large. A predictive illustration of Laputa by J. J. Grandville, made in the wake of Blanchard's balloon travel across the English Channel and beyond (1785), provides an early case in point (see Illustration 28).

There was even more to it than that. Before the advent of air travel, the coast appeared to be the line where one substance (water), in the form of one state of matter (liquid), met another substance (rock), in another state of matter (solid). The idea of that physically defined shoreline had, over the centuries, become a model for thinking about political boundaries. Boundaries and sovereign realms were explained mainly in terms of natural coastlines between sea and land. The shoreline had been defined in terms of a militarily and politically useful horizontal division between land and sea.

The successes of aeronautic technology led to changes in the practical meaning of a "shoreline." Flying nullified some of the implications of the difference between motion on solid land and motion on liquid water (between walking, say, and swimming). Not only did airships move *through* gaseous air, a substance and a state of matter different from both solid land and liquid water; they also moved over *both* land and sea together.

Reactive modifications to the new technology and to the consequent need for a new understanding of the "territory" of an island included novel definitions of *airspace* based on often-tenuous analogies with those of *territorial waters*. In 1908, for example, the *Proceedings of the American Political Science Association* stated: "[When] airspace comes under the domination and control of man, it is embraced within the jurisdiction of law."[90] A 1912 issue of *Political Science Quarterly* typically discussed "legal literature relating to the extent of the rights of the landowner in the airspace above his head."[91]

The larger political consequences were now extreme. Understanding and defending sovereignty in terms of political boundary or natural coast meant also comprehending Earth as a single island, composed of both land and sea and surrounded by air. To deal with the new situation, Lord Northcliffe's publishing empire sponsored aviation events: the Circuit of Britain air race (announced in 1910), the first nonstop transatlantic flight (1913),[92] and the Circuit of Great Britain air race for water planes (1913).

ONCE UPON A TIME, being an "islander" meant relative isolation, at least for the British people, with their "island fortress" *mentalité*. The French historian Élie Halévy thus writes: "The habit of depending on her navy for protection was so deeply ingrained in the British mind that it had become an instinct."[93] A good part of England's sense of itself in the Victorian period depended, in fact, on Norse-style islandology.[94] All that changed—or it seemed to—with the advent of the blimp. In 1908, at the second meeting of the Aerial Navigation Subcommittee, Charles Stewart Rolls, founder of both the Aero Club of Great Britain and the Rolls-Royce car company, said, "England will cease to be an island."[95] H. G. Wells's imaginary *War in the Air* (1908) is a science fiction story that foretells the destruction of cities and navies by airship attack. The *Times of London* (July 13, 1908) warned in "The Conquest of the Air": "England's safety as an island will vanish if not ensured against aerial attack."[96]

Understanding the new world was apparently easier for mainlanders like Americans and Germans. Before World War I, American airship pioneer Walter Wellman published his book *The Aerial Age: A Thousand Miles by Airship over the Atlantic Ocean* (1911), partly about his journey onboard the dirigible *America* with the panoramic-heights photographer Chester Melvin Vaniman.[97] After the war, Wellman published *German Republic* (1916), an upbeat imaginary political postwar history[98] that was meant to match his *Aerial Age*. England, he emphasized, was no longer an island, and he now looked forward to an era of world peace.

The German dirigible, introduced in 1900 as the Zeppelin Company's *Luftschiff*, was put into service partly to explore the northern region—where the German-American movie *S.O.S. Eisberg* (1933) was set. So, too, did Wellman seek out islands in the North. His *Tragedy of the North* (1902) tells the tale.[99] However, he was unacquainted with the mysterious notions that some Germans had of this

region, the "Ultima Thule." In some public topography, for example, German mapmakers included islands in this region where there were none and excluded islands where there were some.

The bellicose ideology was palpable. One case involved the official postage stamp issued by both the liberal Weimer Republic (1928–1931) and the Third Reich (1933–1934), depicting the *Graf Zeppelin* flying over the Atlantic Ocean between "Europe" and "Amerika."[100] When Hitler came to power, the Third Reich adapted the Weimar design for a stamp to commemorate the appearance of the swastika-emblazoned *Graf Zeppelin* at the Century of Progress International Exposition—the World's Fair held in 1933 in the then noticeably German-American city of Chicago (see Illustration 70).[101] The map on this stamp, remarkably, has no standard cartographic projection and leaves entirely blank a good part of the northern region between "Amerika" and "Europe." The island of Iceland and the continental island of Greenland have apparently sunk into the expanse of sea or "the white world."

That blankness is where many Nazi ideologues located their "Ultima Thule." Often members of the German Thule Society, they included leaders like Rudolf Hess (deputy to the führer who had been a pilot during World War I), Alfred Rosenberg (who promulgated the idea that Jesus was an Aryan),[102] Hans Frank (head of the National Socialist Jurists Association), and Dietrich Eckart (to whom Hitler dedicated the second volume of *Mein Kampf* [1926]). The *Graf Zeppelin* depicted on the 1933 stamp hovers over the region where the Aryans presumably lost Atlantis. (The blankness surrounding the area recalls the Sargasso Sea, that "sea within a sea"[103] or "sea without shores," relevant to which is Maurice Tourneur's now lost 1923 movie *The Isle of Lost Ships*[104] as well as Jean Rhys's *Wide Sargasso Sea* [1966], a prequel to Charlotte Brontë's *Jane Eyre* and a response to the young Brontë sisters' make-believe island of Gondal.[105])

Chester Melvin Vaniman was Walter Wellman's collaborator on his early flights across the Atlantic and to the northern regions. The "acrobatic photographer" for the American team, his special forte was taking panoramic images from high above the ground. Vaniman, said the newspapers, was "a photographer with a head for heights."[106] With him around, there would be no such problem with cartographic projection as there was for the *Graf Zeppelin* stamp.

In 1844, the prescient author known as J. J. Grandville published *Another World*, which includes a series of illustrations that project even more overtly imaginary spaces than Thule. Grandville's drawings slyly suggest that air travel would bridge space, conjoining planetary Earths across the vacuum. It would yoke worlds together in the way that the Pont Neuf in Paris bridges the two shores (*rives*) of the Seine (see Illustration 71). In the age of the *Luftschiff*, one would not only bridge island-continents like "Europe" and "Amerika" but also hook up island worlds.

LES MYSTÈRES DE L'INFINI.   139

le courant atmosphérique se mit à l'amble, de sorte que notre voyageur put examiner tout à son aise les objets qui l'entouraient. La trois cent trente-trois millième pile était appuyée sur Saturne. Hahblle put se convaincre alors que l'anneau de cette planète n'était autre chose qu'un balcon circulaire sur lequel les Saturniens viennent le soir prendre le frais.

A l'autre extrémité du pont le courant reprit son essor, et entraîna Hahblle vers des régions plus élevées encore. La mécanique céleste lui fut entièrement dévoilée, grâce à la négligence de son propriétaire, qui ce jour-là avait oublié de fermer ses persiennes de nuages. Ce propriétaire était un vieux magicien qui insufflait des globules de savon et les lançait ensuite dans l'infini. Affaibli sans

ILLUSTRATION 71
The Bridges between the Worlds. From J. J. Grandville (Jean Ignace Isidore Gérard), *Un autre monde* (1844). Reproduced by permission of the collections of the Printing and Graphic Arts Department, Houghton Library, Harvard University, Typ. 815.44.4380.

Just as Francis Bacon, in his utopian novel *New Atlantis* (1624), hypothesizes a fictional sea island called Bensalem that allows for a scientific "experiment solitary touching the super-natation of bodies,"[107] so does Edward Everett Hale, in his prescient novella "The Brick Moon" (1869), imagine such a place on the moon (see Illustration 72).[108] Hale's was one of the earliest literary conceptions of the sort of scientific satellite in the "archipelago" that the astronomer John Herschel called the "Solar System." Hale describes a fictional, artificial, inhabited insular

ILLUSTRATION 72
The Brick Moon.
From Edward Everett
Hale, "The Brick
Moon," *Atlantic
Monthly,* October–
December 1869.
Hale's novella, which
tells the story of an
artificially constructed
island in space, was
the first story of a
satellite-based space
colony in literary
history. Source:
NASA.

STRABON GEOGRAPHE.
Chap. 35.

ILLUSTRATION 73
Strabo Holding the
Globe in His Hands.
From André Thévet,
*Les vrais pourtraits
et vies des hommes
illustres* (Paris, 1584).

'E S T I M E qu'il n'y a celuy qui contem-
plant l'effigie de ce tref-docte Geographe
Strabon, n'admire les perfections & excel-
lence de fon efprit, tant pour auoir efté au-
tant diligent rechercheur de cefte fcience
Geographique, que tref-graue hiftorien &
Philofophe, qui font les deux fingularitez,
qui rendent excellent le Geographe, lequel
traicte les defcriptions, qui fe peuuent apprehender plus aifément &
vniuerfellement, comme des fleuues, des grandes citez, des nations

N iiij

satellite, a planet-like island[109]—the "brick moon"—that orbits the Earth, providing it with navigational and other information.[110] Thanks to gravity's pull, in fact, this "space station" is no less tied to Earth than Ahab's whaler is free of the whale (or vice versa). After reading "The Brick Moon," William Sloane Kennedy wrote in 1869 that, as it seemed to him, Hale as author was "standing" on something like "nowhere":

Give him [Hale] the least bit of a *pou stó* [the Archimedean principle], and by sheer force of genius and fancy, he will project you into the air a full blown romance, which shall keep touch with the base earth of reality by said pivotal *pou stó*, and nothing else.[111]

Geography's elementary concern since the time of Strabo, often called the father of geography, is what "ground," if any, we can take a stand on in order to understand the one Earth, often defined or depicted as an island. It is no wonder that "moderns" wanted to think of the geographer as if he could hold the world "in his hands." After all, Hercules supports the globe on his shoulders; the little boy depicted in *A Child's Geography of the World* observes the Earth (see Illustration 6); and Archimedes hypothesizes that he can move the world (see Illustration 57). It is no surprise that the Renaissance traveler André Thévet should depict the sagacious Strabo holding the whole world in his hands (see Illustration 73), as if Strabo and Thévet himself were not also part of that world.

"He's Got the Whole World in His Hands" is the title of a traditional African American spiritual collected by Frank Warner in 1933—the year that both Hitler and Roosevelt came to power.[112] It became an international hit soon after World War II when a fourteen-year-old white boy of dubious talent made it the best-selling song in Great Britain (for the entire decade of 1950–1960); it continued to be popular thanks to the American civil rights movement of the 1960s.[113] Innocents might believe that one could "[h]old infinity in the palm of your hand"—as the poet William Blake puts it in his *Auguries of Innocence* (1803). Others might believe that a person holds his "self" in his own hands—as Sir Thomas More puts it in Robert Bolt's immensely successful postwar BBC radio play *A Man for All Seasons* (1954), first performed onstage at London's Globe Theatre (1960).[114] And yet there was always the ongoing debate as to whether it would be the spiritual's universal, monotheist God who would hold the new world—or the human aviator circling the globe and exploring the universe. An airplane circling the globe—the cinematic logo, as we have seen, of German-American Carl Laemmle Sr.'s Universal Studios (see Illustration 68)—now became a monotheist, universalist dream presumably transposable to the realm of islandic earthly politics.

Building for a Future

15

O limèd soul, that, struggling to be free,
Art more engaged!

—Shakespeare, *Hamlet*

BEYOND THE PALE
Settlement Archeology

In those climates a vast tract of land, invaded twice each day and night
by the overflowing waves of the ocean, opens a question that is eternally
proposed to us by Nature, whether these regions are to be looked upon
as belonging to the land, or whether as forming a portion of the sea?

—Pliny the Elder, *Natural History*

In the preceding epigraph, Pliny is writing about first-century Emsland in Lower
Saxony. The ancient dwellings in low-lying Emsland were often built on artificial
hills, terps, or *Wurten*, which provided protection during tides and river floods.
Writing as a natural philosopher, Pliny notes that, when the tide was up, the
mounds were more like ships than islands:

Here [on the mounds] they pitch their cabins; and when the waves cover the surrounding
country far and wide, like so many mariners on board ship are they: when, again, the tide
recedes, their condition is that of so many shipwrecked men.

Such dwelling places abound in the Baltic region. Those in Saxony are among the
older and larger such habitations worldwide.[1] However, their inhabitants were
not mound dwellers but lake dwellers living in stilt homes constructed on *pales*,
or "stakes driven into the ground," that attracted the attention of Germans in the
late nineteenth and early twentieth centuries.

The idea of a lake-dwelling community living in pile dwellings excited the European imagination as much as the notion of Greek islandic origins once had. One
historian actually calls the period of German archeological thinking, from 1888 to

1940, the "[European] Era of Pile Dwellings."[2] Europeans were then fashioning their present notion of themselves in terms of an ideal islandic pile-dwelling past.[3]

The purpose of this ideal construction was straightforwardly political. Concerning "settlement theory," Gustaf Kossinna, in his influential *German Prehistory: A Pre-Eminently National Discipline* (1911), states that the main purpose of his work is

as a building block in the reconstruction of the externally as well as internally disintegrated fatherland.[4]

Alfred Rosenberg, a member of the Thule Society, put forward the view that the Nordic-Aryans came from a lost island that was reflected in these dwellings.[5] Hans Reinerth, who excavated stilt houses at the Federsee and was a member of the Nazi Ahnenerbe, also promoted the idea that the Aryans came from a northern Atlantis. When he became Reich deputy of German prehistory, Reinerth promulgated the lake-dwelling theory of European settlement and created idealized architectural reconstructions of Neolithic and Bronze Age pile structures in the village of Unteruhldingen on Lake Constance. His Pfahlbaumuseum was not essentially changed until 1961![6] Moreover, his dwellings were not without influence and effect. The Swiss architect and planner Le Corbusier, for example, was much influenced by seeing them on water when he visited the exhibitions as a young boy. The *piloti*, or stilt-like pylons, of Le Corbusier's buildings reflect the ancient style,[7] and he argued, with these in mind, that dwellings are not only machines for living in (*machines à habiter*)[8] but also vehicles of human ritual and belief.

Building on Water

> . . . He's in a stilt house now,
> The water passing beneath him half the day,
> The other half it's mud. The tides
> Do this. . . .
>
> —Thomas Lux, "He Has Lived in Many Houses"

Human activity is sometimes godlike. The Bible puts it thus: "And God said, Let . . . the dry land appear: and it was so."[9] Creating an island—as Xerxes of Persia did at Mount Athos and King Utopus did in *Utopia*—or producing a stilt building in and above a marsh—as does the peasant in *Alfred*—is a partly ritualistic business.

The practice is worldwide. Stilt buildings in continental Europe range from Lake Constance to Alvastra, a small village in eastern Sweden. There are pile dwellings in South America, floating villages at Tonlé Sap in Cambodia, and islet-home "duns"[10] in Scotland. There are the artificial crannogs that the Irish called *oileán* (island),[11] modern offshore oil rigs, and private artificial island-state experiments with utopian pretensions.[12] Florida's Stiltsville off Key Biscayne has buildings entirely disconnected from the mainland. There are similar dwellings on the Caroline Islands in the western Pacific, Micronesia, and Oceania at large.[13] The *palafitas* on the island of Marajó at the estuarial mouth of the Amazon stand in "a region of islands"[14] where twice-daily tidal flooding and massive annual river flooding saturate the earth or muddy the water.

This way of selecting a locale and building for it has yet to be theorized even in an age of "global warming" in which a wise theory would be useful. Why do people build such buildings and create such places? We often assume that most "environmental migrants"—there are hundreds of millions of them in the world today, or there soon will be—will flee the coast. And yet, at the same time, people want to live on the water, even over it. Why? One kind of answer involves practical conditions: there is not much useable dry land, and the coast provides ready transportation and food; also, there are the many defensive advantages, to which Herodotus draws attention in discussing the Paeonian lake dwellers on Lake Prasias.[15] Another kind of answer involves a different sort of affinity with the water.

First, people have a tendency to go to the water. Maybe it is because we are universally water creatures of a sort, as Rachel Carson has it. (That we are mermen and mermaids of a sort is an old idea in the history of hydrotherapy,[16] and, in some places, the merman is a matter-of-fact creature of the mind, as it is for Newfoundlanders' straightforwardly named "waterman."[17]) One writer about water dwelling points to "people's [presumably essential] interaction with lakes."[18] Second, purposeful accommodation to a place, whether by means of physical construction or intellectual production, creates its own "stilt culture." (Venice is such a place.)

A third explanation for the attractiveness of insular coasts models itself on the argument that Eugene Odum presents in his foundational book *Ecology* (1963). Odum argues, in a quasi-determinist tradition, that the border between forest and plain was a necessary locale for the origin of civilization: "Human civilization has so far reached its greatest development in what was originally forest and grassland in temperate regions. . . . Man . . . tends to combine features of both grasslands and forests into a habitat for himself that might be called *forest edge*."[19] The architectural historian Robert Geddes takes up this argument in the *Forest Edge* (1982). For our present purpose, the place at the border between forest and plain is the coast, or "cut"—the shearing off, or "shore." Languages are full of philological evidence. Consider that the English-language *terp*, which means "pile island" or "pile village," comes to mean "town," as if to suggest the historical and conceptual understanding of a city as an island.[20]

FLOATING THE BRITISH EMPIRE

We do not believe that a tide rises and falls behind every man which can
float the British Empire like a chip, if he should ever harbor it in
his mind.

—Thoreau, *Walden*

The German story of discovering a German self in the northern islands ends with
World War II. The most important German expression of islandological propaganda during that period is the Nazi Carl Schmitt's *Land and Sea* (1942), from
which I have quoted several times in the present volume. Schmitt's work shows
an intellectually neurotic and doubly infirm purpose: both to emulate and to put
down the presumably insular culture of the Norse as well as that of the British Empire and its American cousin. Both Englishmen and Anglo-Americans were much
aware of this aspect of German culture. (No wonder Ernst Udet killed himself.)
Two major British-American examples will here suffice: Eliot and Auden.

Eliot's Defence and Auden's Stranger

On the one hand is T. S. Eliot, whose "Defence of the Islands" (1940), a blandly
disturbing poetic call for a specifically nationalist English verse, was written by
this American immigrant in order to aid the British war effort. Two poems are
counterpart to "Defence." First is "East Coker" (1940), which partly expounds
war as a spiritually cleansing ordeal. The other is the quartet "The Dry Salvages"
(1940), written during the Luftwaffe's sustained, strategic bombing of London
during the Blitzkrieg. This work starts with an epigraphic reference—complete
with scholarly annotation—to an archipelago of three small granite islands off
the coast of Massachusetts: the so-called Dry Salvages,[21] which Eliot's ancestor
Andrew Eliott passed during his emigration from East Coker, England, to the
Massachusetts Bay Colony in 1649.[22] The Salvages usually show only at low tide;
near their middle stands the assuaging beacon to which Eliot famously refers.

The text proper of "The Dry Salvages" starts with an expression of how the sea
sounds as it meets the rocky land—"the distant rote in the granite teeth." It then
moves on to how the sea surrounds human beings, who are in that sense islands—
"the sea is all about us." For Eliot, finally, the sea and the land are one and the same:

> The sea is the land's edge also, the granite
> Into which it reaches, the beaches where it tosses.

Charles Darwin–like imagery pervades Eliot's writing.[23]

Apposite to Eliot's "Defence" and marine poetics is W. H. Auden's "Look,
Stranger," published in *On This Island* (1937),[24] in which Auden approaches otherwise the "rote"—as Eliot calls the stormy music that tide and rock produce

together. Auden seeks out the strange "channels of the ear," focusing on the river-
ine English Channel that separates island Britain from mainland Europe and also
binds the two.

> Look, stranger, on this island now
> The leaping light for your delight discovers,
> Stand stable here
> And silent be,
> That through the channels of the ear
> May wander like a river
> The swaying sound of the sea.[25]

Where I live, on Grand Manan Island in the Bay of Fundy, the fishermen really know their distance from the circle of land in foggy weather by "the rout of
the sea"[26]—the voice of the water and land, taken as bow and lyre. Interpreting
that signal music is the work especially of island peoples. The ancient Greek poet
Aratus, in his astronomical poem *Phaenomena* (third century BC), refers to "the
far sounding beach"[27] as a sure sign of storms to come. One reads in the *Newfoundland Quarterly* (1909) that

according to the nature of the shore, whether sandy beach, gravel, rocky caves, and so forth,
a different rote is made, and the fishermen are wonderfully expert in detecting their whereabouts by this sign. Sometimes the rote is deep and hollow, like the bellowing of distant
thunder or of artillery, as the water rushes into deep caves, again sharp and shrill as it rolls
over moving pebbly beaches; again hissing and seething as it creeps over a sandy shore.[28]

The "rote," rightly understood, informs the thinking of those island fishermen
with ears to hear it both where the right *land* is (a harbor) and where the right
water is (a fishing ground). A once well-known Newfoundland debate about the
etymology of the hydronym *Double Road Point* serves to illustrate the doubleness
of the sign. The toponym *Double Road Point* contains a word, *road*, which Newfoundlanders understand to be a local dialectal pronunciation of *rote* and which
they agree names some spot in the waters in the southern portion of the entrance
to St. Mary's Harbour (on the southern shore of Newfoundland's Avalon Peninsula). But what vaguely Archimedean spot is it? One group of Newfoundlanders
claims that it is the small bit of *land* surrounded by water (an *insula*) where one
can hear two differently tuned rote sounds *in alternate sequence*:

This point at St. Mary's has a sort of cave or gorge or split in the rock, so that after the sea
strikes and makes its first rote, it then rushes into the fissure of the rock and again striking,
it produces a second rote.

Another group says that the spot is a small bit of *water*, a "fishing ground" where one can hear two rote sounds issuing from two different places *simultaneously*:

In about the middle of the entrance to St. Mary's Harbour, half-way between Double Rote Point and Crapeau Point on the north, there is a very good fishing "ground." It is called the Double Rote Ground, and the way to find it in a fog is to row out from the shore till you hear the two rotes one from Double Rote Point on the south, and one from Crapeau Point on the north, then you are on the "ground."[29]

When I sailed to this latter ground in 2009, the effect was stereophonic: the one voice singing with the other, each to each and reef to reef, like the modulated music of Wagner's "Rhine maidens" singing mermaid-like at the beginning of *The Ring of the Nibelung*. T. S. Eliot, in his *Wasteland* (1922), twice quotes them thus: "Weialala leia / Wallala leialala." It is the sound of the beach speaking to humankind, or trying to.

And yet one never knows how well one understands the rhetoric of the seas. The English navigator Thomas James, in *Strange and Dangerous Voyage* (1633), writes about the music that he and his crew heard while searching for the Northwest Passage: "We heard the rutt of the shoare, as we thought: but it prooued to be the rutt against a banke of Ice."[30] That, as we saw in *S.O.S. Eisberg*, is the littoral voice: the speaking of the beach.

Schmitt's Emergency and Heidegger's Thing

> . . . The play's the thing
> Wherein I'll catch the conscience of the king.
>
> —Shakespeare, *Hamlet*

German writers, painters, and composers of the nineteenth century were already transforming the presumably democratic politics of the Norse island *ting* into variously aesthetic formats. During the Nazi era, the legislative political forum of the *Thing* found architectural and dramatic expression throughout Germany. Theater propagandists for the *Thingspiel* (*ting* drama), including "Hitler's muse" Eberhard Wolfgang Möller,[31] found *Thing* ready-to-hand in the German language.[32] Architects like Fritz Schaller[33] constructed forty *Thingstätten* or *Thingplatzen* throughout Germany and had plans for eleven hundred more.[34] The *Thingstätte* on Rügen itself was completed in 1937. Propaganda minister Joseph Goebbels opened the Heidelberg *Thingstätte* in 1935. (His 1922 Heidelberg PhD thesis had been about the dramatic work of the early Romantic dramatist Wilhelm von Schütz. Its original subtitle was "A Contribution to the History of Romantic Drama," but Goebbels now repressed it and substituted the more apparently political "The Spiritual and Political Undercurrents of the Early Romantics.")[35] At his dedication for the

Heidelberg Thingstätte, he called the place a "holy mountain"[36] and argued that "these sites [*Stätten*] are the state diets or parliaments [*Lagtagen*] of our time." "There will come a day," he claimed, "when the German people will walk on these stone sites and marvel at the ritual plays [*kultischen Spielen*]."[37] Nazi playwrights wrote dozens of plays[38] about *Things*, hoping thus to "stage [the] national community."[39] Most influential among these was Möller's own *Frankenburger Würfelspiel* (Frankenburg Dice Game) written in 1936[40] and set during the Thirty Years' War (1618–1648).[41] Another popular writer of the Nazi genre was Heinrich Zerkaulen, who wrote such plays as *The Island of Thule*,[42] *Leider auf Rhein*,[43] and *Tage auf Rügen*.[44] Martin Heidegger, a member of the Nazi Party since 1933, joined the fray only obliquely, as in his Freiburg seminar "What Is a Thing [*Ding*]?" (1935–1936).[45] In "The Thing" (1950), written after Germany's defeat, he makes the *Ting* movement sound more philological than political: "[T]he Old High German word *thing* means a gathering, and specifically a gathering to deliberate on a matter under discussion, a contested matter. In consequence, the Old German words *Thing* and *Ding* become the names for an affair or matter of pertinence. They denote anything that in any way bears upon men, concerns them, and that accordingly is a matter for discourse."[46] It is as though Heidegger wanted to forget about the *Dingpolitik* or *Thingpolitik*, which he himself had touted.[47]

Carl Schmitt, a German of Heidegger's generation and religious background, began his professorial career at the University of Greifswald, on the banks of the Ryck River near where it empties into the Baltic Sea. The city of Greifswald is equidistant from the islands of Rügen and Usedom, cultural projections of which were important to nineteenth-century German nationalism, and it was the hometown of the influential coastal painter Caspar David Friedrich. At Greifswald, Schmitt wrote or conceived works involving politics and the sea that informed his entire career: "Die Diktatur: Von den Anfängen des modernen Souveränitätsgedankens bis zum proletarischen Klassenkampf" (1921), with its incipient theory of "a state of emergency and exception" (*Ausnahmezustand*), and "Politische Theologie: Vier Kapitel zur Lehre von der Souveränität" (1922).[48]

The historical issue informing such work on political emergency and exception was the role of "ship money" in island Britain in the sixteenth and seventeenth centuries. In fact, there were two periods when the English sovereign raised ship money. First, the Tudor queen Elizabeth I urgently needed funds to prosecute the war with Spain (1588), and so, exceptionally, she raised special taxes—"ship money"—from people who lived on the coast. This is the period that Schmitt reconsiders in *Hamlet or Hecuba: The Intrusion of Time into the Play* (1956). Second, the Stuart king Charles I, calling up the same sort of urgency (1634, 1635, 1636), raised ship money, but from the inland shires as well as from the coast. The Whig historian and English secretary of war (1839–1841) Thomas B. Macaulay, in *History of England* (1849), writes that "former princes . . . had raised ship-money only along the coasts: it was now exacted from the inland shires."[49] This time Britain was apparently at relative peace, albeit the Dutch commercial fleets and navy were besting the British around the globe and nearer to home.[50] The people were not happy with ship money. Charles used his to build such vessels as the HMS *Sovereign of the Seas*, but in 1649 he was beheaded.

For Schmitt, the political issue involved the old debate about the law of the sea as well as the Hobbesian views of sovereignty and urgent political "exceptionalism." This is clearest in Schmitt's *Land and Sea* (1954). In his *Mare Clausum* (1618), Thomas Hobbes's friend John Selden, who hailed from Salvington on the south coast of England, argued that the seas were potentially closed. His view so countered the argument of the Dutch jurist and scholar Hugo Grotius, in *Mare Liberum* (1609), that the sea was free, that the English king James I suppressed the manuscript. But King Charles I, now seeking to protect English waters, had it published as an anti-Dutch salvo, compelling the Dutch herring fishermen off the east coast of England to purchase "licenses" in order to have "protection" from what the English said were "Channel pirates."

King Charles I's contemporary Thomas Hobbes says that the king's raising of ship money was the trigger for the revolution. In *Dialogue between a Philosopher and a Student of the Common Laws of England*, written in 1666, he pretends to wonder whether "a King should have the Right to take from his Subjects, upon the pretense of Necessity what he pleaseth." In *Behemoth: The History of the Causes of the Civil Wars of England, and of the counsels and artifices by which they were carried on from the year 1640 to the year 1660* (published abroad beginning in 1668), Hobbes says outright that ship money was the first event leading to the English Civil War and the Commonwealth (the so-called English Revolution).[51] No wonder Schmitt published *Der Leviathan in der Staatslehre des Thomas Hobbes* the same year (1938) that Hitler, declaring a state of emergency, announced that he was now supreme commander of the armed forces and began promoting the geopolitical idea of *Lebensraum*.

The doctrine of *Lebensraum* had its geopolitically islandic aspect even at the level of the national anthem. It was on British Heligoland that August Heinrich Hoffmann von Fallersleben wrote the "Lied der Deutschen" (1841), which in 1922 became the German national anthem. Its first line, "Deutschland, Deutschland über alles / Über alles in der Welt" (Germany, Germany above all, above all in the world), is a response to the British national anthem "God Save the King" (watchword of the British Royal Navy) as set by Thomas Arne (composer of "Rule, Britannia!"). The first stanza of Hoffmann's "Lied" refers to the "Belt" as the proper borderland of the German nation.

Von der Maas bis an die Memel,	From the Meuse to the Memel,
Von der Etsch bis an den Belt.	From the Adige to the Belt.

Belt can refer to the Little Belt (or strait), between the now Danish islands of Funen and Jutland, to the Great Belt, between the Danish islands of Funen and Zealand, and/or to the Fehmarn Belt, between the presently German island of Fehmarn and the presently Danish island of Lolland. All of these straits and some of their defining islands have been disputed boundaries. Indeed, Eric of Pomerania—the first king of the Scandinavian Kalmar Union—occupied Fehmarn in the fifteenth century. The Danish king Frederick I gave Fehmarn to Schleswig in 1580, although the Danes often reoccupied it, as in the 1680s, 1850s, and 1860s.

The present version of the German national anthem represses the lines about the "Belt." Still, Nietzsche's warning about the popularity of the "Lied" rings true. "Deutschland, Deutschland über alles—I fear that was the end of German philosophy," he writes in *Twilight of the Idols, or, How to Philosophize with a Hammer* (1889).[52] Following World War II, many Germans argued that they would do well to adopt instead Bertolt Brecht's rewriting of the anthem. After Hitler came to power, Brecht left Germany to live on Danish Funen Island (1933–1939), at the shoreline of the Svendborg Sound, which runs between the islands of Funen and Tåsinge.[53] His eventual rewriting of the national anthem, called "Kinderhymne" (1950),[54] seems to call for an insular geopolitics with no "Belt":

Von der See bis zu den Alpen	From the Alps to the [North] Sea,
Von der Oder bis zum Rhein.	From the Oder to the Rhine.

Britten's Britain
Before Britten

The British composer Benjamin Britten, who often wrote of islands, was influenced by two sorts of earlier English music about the sea. One was the utopian, satirical, and lighthearted British operettas with island settings—latter-day variants of Arne's *Alfred* (1740). These he reacted against. The second was the new sea music of the early twentieth century, which he partly adopted as his own.

Concerning the operetta's prehistory, there had been for centuries an onstage playing-out of political issues in the form of fictions projected onto colonized islands. Shakespeare wrote *The Tempest* (1611), for example, at a time when there were two relevant British attempts at island colonization in the New World. First was Sir Walter Raleigh's colonizing mission on Roanoke Island, located between the mainland and barrier islands of what is now North Carolina.[55] This island colony failed—the settlers mysteriously disappeared in 1590. The second attempt, in 1606, was the establishment on Jamestown Island in the tidal estuary of the James River in present-day Virginia. One of the ships going to Jamestown, the *Sea Venture*, came into a tempest and beached at Bermuda,[56] an event described in *A Plaine Description of the Barmudas, Now Called Somer Ilands* (1613) by "W. C." When the survivors of that wreck finally arrived at Jamestown, they discovered that many of their fellow would-be colonists had died. (Shakespeare probably knew Sylvester Jourdain's *A Discovery of the Barmudas*, written in 1610; Robert Johnson's publicity pamphlet *Nova Britannia*, written in 1609 and published in Amsterdam; the advertising flyer *A True Declaration of the Estate of the Colonie in Virginia*, written in 1610;[57] and/or William Strachey's works on voyages to Virginia, written between 1609 and 1610.)[58] When Shakespeare died, in 1616, the colonists at Jamestown were thriving, but in 1622 Indians killed a third of them.[59]

That same year, the dramatist John Fletcher put out *The Island Princess*, setting the story on the East Indian islands of Ternate and Tidore (in the present-day Malaku Archipelago in Indonesia). Fletcher and Philip Massinger then brought out their short play *The Sea Voyage* (1622), set on a geographically ambiguous island populated by Amazonian women. A later version of *The Sea Voyage*, called *The Storm*, competed directly for audiences with William Davenant and John Dryden's *Tempest, or The Enchanted Island* (1667), a reworking of Shakespeare's play. When Thomas Shadwell arranged for the Davenant-Dryden version to be set to music by Henry Purcell's mentor, Matthew Locke (1674),[60] this *Tempest* became a box office success. A century later, in 1770, David Garrick and Thomas Arne revived Dryden and Purcell's *King Arthur; or The British Worthy* (1691), in whose final act "an island arises, to a soft tune." According to stage directions, "Britannia [is] seated in the island, with fishermen at her feet," and Venus sings "Fairest Isle, All Isles Excelling"—as if Britain were both an isle and, insofar as it excelled *all* (other) isles, more than a regular one.

An example of such seventeenth-century island drama providing the historical backdrop to W. S. Gilbert and Arthur Sullivan is Thomas d'Urfey's *Commonwealth of Women* (1685), which was another adaptation of Fletcher and Massinger's *Sea Voyage*. D'Urfey added the "Honest Pirate" to the earlier story line, which much influenced Gilbert and Sullivan when they wrote *The Pirates of Penzance* in 1879.

Gilbert was an English island librettist whose tales, written at the height of the British Empire, usually couch a utopian social critique. With composer Thomas German Reed, for example, he created the one-act *Our Island Home* (1870), set on an island in the Indian Ocean.[61] With composer Arthur Sullivan, he wrote *The Gondoliers* (1889), set in Venice (a real island-city) and Barataria. (Barataria [cheapness] recalls the imaginary *insula* promised to Sancho Panza in Miguel de Cervantes's novel *Don Quixote* [1604, 1615].)[62] The successful *H.M.S. Pinafore* (1878), which partly revises Arne's musical work "Rule, Britannia!" (1740)—"Shall [we/they] submit? Are [we/they] but slaves? /. . . / Britannia's sailors rule the waves"[63]—takes place shipboard at Portsmouth on Portsea Island. *The Mikado* (1885) represents the island nation of Japan even as it presents the island nation of Britain: G. K. Chesterton, comparing *The Mikado* with Jonathan Swift's *Gulliver's Travels*, argues that "the satire of the *Mikado* is not at all directed against Japanese things, but exclusively against English things."[64] The apogee of Gilbert and Sullivan's work, in this context, is probably the politically controversial *Utopia Limited* (1893), with its South Seas island of Utopia and its satire of limited liability companies and British cultural imperialism. So politically sensitive were some of the lines in *Utopia Limited* that they had to be cut after the first few performances. For example, consider the lines spoken by the governess Lady Sophy to Paramount, king of Utopia, the island paradise: "As there is not a civilized king who is sufficiently single to realize my ideal of Abstract Respectability, I extended my sphere of action to the Islands of the South Pacific—only to discover that the monarchs of those favoured climes are at least as lax in their domestic arrangements as the worst of their European brethren." The island of the South Seas begins as paradise, but thanks to British imperialist takeover, ends as ruin.

Britten

More than reacting against such work particularized on Britain, Benjamin Britten moved in accord with the classical music of the first decades of the twentieth century dedicated to the universalist sea. Such work included Charles Villiers Stanford's *Songs of the Sea* (1904) and *Songs of the Fleet* (1910); Edward Elgar's *Sea Pictures* (1899); Claude Debussy's *La mer* (1905) and *L'isle joyeuse* (The Happy Island); and Ralph Vaughan Williams's choral work *A Sea Symphony* (1910), which was influenced by Walt Whitman's universalist poem *Leaves of Grass* (1855).[65] Britten's first important teacher, Frank Bridge, had written an orchestral poem called *The Sea* (1911), and when young Britten first heard it at a festival, it caused him to be "knocked sideways." An older Britten would go on to write his *Variations on a Theme of Frank Bridge* (1937)[66] and an unfinished *Sea Symphony* (1976).

Britten himself was from Lowestoft, on the East Anglian coast, and the storms of that area, with which the local oral and written history is replete, inform his music. A stormy shift of wind during the Battle of Lowestoft (June 3, 1662), for example, favored the British. (The Dutch still remember this battle as their greatest naval defeat.) Stormy weather also meant the defeat of the Spanish Armada (1588) in the same general region. The Lowestoft ship *Elizabeth* played a key role. Wars can be won thanks to storms, but, at the same time, wars are themselves storms—or so goes the common English "metaphor," especially during World War II.[67] The notion that war is storm informs, for example, Britten's opera *Peter Grimes* (1945),[68] which is based on an early nineteenth-century poem, "The Borough,"[69] by George Crabbe, who was from Aldeburgh, just down the coast from Lowestoft, where Britten founded the Aldeburgh Festival (1948) and from whose port, Slaughden, on the banks of the Alde River, four ships were launched to fight against the Spanish Armada.

Likewise, there is Britten's shipboard *Billy Budd* (1951),[70] whose libretto E. M. Forster and others modeled on Herman Melville's novella and whose historical grounding is the 1797 mutinies at Spithead, near Portsmouth, and at the Nore, the sandbank island in the Thames Estuary where the river meets the sea.[71] There the sailor Richard Parker became "president" of the rebellious "Floating Republic"; he was eventually hanged on nearby Sheppey Island.[72] At the beginning of *Billy Budd*, in which French-style republicanism confronts British-style monarchy, Captain Edward Fairfax Vere brings up the French-inspired "Mutiny at the Nore" of 1797 in the Prologue.[73] In act 1, scene 2, the Sailing Master recalls darkly "Spithead, the Nore, the floating republic"—to which First Lieutenant and Vere respond.[74] Toward the end of the opera, in the second act, the Captain recalls sadly that, as captain of the ship, he is the forlorn "king of this fragment of earth, of this floating monarchy."[75] The man-made ship is no less earthen than island Britain.

Two Britten works specifically about islands and politics stand out from the others: *On This Island* (1937) and *Noye's Fludde* (1957). *On This Island*,[76] which sets to music poems from W. H. Auden's *On This Island* (1936),[77] was written at a time when Britten and Auden were both working on the documentary film *The Way to the Sea* (1936), partly about Portsea Island, the seat of the Royal Navy and the shipboard setting for the popular *H.M.S. Pinafore* (1878) by Gilbert and Sullivan. Britten provided the score and Auden provided the much repeated line "We seek an island."[78] In the title poem of the collection *On This Island*, the speaker addresses a "stranger," meaning "a person unfamiliar to the speaker" or "a person unfamiliar with the island." (Auden wrote the work while headmaster at Larchfield Academy[79] on the Firth of Clyde.[80]) Or perhaps the speaker addresses a person who is actually long familiar with the island but whom the speaker asks *now* to look on the island anew, as if it were, or had now become, strange to him.

> Look, stranger, at this island now
> The leaping light for your delight discovers,
> Stand stable here
> And silent be,
> That through the *channels of the ear*
> May wander like a river
> The swaying *sound of the sea*.
>
> (emphasis mine)

The place is sublime, like the island in the Øresund to which the presumed Ghost of Old Hamlet beckons Young Hamlet. It is an "ending pause / Where the chalk wall falls to the foam." Even as Horatio would "assail" the channels (of Hamlet's ears),[81] asking Hamlet to "Season your admiration for awhile / With an attent ear, till I may deliver, / . . . / This marvel to you" (act 1, scene 2), so the speaker requires heed to the swaying sound of the sea.

The second of Britten's major island works is *Noye's Fludde*. The historical counterpart is clear: the North Sea Flood of 1953, when large parts of the Netherlands, Belgium, England, and Scotland were inundated and thousands died.[82] The English island was no longer the safe haven of old, if indeed it had ever been. Britten's house in Lowestoft was flooded. Houseboats were used as "saviour arks."[83] People recalled the old notion that Britain was an insular ark—as with the Armada Jewel (see Illustration 62, Color Plate 6). Winston Churchill reminded people that Britain was a safe and glorious island, as in his speeches "I Have Watched This Famous Island" (1938)[84] and "Renewing the Glory of Our Island Home" (1951).[85] *A Floating Home* is the name of Cyril Ionides and J. B. Atkins's history of British yacht conversion (1918); in one example that the authors give, a yacht is christened *Ark Royal*—recalling the names of the flagship of the English fleet during the Spanish Armada and of the British aircraft carrier during World War I.[86]

Britishers well knew the biblical story of Noah's Ark, an artificial floating island. That floating home becomes a grounded island when stranded atop a submarine mountain, as in the illumination from the thirteenth-century *North French Hebrew Miscellany* (see Illustration 74). Some commentators say the ark landed at Mount Judi (*Judi* meaning "high hill or island").[87] Others say it ran aground near the city of Cizre (*Cizre* meaning "island").[88] Still others say Noah's ark landed on Mount Ararat. The New York politician and playwright Mordecai Manuel Noah adopted the Ararat hypothesis when he named "Ararat" the Jewish refuge state that he founded on Grand Island in the Niagara River between Canada and the

זה עץ בתוך היתבה ׳ והיונה עליו בחה ״

United States. Israel Zangwill tells the story in his "Noah's Ark" (1899).[89] One response to such island-refuge hypotheses and projects was the German scholar and orientalist Paul de Lagarde's anti-Semitic plan (1885) to relocate European Jews to the French-held island of Madagascar and its subsequent adoption by the Nazis.[90] No matter the landing place for the biblical Noah, for Britishers the ark and the island were one: England.

Britten's *Noye's Fludde* takes as its general text one of the medieval Catholic "Chester Mystery Plays" banned by the Protestant Elizabeth I.[91] (Igor Stravinsky's twenty-four-minute work for CBS television, *The Flood: A Musical Play* (1962) with choreography by George Balanchine, likewise took its material from the Chester plays, except that it overtly conflates the relatively historical story of the Flood with the partly prehistorical story of the Creation. Stravinsky thus writes: "The music imitates not waves and winds, but time.")[92] Britten notably adds to his *Noye's Fludde* William Whiting's naval hymn "Eternal Father, Strong to Save"

(1860) as set to John B. Dykes's melody "Melita." "Melita," meaning "island of Malta," names the coast where Saint Paul, from the port city of Tarsus, warns his followers:

Howbeit we must be cast upon a certain island. Except these abide in the ship, ye cannot be saved.[93]

The first verse of *Noye's Fludde* refers to God's forbidding the waters to flood the Earth: "Eternal Father, strong to save, / Whose arm hath bound the restless wave, / Who bidd'st the mighty ocean deep / Its own appointed limits keep."[94] The second refers to Jesus's miracles of stilling the storm and walking on the Sea of Galilee: "O Christ! . . . / Who walkedst on the foaming deep." The third invokes the separation of water and earth in Genesis: "Most Holy Spirit! Who didst brood / Upon the chaos dark and rude."[95] The final verse prays for protection from the storm: "Our brethren shield in danger's hour; / From rock and tempest, fire and foe."[96] Never had the words of this hymn been more carefully scanned than in August 1941 aboard the Royal Navy battleship HMS *Prince of Wales*. At that time, Winston Churchill (first lord of the Admiralty during the Great War and now prime minister) and Franklin D. Roosevelt (assistant secretary of the navy and now president) met on this floating island and created the Atlantic Charter so as to bring security to the Allied Powers.

For Britten, it was important that the opera be attractive to children as well as adults. The story itself allows for that. After all, the Hebrew Bible uses the word *teyva* twice, both involving floating artificial islands: first, to refer to the savior "ship" that Noah, whose name in Hebrew can mean "comfort" (*naham*) or "peace" (*nuah*), built for all nonmarine creatures; second, to indicate the "basket" that Miriam made for the child Moses and set out on the Nile.[97] Children know that, although Noah saved only two of each land animal, he saved his own nuclear family, including the children. The old African American spiritual "Who Built the Ark?" notes the fact thus:

> Old man Noah built the Ark,
> He built it out of hickory bark.
> He built it long, both wide and tall.
> With plenty of room for the large and small.

Wes Anderson's movie *Moonrise Kingdom* (2012) features a violent storm on make-believe island New Penzance—a reference to Gilbert and Sullivan's *Pirates of Penzance* (1879), set in the real Cornish port village of Penzance, which was a favorite spot for pirates.[98] Penzance is the location, says Harry Morris, where the adult Britten made sexual advances to him when Morris was a child.[99]

Noye's Fludde was part of an ongoing twentieth-century debate about musical pedagogy for children initiated in the 1920s by the politically opportunistic German composer Carl Orff[100] in his *Schulwerk* and *Music for Children*.[101] Britten's response included such works as *The Young Person's Guide to the Orchestra* (1946),[102] *The Little Sweep* (1949)[103]—a child's "rescue opera"—and *The Chimney Sweeper* (1949).[104] The last two, based on William Blake's child labor poems, reference also Charles Kingsley's children's novel *The Water-Babies: A Fairy Tale for a Land Baby* (1862–1863). Britten's librettist, Eric Crozier, reminds us that the three-year-old Britten acted in a dramatic version of *The Water-Babies*,[105] which was set on the banks of the tidal river Alde, where Britten ended up living most of his life. He died at his house in Aldeburgh as "Baron Britten of Aldeburgh." He had purchased the house in 1957, the year of *Noye's Fludde*.

Postamble

He quietly lifts his pilgrim-staff, and begins a perambulation and
circumambulation of the terraqueous Globe.

> —Thomas Carlyle, *Sartor Resartus: The Life and Opinions
> of Herr Teufelsdröch*

SPEAKING ON THE SHORE

Who wishes to speak? (Τίς ἀγορεύειν βούλεται;)[1]

> —Aeschines, *Against Timarchus*

The ancient Athenian speaker Demosthenes (384–322 BC) had a speech dysfluency and talked too softly.[2] Yet he became the great orator of the Athenian assembly (*ekklēsia*)[3]—the constitutionally established democratic body regularly "convoked" and open to all citizens at the islandic Pnyx. Why? The way Demosthenes managed to learn to speak "fluently" is described by the first-century Greek writer Plutarch, who says that the young Demosthenes filled his mouth with pebbles from the seashore. He then practiced talking—through the combination of salivous[4] water and small rocks. Precisely how Demosthenes learned to speak "loudly" is reported in *The Lives of the Ten Orators*, one of the later writings sometimes attributed to Plutarch, which says that he used to go for walks along the seashore. There he practiced talking so loudly as to be heard above the *rut*[5]—the racket that is made when the waves and tides grind the rocks at a shingle beach against one another.[6] Just so, Demosthenes developed, from working the two defining elements of the beach—rock and water—a politically powerful hortatory talent that issued forth, fluently and loudly, from his body (the mouth where fluid water comes up against rocky impediment), in such a way as to be heard above the natural ruckus of the body politic (the shingle where rock and water together create the rut). (See Illustration 75, Color Plate 7.)

Demosthenes managed to turn his speaking impediment to rhetorical advantage just at the time when Athenian democracy needed it most. He used his special rhetorical skill, deriving from land and sea, to encourage his compatriots

ILLUSTRATION 75
*Demosthenes Practicing
Oratory.* From
Jean Jules Antoine
Lecomte du Nouy,
*Démosthène s'exerçant
à la parole* (1870).
Private Collection /
© Look and Learn /
The Bridgeman Art
Library.

to build up a navy for defense against the powerful Persians. The Persians had turned mainland Mount Athos into a "European" island; their tyrant, Xerxes, had challenged the tide itself at the Hellespont.[7] Greek naval preparation, land and sea, was therefore the subject of Demosthenes' first political oration, *On the Navy*.[8] Demosthenes also used his rhetorical skills to decry Philip II of Macedon, whom he characterized as being as dangerous to Athens as any Persian tyrant, whose assassination he celebrated, and whose son, Alexander the Great, turned island Tyre into part of the "Asian" mainland.[9]

Historians often recall Athens as a fount of Western democracy, which they define in terms of the constitution of the assembly (*ekklēsia*, described by Aristotle in his *Constitution of the Athenians*). They characterize the highest period of Athenian democracy as the "Age of Demosthenes."[10] This characterization makes sense, on the political level, when we recall Demosthenes' critiques of antidemocratic Persian and Macedonian factions in Athens. It seems especially reasonable, on the personal level, when we recall his own struggle to define the principle of *isēgoria* (equal speech) for all citizens, as guaranteed by the *ekklēsia*,[11] and then to make that principle practicable in his own case. Demosthenes, who spoke out against Alexander of Macedon, eventually went into self-exile from Athens on the island of Calauria (Poros)—whose legendary amphictyony had been one model for the Athenian Empire.[12] When Athenian authorities sought him out there, he committed suicide.

In a tyranny one must hold one's tongue—or even pretend, more or less "esoterically," that one is dumb, stupid, or brutish. Otherwise, one will lose one's life. The play *Hamlet* suggests several examples of persons who pretend to be stupid or unable to speak their minds. One is the Roman Lucius Junius Brutus, referenced also in Shakespeare's *Julius Caesar*. This Brutus counterfeited himself as a dullard to avoid being killed by his uncle, the last tyrant of Rome, Lucius Tarquinius Superbus.[13] Another example is the stuttering Roman emperor Claudius, who likewise counterfeited stupidity to be safe. This Claudius was the "uncle-father" (according to Hamlet) of Nero in much the same way that Claudius is Hamlet's uncle-father. The Roman Claudius became Nero's stepfather when he married

Agrippina, Nero's consanguineous mother.[14] He actually wrote a defense of the republican oratory of Cicero—whom Asinius Gallus had attacked[15] and whom Plutarch compares with Demosthenes. Eventually Claudius managed his empire from what he hoped would be the secure shores of the island of Capri in the Tyrrhenian Sea near Naples,[16] though Agrippina (and Nero) managed to poison him.[17]

Something like the same fate had awaited Demosthenes, who eventually escaped from Athens to Calauria. Finally he was able to elude his captors only by taking poison from a reed with which he pretended he wanted only to write.[18] No wonder, then, that Hamlet, who still wants to live, says, "I must hold my tongue." The name *Hamlet* (Amleth) means "dumb."[19]

The sea, whose sound Demosthenes learned to speak with, has a voice of its own, especially in *Hamlet*, where the tyrannical king is eventually overthrown only by the islandic *ting*. The "rut" of the sea plays a distinct role in earlier Hamlet legends. These legends come to the fore again during the later nineteenth century, with the developing interest in the northern isles and "Thule societies" in the European countries, especially islandic Britain and mainland Germany.

In the twelfth century, for example, the Icelandic Snorri Sturluson began to preside as lawspeaker over the islandic *Alþingi* in the same year that the barons of Britain forced King John to sign the Great Charter (1215) at "Magna Carta Island." Sturluson quotes in his *Skáldskaparmál*[20] a lengthy dialogue between the god of the sea and the god of poetry,[21] a work of the Arctic "sailor-poet" Snæbjörn,[22] who had it that "Hamlet's Mill" (*Amlóða mólu, Amlóða kvern*) is the great marine whirlpool or maelström, a central subject of Norse mythology.[23] Once upon a time, according to Snæbjörn, Hamlet (Amlodhi, "Frohdi") had "the ownership of a fabled mill which, in his own time, ground out peace and plenty." In later days, however, the mill came to churn out only salt. Moreover, nowadays, he says, "it is grinding rock and sand, creating a vast whirlpool, the Maelström."[24] That is, in effect, the turning point of Shakespeare's play. *Hamlet in Iceland*—written in 1898 by the Shakespearean scholar Israel Gollancz, a member of various Thule societies[25]—was a critical sensation.[26]

For the Norse Hamlet, the relevant democratic assembly was not Demosthenes' Athenian *ekklēsia*, which met at the landed Pnyx, a "hill that rises above the surrounding land" in Athens and conveys the principle of *isēgoria* (equal speech). The democratic assembly of the Norse was, instead, the *ting*, which met on an island, or "a hill that rises up out of the sea," and compared to which the king was as nothing, as presumably King John himself felt at the "water-meadow" that was Runnymede, one of the ancient islandic meeting places of the Britons.

Must the *ting* be islandic? The Dutch Caribbean writer Frank Martinus Arion suggests one answer to that question in his novel *De laatste vrijheid* (The Ultimate

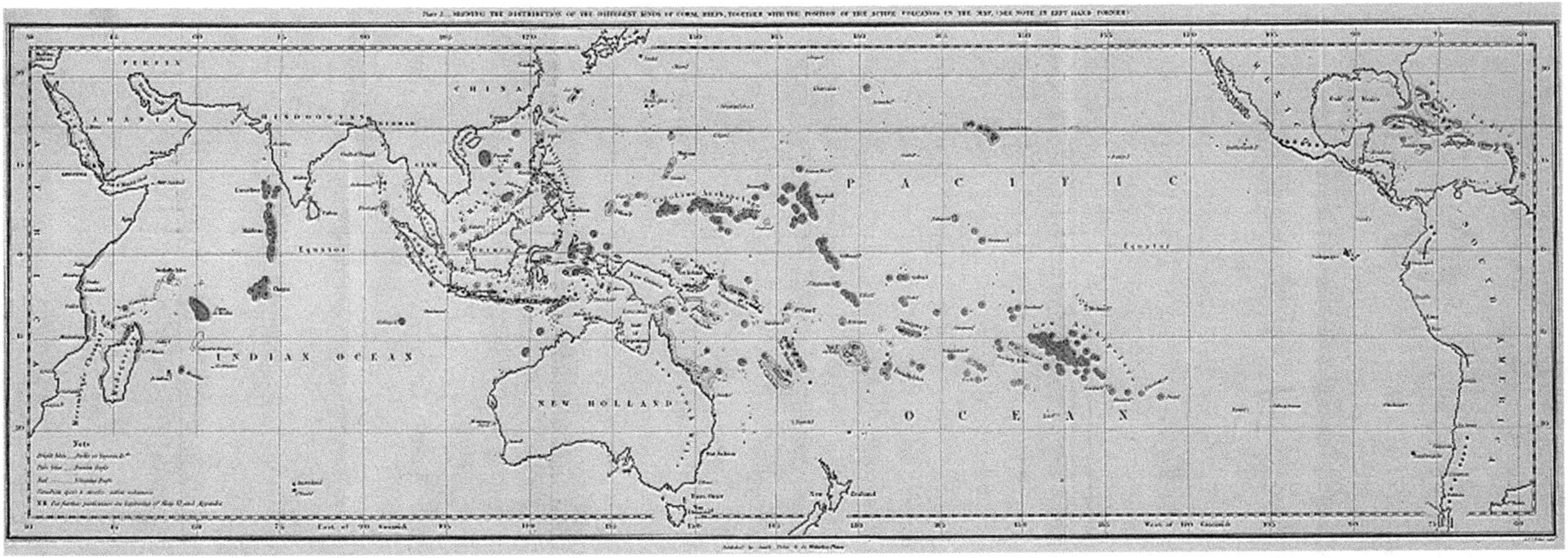

ILLUSTRATION 76 World Tectonic Plates as Islands. From Charles Darwin, *Coral Reefs with Reference to Their Formation* (1842). Reproduced with permission from John van Wyhe, ed., *The Complete Work of Charles Darwin Online*, http://darwin-online.org.uk/. Harvard College Libraries.

Freedom; 1995).[27] His husband and wife heroes—one comes from Curaçao in the Dutch Antilles, like Arion himself, and the other from Dutch Saint Martin Island—discover freedom on Suriname, which, though mainland, achieved independence from the Netherlands and is thus the only (formerly) Dutch-held Caribbean colony that has its own political constitution. Suriname is where the wife can do "*haar eigen ding*" (her own thing);[28] her *own* thing, however, turns out to be less democratic than individualistically liberal.

As ancient as the Greek hills is the vaguely determinist question of whether island places, natural or artificial, encourage the development of democratic political cultures. Nineteenth-century geography took up this sort of question with particular fervor. In that regard, Part 4 addressed the Scandinavian *ting* and its latter-day reformulations in early twentieth-century Germany.

TERRA INFIRMA AND CRITICAL TOPOGRAPHY

One feller said, "Well," he said, "we shall never see land nor stand no more."

—*Dictionary of Newfoundland English*, s.v. "land"

Charles Darwin's theories about the workings of coral (animate) islands, as in *Coral Reefs* (1842), and volcanic (inanimate) ones, as in *Volcanic Islands* (1844), work toward both a theory of geological subsidence and uplift (see Illustration 76) and a wide-ranging argument about biological evolution around the world, as in *The Origin of Species* (1859). His last major book, *Earthworms* (1881), like the first, on corals, homes in on telling links between animate and inanimate evolution on Earth.

Darwin, as we have seen, believed geological and hydrological subsidence and uplift to be a "balancing process." "As [the island of] Sumatra rises . . . the other end of the lever [Cocos (Keeling) Atoll, which is six hundred miles away] sinks down." Those are words he spoke in a lecture delivered in 1837 at age twenty-eight—in a mainly deductive scientific paper that, given the hindsight of history, must count as one of the most significant ever.[29] The up-and-down isostasy recalls Archimedes' relatively isostatic seesaw: "Give me a place to stand and I will move the Earth."[30]

Darwin's idea, that rising land is balanced by subsidence in ocean areas, was soon superseded by theories of continental drift and modern plate tectonics that he did not anticipate. These actually make *both* the oceans and the land into islands floating on molten rock. Theories of continental drift (*die Verschiebung der Kontinente*),[31] in fact, tend to explain the Pacific Ring of Fire by redefining *continents* as "floating islands"[32] and looking anew at older theories of a hollow Earth. These theories date to the first decade of the twentieth century, when the astronomer Alfred Lothar Wegener, whose disappearance on a polar expedition inspired the movie *S.O.S. Eisberg*, articulated a new explanation for why scientists were discovering, on opposite sides of the Atlantic Ocean, fossils of identical plant and animal species. Wegener rejected the old hypothesis of a "land bridge" (or dry isthmus) that had sunken, Atlantis-like, beneath the surface of the water. In his *Origin of Continents and Oceans* (1915),[33] he put forward instead the thesis that long ago there had been only one island, or landmass, in the world, which he called *Pangaea* (all earth), and which eventually split into segments. The segments drifted far apart from one another to form today's "continents."

Most scientists nowadays say that the oceans and continents, taken together, are floating "archipelagoes." The water and land, en masse, float on an asthenosphere—a "sphere of weakness" or "realm of infirmity." The shell-like firmament of Earth[34] thus rests on an infirmity—whence, says the perspicacious book of Genesis, the firmament arose in the first place. The islandic lithosphere (sphere of stone) seems firm enough,[35] but it consists of ever-shifting plates on which the continents and the oceans rest. The plates lie on the infirm, magma asthenosphere.

Just so, the entire surface of the Earth, both water and earth, is a single archipelago of floating islands (see Illustration 77). The island of Thwart-the-Way stands here at a special strait, a meeting place of world archipelagic and seismic ridges or rings.[36] Likewise, the Intermontane Islands of Canada's British Columbia make up an arc of eastward-moving Pacific Ocean islands which, since they are "too big to sink" beneath the continent against which they collide, weld with the mainland to form the "insular mountains," "insular islands," and "insular belts" of the northern part of western North America.

Similar to the action of *plate tectonics*, understood as "the motion that sometimes cracks the solid rocky crust floating above the liquid rock below," is *tidal flexing*, understood as "the vertical motion that sometimes cracks the solid icy crust that floats above liquid water." Consider the tides of Antarctica's *entirely* subglacial Lake Vostok. (Vostok is large: its "surface" area matches, and its average depth is double, that of Lake Ontario.) Lake Vostok lies two miles beneath a "permanently" solid ice sheet that covers it with a hydrostatic seal that has kept it isolated from the world above for fifteen million years. That ice and water are subject to tidal forces that, together with geothermal energy, help keep the water of Lake Vostok in a liquid state.[37] Writes Heraclitus, "Water lives the death of earth, earth that of water."

To say that the geology and hydrology of Vostok, with its various subglacial islands of earth,[38] is otherworldly means no more than to note the surface ice on heavenly bodies in the solar system other than the planet Earth: Europa (a moon

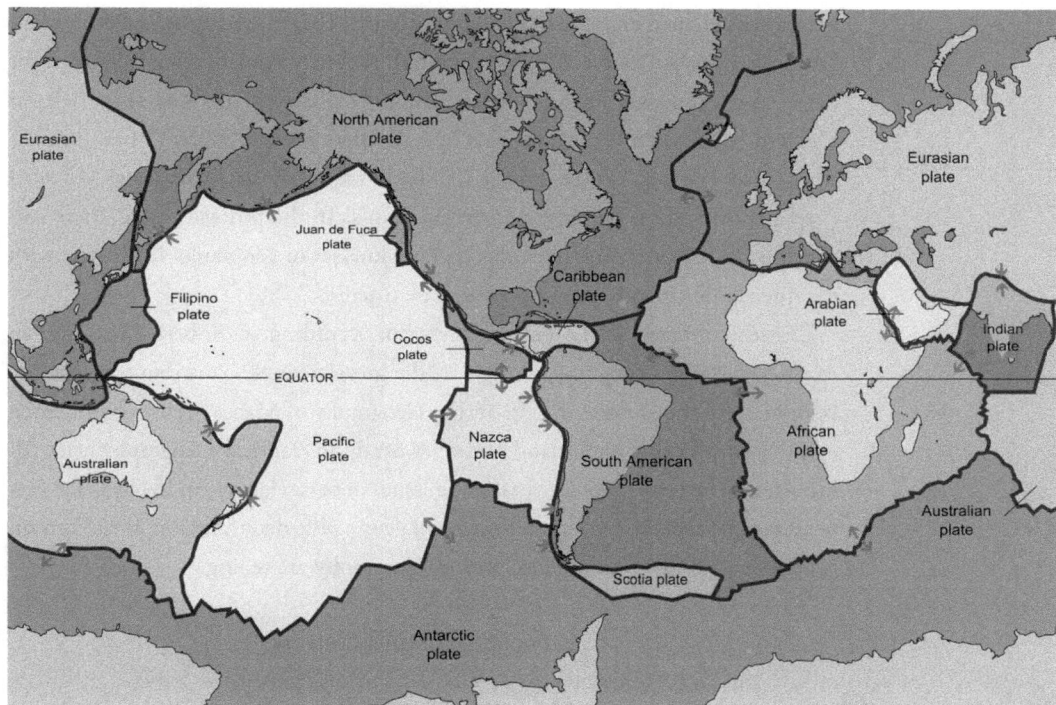

ILLUSTRATION 77
Continental Drift
or Plate Tectonics.
Source: U.S.
Geological Survey.

of Jupiter)[39] and Enceladus (a moon of Saturn) may both have subsurface heat-generating tides.

Here speculative earth science becomes planetology. Hercules takes his place alongside Archimedes.

In recent years, the term *critical topography*, once reserved for the geological sciences' description of orogens—"the great mountain systems . . . zones of compression of the earth's crust weaving a complex pattern of majestic sweeps around the world"[40]—has been making apparent incursions into such apparently disparate fields as literary study,[41] contemporary philosophy,[42] and social geography.[43] These fields await the islandological explanation for their seeming importation of a term that was in any case always there from the beginning.

GEOGRAPHY AS INSTITUTION

There is very little ground, either from reason or observation, to
conclude the world eternal or incorruptible.

—David Hume, "Of the Populousness of Ancient Nations"

Rationalizations pretending to explain the demise of geography often include stories about President James B. Conant of Harvard University, who was so much influenced by Marland Pratt Billings and Isaiah Bowman[44] that he decided to permanently close Harvard's geography department. "Geography," said Conant, "is not a university subject."[45] Derwent Whittlesey, whom Bowman detested for personal reasons, was for a few years "the only professor of Geography [left] at Harvard," and when Whittlesey died in 1956, there were no more. Other universities copied Harvard, so the story goes. There are many such anecdotes. All skirt deeper reasons for geography's American demise in the postwar years. These reasons involve, at once, the strengths and weaknesses of geography in dealing with the question of islands and in serving state interests.

Friedrich Ratzel is the geographer who, according to Robert Dickinson in *Makers of Modern Geography* (1969), is "the greatest single contributor to the development of the [academic field of the] Geography of Man."[46] Ratzel's influential essay "Island People and Island States: A Study in Political Geography" (1895)[47] introduces the term *geopolitics*, takes up issues of sovereignty and living space (*Lebensraum*),[48] and aims to fix a definition of geographic determinism. The effects on geopolitics and then on the discipline of geography are telling.

Environment and Culture

The sea is History.

—Derek Walcott, "The Sea Is History"

Ratzel had a distinctive bias toward "geographic determinism": the view that their physical environments (climate, water, land) help to determine particular human cultures. Likewise, he had a characteristic prejudice for "social Darwinism," the view that groups of human beings, defined in terms of culture or politics, struggle to survive in particular environments in much the same way that individual organisms struggle.[49] These views, taken together, are hardly Darwinian. Darwin's thinking had begun with geology and then worked toward an open-ended process of evolutionary biology and a far-reaching theory of animate and inanimate geography. Ratzel's ideas, on the other hand, started with biology, as in his Darwin-inspired *Being and Becoming in the Organic World* (1869)[50]—a popular history of creation—and then, working within a tradition holding that landscape is destiny, aimed toward particularized cultural geographies. His American works—including *Sketches of Urban and Cultural Life in North America* (1876)[51] and *The United States of America* (1878, 1880)[52]—focused on the linkage of physical geography with "natural characteristics of human beings in society" (which was a special subject of the first volume of his book about the United States). His German works did the same. Thus *Snow Cover; with Special Reference to the Mountains of Germany* (1889)[53] was a special interest of the geographer turned cinematographer Arnold Fanck, inventor of German land-mountain movies, like *The Holy Mountain* (1926) and *The Miracle of Snowshoes* (1919), and ice-mountain movies, like *S.O.S. Eisberg*

(1933). Ratzel's chorological theories of locations' effects on cultures appear most generally in his massively influential *Anthropogeography; or an Introduction to the Application of Geography* [Erdkunde] *to History* (1882, 1891).[54]

Geographic determinism, intellectually questionable and politically dangerous, has much affected the history of geography, as it did in the case of the American geographer Ellen Churchill Semple. Geographic determinism affects other fields as well. The literary critic Elizabeth DeLoughrey, for example, argues for the distinctiveness of literary production that "originates" from island peoples and the waters that surround them.[55] The field of "psychogeography" provides another example. According to its founder, Guy Debord, this field concerns "the precise laws and specific effects of the geographical environment, consciously organized or not, on the emotions and behavior of individuals."[56] When it touches on issues of nature and politics, the effects of geographic determinism can be grim—as we have seen.

Geographers following in Ratzel's wake, and more influenced than he was by the introduction of air travel, reconfigured the position of Strabo (see Illustration 78) and Pliny the Elder, as they understood it, that the world was a single island. In the *political* realm there now developed the concomitant thesis that there was only one world island, a theory of Halford Mackinder (a "conservative" position, as we will see), or effectively only one archipelago.

These reasons include the questionable place of "geographic determinism" in the American academy alongside racialist "genetics" and linguistic and climatic

ILLUSTRATION 78 Map of the World Island According to Strabo. Engraving by J. Bye. From Principal Playfair, *A New General Atlas, Ancient and Modern* (Edinburgh: P. Hill, 1814). Courtesy of Harvard Map Collection.

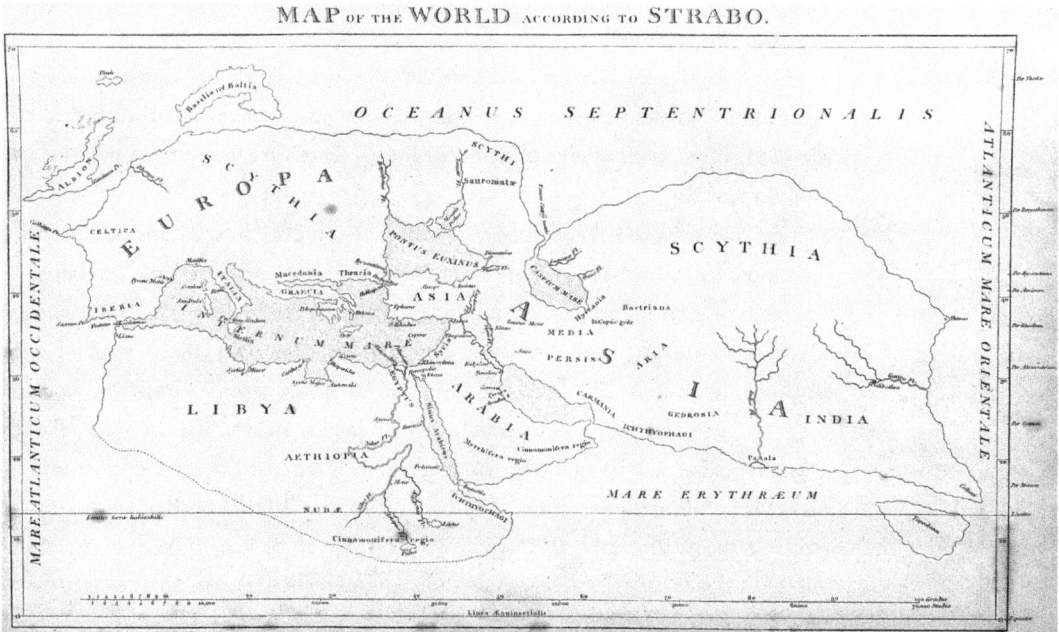

MAP OF THE WORLD ACCORDING TO STRABO.

determination.[57] Relevant here is the role that German geographers played during World War II,[58] the rise of the so-called value-neutral social sciences (including political science and economics) and geological sciences, and the eventual rise of American "world hegemony" in the wake of World War II.[59]

Living Space

A universal activity . . . which forms an all-comprehending *Lebensraum*
in which the manifold may meet and enter into relation.

—David Morrison, *Mind*

Ratzel is remembered for having introduced the term *Lebensraum* (living space) in an essay of the same name published in 1901.[60] The idea of *Lebensraum* took on a psychological or spiritual dimension for philosophers like Rudolf Eucken,[61] as well as a political one, as when the English straightaway defined the term in English as meaning the "territory which the Germans believed was needed for their natural development." Within four years of its publication, *Lebensraum* was a standard English-language word.

The term *Lebensraum* touched on the islandological sense in which "political borders" were not "natural borders" and, in definite national circumstances, required expansion by whatever means necessary.[62] Ratzel's Scandinavian student Rudolf Kjellén teased out of Ratzel's *Political Geography* (1897) a second term that soon entered dozens of languages: *geopolitics*.[63] This gave a general direction to notions of globalism and strategy in the age of big guns on ships[64]—and eventually the age of aviation, which, as we have seen, gave new impetus to defining the "horizon" or "shoreline."

Here, understanding the "difference" between land and sea was crucial: "geopolitics" as such arose in part from the landed German attempt, during the age of airships, somewhat anachronistically to define itself as a sea power in the modern world.

The notion, as Ratzel expressed it, was originally American.[65] The old idea that there were basically two kinds of power in the world—land-based and sea-based—had been pressed home by the American Alfred Thayer Mahan in his internationally significant *Influence of Sea Power upon History, 1600–1783* (1890), which found its way into military libraries and commercial headquarters across the world.[66] Mahan's historical and global purview, which begins around the time that Shakespeare's *Hamlet* saw its first performance in the Globe, stresses the importance of particular geographic locations. These include natural naval bottlenecks or choke points, like the Øresund, through which German ships had to pass on their way into or out of the Baltic; they also include commercially consequential "man-made straits," like the Suez Canal (1869) and the soon-to-be-completed Panama Canal

(1914), about whose geopolitical significance Ratzel had written decades earlier (1880).[67] Mahan's works were read widely in Germany and Japan and helped to shape modern naval theory. Ratzel followed the rhetorical lead of Mahan and brought it into the geopolitical sphere in his "Sea Power: A Geopolitical Study" (1896)[68] and *The Sea as a Source of National Greatness: A Geopolitical Study* (1900).[69]

The "new" field of geopolitics, understood in terms of land and sea (Mahan), was now established throughout the world. Two examples will suffice here: the British geopolitician Halford Mackinder and, following him in the wake of Ratzel, the German geopolitician Karl Haushofer.

Mackinder was the author of the geomorphological *Britain and the British Seas* (1902) and *Our Own Islands* (1907)[70] and a special student of German notions of space (*Raum*) and German places, which interest shows in his work on the Rhine River and its human cultures.[71] Mackinder's most influential essay, however, is "The Geographical Pivot of History" (1904),[72] which put forth a startling global thesis with a geostrategic head (see Illustration 79). The *pivot* in the title means "that on which everything depends"[73] and suggests two sorts of axes. First is the sort of *fulcrum*[74] that Archimedes knew would help him move the Earth, if only he had a place to stand, floating out in space. Second is the sort of *pivot ship*, or artificial floating island, "used as a reference point by a [military] fleet when moving or changing course in formation."[75]

To understand the world of the twentieth century, said Mackinder, one needs to see that there is now one "World Island"—Europe, Africa, and Asia taken

ILLUSTRATION 79
World Island and Natural Seats of Power. From Halford J. Mackinder, "The Geographical Pivot of History," *Geographical Journal* 23 (1904): 421–437. Source: Harvard College Libraries.

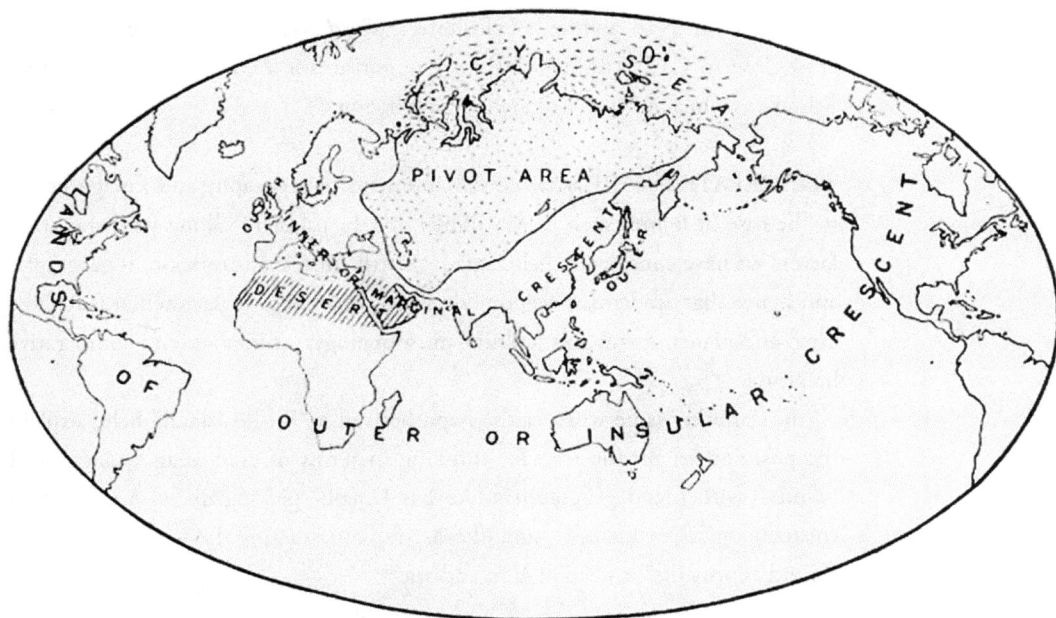

together as one, much in the fashion of Strabo. This World Island has "Offshore Islands" at both ends (Britain and Japan); likewise, it has "Outlying Islands" (Australia and the Americas). Mackinder's *Democratic Ideals and Reality: A Study in the Politics of Reconstruction* (1919) presented this idea at the Paris Peace Conference (1919), convened by the Allied Powers following the end of World War I in order to set the peace terms for defeated Germany. At this convention, Mackinder contrasted his realistic geopolitics with what he believed to be Woodrow Wilson's unworldly idealism. The main requirement for the Allies to rule the entire world, says Mackinder, was to rule the Heartland of the World Island. Here is how he put the argument:

Who rules East Europe commands	The Heartland
Who rules the Heartland commands	The World Island
Who rules the World Island commands	The World[76]

"When united by overland communications, the World Island [as Mackinder often calls the "Heartland"] is in fact possessed potentially of the advantages of both insularity and incomparably great resources."[77]

In twentieth-century Germany, Karl Haushofer, evincing the geopolitical tendency that David Harvey dislikes to the dangerous point of ignoring it, took up the geopolitical mantle from Ratzel.[78] He taught at the Bavarian War Academy (1903–1908), linked German and Japanese war powers,[79] served as general during World War I, became a professor of "political geography" at Munich in the same year that Hitler came to power (1933), joined the Thule Society (probably),[80] and participated in Nazi politics.[81] Haushofer's *World Seas and World Powers* (1937)[82] is so close to Carl Schmitt's obviously opportunistic *Land and Sea* (1954) that Schmitt's claims to "mere originality" ring hollow.[83]

THE RELATIONSHIP between developments in geography and foreign policy in the case of figures such as Mackinder and Haushofer,[84] along with the other factors we have considered, help explain North American suspicion of geography and hence that academic discipline's splitting apart into departmental fields like earth and planetary sciences, cultural anthropology, economics, and comparative literature.

It is another issue whether the hypothesis of a "World Island" helps explain the postmodern predilection for thinking in terms of archipelagoes instead of islands—with island geographers like Bas Umali[85] picking up on anti-Kantian epistemological analogies[86]—and likewise for constructing the world in terms of "liquid continents" instead of land and sea.[87]

The contemporary geographer David Harvey correctly says that Mackinder is "a very clear example of someone who puts his work in service of the territorial logic of power," even as he notes that Mackinder's view of the Heartland was "at once [Mackinder's] greatest fear and Germany's greatest dream."[88] However, Harvey's narrowly economic viewpoint means that he uses *island* only to indicate "a region of wealth or poverty," excluding "a natural entity having to do with land and water"—which is how Mackinder and his generation had it. Harvey empties *island* of its meaning when, for example, he writes:

A serious social danger attaches to creating an island of affluence and power in the midst of a sea of impoverishment, disempowerment, and decay.[89]

Likewise, he argues that "factions of labor that have through struggle or out of scarcity managed to create islands of privilege within a sea of exploitation will also just as surely rally to the cause of the alliance to preserve their gains."[90] He describes an effort to "divide up the urban realm into a patchwork quilt of islands of relative affluence struggling to secure themselves in a sea of spreading squalor and decay."[91] In all such cases, Harvey disposes of the geopolitical term in favor of a merely economic one, as if the former were of no consequence. The unfortunate result is that he throws out the baby (traditional geography) with the bathwater (the geopolitical gist of the traditional geography that he detests). That is, anyhow, how Kepler might have put it.[92]

A Large World Writ Small

> When taken to his own garden [which the infirm elderly Kant "had not done for more than two years"], he declared he felt himself as nonplussed as if he were on a desert island.
>
> —William Wallace, *Kant*

Immanuel Kant, who set the measures for the modern geography of the world, never left his hometown of Königsberg. An excellent "armchair" traveler who invented the modern notion of "the world," he was himself a remarkably poor traveler and tourist. He spent almost all of his life on one small island near Königsberg, Kneiphof, which stands in the Pregel River a little way before it empties into the freshwater Vistula Lagoon.[93] Kant's university (the "Albertina"; see Illustration 80) was on Kneiphof, and so were most of his dwellings and lecture halls, including those on Magisterstrasse and Neustadt Street (see Illustration 81).

The area of Königsberg, a trading center, was multilingual and cosmopolitan. Kant took an interest in languages—his last publication was a short introduction to a Lithuanian dictionary. He knew that the name *Kneiphof* means "area flushed by water"[94] in Old Prussian, as suggested in Christian Gottlieb Mielcke's

ILLUSTRATION 80 Collegium Albertinum on Kneiphof Island. Source: Archiv Corps Masovia.

ILLUSTRATION 81 Map of Königsberg. "Grundriss der Haupt und Residenz Stadt Königsberg nach der neuen polizeilichen Eintheilung und den jetzt revidierten Stassenbenennungen zusammengetragen im Jahr 1809." Courtesy of the Harvard Map Collection.

Königsberg-published German-Lithuanian dictionary, for which Kant wrote the postscript.[95] (Old Prussian is a non-Germanic Baltic language,[96] not to be confused with Low German and High Prussian, which are German dialects.) The term *ape* in *Knypabe* (*Kneip-ape*) is cognate with the English term *over*, which means "the bank of a river" or "the shore of a sea,"[97] and it is cognate with the English word *over*, as when one says, "I am thinking *over* the logic of islandology."

Kant's island Kneiphof, as he knew, marks both a sea passage (the Baltic) and a river passage (the Pregel). He describes it thus:

A large town, the centre of a kingdom, in which are situated the ministries of the local government, which has a university [Albertina] (for the culture of the sciences), and which, moreover, possesses a site suitable for maritime trade,—which by means of rivers [Pregel along with its tributaries and canals] favors intercommunication with the interior of the country not less than with the remote lands on the frontier lands of different languages and customs—such a town, like Königsberg [in the kingdom of Prussia, with its Albertina] on the river Pregel [which empties into the Baltic], may be taken as a suitable spot for extending not merely a knowledge of men, but even a knowledge of the world, so far as it is possible to acquire the latter without travelling.[98]

What happened to Kant's island, whence came so much modern thinking about "the world," is terrible to report.

The Royal Air Force flattened almost all of the island's buildings in 1944; then, in 1945, the Soviet occupation expelled the entire German population. Little remains from the olden days.[99] The toponym *Königsberg* has become *Kaliningrad*. Kant had argued, on grounds both scientific and nationalistic, for the preservation of the "insular" and "isolated" Lithuanian and Baltic languages, but out of misplaced respect for Kant and understandable desire for high-spending tourists, the old hydronym *Kneiphof* has become Kant Island.

WHERE TO SIT

"Give me whereon to stand," said Archimedes, "and I will move the earth." The boast was a pretty safe one, for he knew quite well that the standing place was wanting, and always would be wanting.

—Mark Twain, "Archimedes," *Australian Standard*

Critical topography needs always to return and tune itself to the problems and opportunities of an always-changing planet, now liquid, now solid, and always Earth. In the circumstances, the actual mnemonic *SOS*, understood as a de facto acronym for "*Save Our Ship*," seems a mild message for tellurians hurtling ever more perilously through space. It is easier, after all, to move some things on Earth than to move the Earth itself. As regards the former task, Plutarch tells how

Archimedes thus effaced the difference between water and land at the shore when the Syracusan tyrant Hieron challenged his claim to be able to move the Earth:

> [Archimedes] had stated [in a letter to Hieron] that given the force, any given weight might be moved, and even boasted, we are told, relying on the strength of demonstration, that if there were another earth, by going into it he could remove this. Hiero being struck with amazement at this, and entreating [Archimedes] to make good this problem by actual experiment, and show some great weight moved by a small engine, [Archimedes] fixed accordingly upon a ship of burden out of the king's arsenal, which could not be drawn out of the dock without great labour and many men; and, loading [the ship] with many passengers and a full freight, sitting himself the while far off, with no great endeavour, but only holding the head of the pulley in his hand and drawing the cords by degrees, he drew the ship in a straight line, as smoothly and evenly as if she had been in the sea.[100]

Thanks to Archimedes' systems of levers and pulleys, the old difference between land and water might become naught (see Illustration 82). Thanks to other devices, in fact, "a ship was frequently lifted up to a great height in the air (a dreadful thing to behold), and was rolled to and fro, and kept swinging, until the mariners were

all thrown out, when at length it was dashed against the rocks, or let fall."[101] Archimedes knew Plato well enough, however, to figure that these machines were toys for boys and weapons for soldiers. ("These machines [Archimedes] had designed and contrived, not as matters of any importance, but as mere amusements.")[102] It was another matter entirely to find what Plutarch here calls "another Earth" from which to move this one—unless, of course, this Earth were itself a different sort of ship, or floating island. Whatever else Prospero could do on his island, his was no brave new world.

Islandology, as we have considered it in *Islandology*, involves less the fact of being an island—since what landmass is not an island?—than the hypothesis that there are different ways of conceiving the natural insular condition in terms both political and individual. That much Carl Schmitt got right. He writes in *Land and Sea* (1942): "It goes without saying that England is an island. But the simple affirmation of this geographical fact does not mean much of anything."[103] The island the mind creates is sometimes something else, only *almost* entirely.

AT AROUND THE SAME TIME that aeronautics was first changing our perception of islands and our understanding of Earth as an island, the influential American geographer and climatologist Ellsworth Huntington wrote: "At the beginning of their volumes, the historians speak respectfully of the influence of geographical factors, but that is usually all."[104] Many people argued that he erred in stressing the effects of geographic determinism. And yet the question of determinism (nature and politics) usually comes to the fore mainly so that one can dismiss it. Thinking about general issues of climatic and geographic determinism in terms both natural and political gradually went out of fashion in the twentieth century, especially after the excesses of World War II, which, as we have seen, seemed fueled by ideas of determinism and their unbearable political consequences. Determinism as such exited the intellectual stage along with most other aspects of nineteenth-century causality-oriented geography, especially in America. There entered instead "value-neutral" fields, including the "political sciences" (as well as the various social sciences) and "natural sciences" (including geology and planetary sciences), which represented the then politically useful schism in geography, as well as the apparently "humanist disciplines" (including comparative literature).

Urgent practical considerations—unending war and global warming—encourage a vigilant, moderate updating of the Kantian discipline of geography along with other hard-pressed disciplines at issue in *Islandology*. Not only do we think *by means of* islands, every last one of us; we also live on one. If there were not islands, the inevitable infirmities of language and contingencies of politics would make it necessary for human beings to invent them.

The Archimedean supposition in *A Child's Geography of the World* (1929) is a heuristic hypothesis:

Just suppose you could go way, way off in the sky, sit on a corner of nothing at all, and look down at the World through a spyglass. [See Illustration 6.]

In the end there is nowhere to sit.

Acknowledgments

Many institutions supported the writing of this book. I am grateful to Harvard University (United States), the United Arab Emirates University (United Arab Emirates), Memorial University (Newfoundland), Trent University (Ontario), the University of New Brunswick (Fredericton), Ca'Foscari University (Venice), Católica University (Lisbon), and the University of Tel Aviv (Israel). Discrete sections of the present work were delivered in earlier versions as lectures at the Weatherhead Center (Cambridge, Massachusetts), the Grand Manan Museum and the Seven Days Work Educational Foundation (Grand Manan), and the Zayed Library (Al Ain). Students who attended my islands graduate seminar in the Comparative Literature Department at Harvard University asked creative and demanding questions. Especially helpful were Kristen Roupenian, Vincent Gélinas-Lemaire, and Sarah Moon.

Island associations and their publishers have provided assistance. Journals devoted generally to island studies include *Shima: The Journal of Island Cultures* (Southern Cross University, Coff's Harbour, Australia), *Island Studies* (University of Prince Edward Island, Charlottetown, Prince Edward Island, Canada), *Island Journal* (Island Institute, Rockland, Maine), and *Journal of Island Studies* (Japanese Society of Island Studies, Hitotsubashi University, Tokyo, Japan). Other helpful institutions include the Scottish Centre for Island Studies (University of the West of Scotland, Scotland), Webbing the Islands (University of Tasmania, Hobart, New Zealand), the International Geographic Union's Commission on Islands (Brisbane, Australia), the Island Resources Foundation (Tortola, United States Virgin Islands), and the Small Cultures Research Initiative (Southern Cross University).

People to whom I am grateful for their support include Ateeq Jakka (Ras Al Khaimah), Donald and Heather Baker (Al Ain), Manfred Malzahn (Al Ain), Peter and Marilyn Cronk (Grand Manan), Jonathan Graves and Nancy Estle (Grand Manan), Daniel Albright (Cambridge, Massachusetts), Werner Sollors and Alide Cagidemetrio (Venice), Richard Rice and Martha Ballantyne (Grand Manan),

Pauline Cox (St. John's, Newfoundland), and Jonathan Bordo and Doreen Small (Peterborough). My wife Susan Lisa Meld Shell has helped in more ways that I can say.

Above all, I am grateful to my intellectually perspicacious and intelligently good-humored son, the geographer Jacob Adam Shell, now teaching in Philadelphia. His encouragement throughout has been the sina qua non.

Marc Shell
Telegraph Island, Musandam Peninsula, Omani Exclave
November 2013

Notes

PREAMBLE

1. Rachel Carson, *The Sea around Us: An Illustrated Commemorative Edition*, with introduction by Robert D. Ballard and afterword by Brian J. Skinner (1951; New York: Oxford University Press, 2003), 110. The chapter titled "The Birth of an Island" was published in the *Yale Review* and won the George Westinghouse Prize for science writing; the film version of the book won the 1953 Oscar for best documentary; and the original edition won the 1952 National Book Award for nonfiction and the Burroughs Medal for nature writing.

2. "Attention Salesmen, Sales Managers: Location, Location, Location, Close to Rogers Park," 1926 classified advertisement in the *Chicago Tribune*; cited in William Safire, "On Language: Location, Location, Location," *New York Times*, June 26, 2009. On the term *critical topography*, see the Postamble.

3. Friedrich Theodor Rink, ed., *Kant's Gesammelte Schriften*, Königliche Preussiche Akademie der Wissenschaften (Berlin: G. Reimer, 1902), 9:159–162.

4. On this, see Onora O'Neil, "Orientation in Thinking: Geographical Problems, Political Solutions," in *Reading Kant's Geography*, ed. Stuart Elden and Eduardo Mendieta (Albany: State University of New York Press, 2011).

5. See Paul Richards, "Kant's Geography and Mental Maps," *Transactions of the Institute of British Geographers* 61, no. 1 (1974): 1–16.

6. See Chapter 6.

7. The Greek text, *Dos moi (phēsi) pou stō kai kinō*, is from *Pappi Alexandrini Collectionis*, ed. Friedrich Otto Hultsch (Berlin: Apud Weidmannos, 1878), 1060. This translation is taken from T. L. Heath, ed., *The Works of Archimedes with the Method of Archimedes* (New York: Dover, 1953), xix.

8. See Chapter 6.

9. Compare the Celtic-French word *nadio*.

10. Rebecca Weaver-Hightower, *Empire Islands: Castaways, Cannibals, and Fantasies of Conquest* (Minneapolis: University of Minnesota Press, 2007); and Diana Loxley, *Problematic Shores: The Literature of Islands* (Houndmills, UK: Palgrave Macmillan, 1991).

11. Jill Franks, *Islands and the Modernists: The Allure of Isolation in Art, Literature and Science* (Jefferson, NC: McFarland, 2006).

12. Jill Franks, "Men Who Loved Islands: D. H. Lawrence and J. M. Synge in Sardinia and Aran," *Études Lawrenciennes* 28 (2003): 133–147.

13. Godfrey Baldacchino and David Milne, eds., *Lessons from the Political Economy of Small Islands: The Resourcefulness of Jurisdiction* (London: Macmillan / Institute of Island Studies, 2000); and Stephen A. Royle, *A Geography of Islands: Small Island Insularity* (London: Routledge, 2001).

14. Stephen A. Royle, "Inseltourismus: Inseln der Träume?" in *Trauminseln? Tourismus und Alltag in "Urlaubsparadiesen,"* ed. Heidi Weinhäupl and Margit Wolfsberger (Vienna: Lit Verlag, 2006), 13–36.

15. Janis Frawley-Holler, *Island Wise: Lessons in Living from the Islands of the World* (Louisville, KY: Broadway, 2003).

16. Tom Conley, *The Self-Made Map: Cartographic Writing in Early Modern France* (Minneapolis: University of Minnesota Press, 1996), esp. 178–182.

17. Islands were "in practical environmental as well as in mental terms, an easily conceived allegory of a whole world." Richard Grove, *Green Imperialism: Colonial Expansion, Tropical Island Edens, and the Origins of Environmentalism, 1600–1800* (Cambridge: Cambridge University Press, 1991), 32.

18. Marshall Sahlins, *Islands of History* (Chicago: University of Chicago Press, 1985).

19. Fernand Braudel, *The Mediterranean and the Mediterranean World in the Age of Phillip II* (New York: Harper & Row, 1972), 2:154.

20. Relevant here is Gillis's anthropologically directed *Back to the Sea: Coasts in Human History* (Chicago: University of Chicago Press, 2012) and his contribution to *Seascapes, Littoral Cultures, and Trans-Oceanic Exchanges*, Library of Congress Conference, Washington, DC, February 12–15, 2003.

21. For example, *O*.

22. *Oxford English Dictionary* (hereafter, *OED*), s.vv. "brachyology," "tautology," n.f. Other examples include: *acrylogia, aetiologia, analogy, battology, bomphilologia, dialogismus, dicaeologia, dissoi logoi, homiologia, hysterologia, ideology, leptologia, neologism, palilogia, paromologia,* and *syllogism*.

23. Thomas Hardy, *Tess of the d'Urbervilles* (London, 1891), 1:249.

24. "Continent vs. Island," *Spokane Daily News*, August 21, 1945. See also *Far Eastern Review* 14, no. 17 (1945): 246.

25. Rosemary G. Gillespie and David A. Clague, eds., *Encyclopedia of Islands* (Berkeley: University of California Press, 2009).

26. Eduard Suess, *Die Entstehung der Alpencow* (Vienna: W. Braunmüller, 1875).

27. VladimirVernadsky, *Biosphere*, ed. Mark A. S. McMenamin, trans. David B. Langmuir (New York: Copernicus, 1998).

28. Among conferences in the last few decades would be "Models, Metaphors, Networks, and Insular Biosphere Reserves: The Virgin Islands Case," reported on by Edward L. Towle and Caroline Sutherland Rogers in *Contribution to the UNESCO/IUCN Workshop on the Application of the Biosphere Reserve Concept to Coastal Marine Areas*, San Francisco, August 14–20, 1989. Another conference would be "Island and Coastal Biospheric Reserves in the Mediterranean: Models for Sustainable Development," Syracuse, Italy, November 10–12, 2009, sponsored by the United Nations Educational, Scientific, and Cultural Organization. The focus on "conservation and the sustainable use of insular biosphere reserves" was the main topic at the ninth meeting of the East Asian Biospheric Reserve Network, Jeju Island, Korea, September 2005.

29. See the Postamble.

30. Edmund Burke, *A Philosophical Inquiry into the Origin of Our Ideas of the Sublime and Beautiful*, 2nd ed. (London: R. & J. Dodsley, 1759), 4.

CHAPTER 1

1. Shakespeare, *Hamlet*. References are to the standard edition except where noted otherwise as First Quarto (1603), Second Quarto (1604), First Folio (1623), Second Folio (1632), Fifth Quarto (1637), Third Folio (1664), or Fourth Folio (1685).

2. "An island is defined by its shore." Joël Bonnemaison, "The Tree and the Canoe: Roots and Mobility in Vanuatu Societies," *Pacific Viewpoint* 26, no. 1 (1985): 30–62.

3. *OED*, s.v. "Venn diagram."

4. Godfrey Baldacchino claims that many people, when asked to draw an island, produce a circle; he concludes that there is "an obsession to control, to embrace an island, as something finite. . . . Being geographically defined and circular, an island is easier to hold, to own." Baldacchino, "Editorial Islands: Objects of Representation," *Geografiska Annaler: Series B, Human Geography* 87, no. 4 (2005): 247.

5. We will turn to this discovery by Johannes Kepler—assistant in 1600 to the Danish astronomer Tycho Brahe on the island of Hven near Elsinore—in Chapter 11.

6. John Venn, *The Principles of Inductive Logic*, 2nd ed. (London: Macmillan, 1889), 355n.

7. Sebastian Münster, preface to *Treatise of Newe India* (1553), sig. Aij.

8. Willard Van Orman Quine, *Methods of Logic* (London: Routledge & Kegan Paul, 1952), 70.

9. Friedrich Ludwig Gottlob Frege, *Grundgesetze der Arithmetik*, vol. 2, para. 56, trans. 139; cited in Geoffrey Bennington, *Frontiers: Kant, Hegel, Frege, Wittgenstein* [University of Sussex Seminars 1989–1992] (New York: Create Space, 2003).

10. Ludwig Wittgenstein, *Philosophical Investigations*, ed. and trans. G. E. M. Anscombe (Malden, MA: Blackwell, 2001), para. 71.

11. For example, Wolfgang Welsch, "Transculturality: The Puzzling Form of Cultures Today," in *Spaces of Culture: City, Nation, World*, ed. Mike Featherstone and Scott Lash (London: Sage, 1999), 194–213.

12. See Chapter 8.

13. http://en.wikipedia.org/wiki/File:EulerDiagram.svg.

14. John Venn, "On the Employment of Geometrical Diagrams for the Sensible Representation of Logical Propositions," *Proceedings of the Cambridge Philosophical Society* 4 (1880): 47–59.

15. John Venn, *Symbolic Logic* (London: Macmillan, 1881).

16. Ruth Heller, *A Sea within a Sea* (New York: Grosset & Dunlap, 2000).

17. Waldon R. Porterfield, "The Sea with No Shores: It's the Sargasso, the Mysterious Oceanic Desert That Surrounds Bermuda," *Milwaukee Journal*, January 9, 1980.

18. "Sargasso Sea without a Coastline," *Island Times* [Malakal, Koror, Republic of Palau], March 24, 2011.

19. Jules Verne, *Twenty Thousand Leagues under the Sea* (1879), pt. 2, chap. 11. For an example of a lesson for grades 3 through 8 based on this assignment, see "Islands in the Stream," *The Lesson Corner* (2002), http://www.lessoncorner.com/Science/Earth_Science/Geography?page=98.

20. John Robert Ross, "Constraints on Variables in Syntax," PhD diss., Massachusetts Institute of Technology, 1967.

21. Benoit Mandelbrot, "How Long Is the Coast of Great Britain? Statistical Self-Similarity and Fractional Dimension," *Science* 156, no. 3775 (May 5, 1967): 636–638.

22. Benoit Mandelbrot, *Fractal Geometry of Nature* (San Francisco: W. H. Freeman, 1982).

23. See Carl L. Amos, M. Brylinsky, T. F. Sutherland, D. O'Brien, S. Lee, and A. Cramp, "The Stability of a Mudflat in the Humber Estuary, South Yorkshire, UK," *Geological Society Special Publications* 139 (1998): 35–43.

24. Carl Schmitt, *Land und Meer: Eine weltgeschichtliche Betrachtung* (Leipzig: Reclam Verlag, 1942); translated by Simona Draghici in 1954 as *Land and Sea*, rev. Greg Johnson (Washington, DC: Plutarch Press, 1997), chap. 17.

25. William Shakespeare, *Richard II*, act 2, sc. 1.

26. Stephen J. Lee, *Aspects of British Political History, 1815–1914* (London: Routledge, 1994), 254–257.

27. The phrase "Empress Island" is used in Robert Cooney, *A Compendious History of the Northern Part of the Province of New Brunswick and of the District of Gaspe in Lower Canada* (Halifax: J. Howe, 1832). It found its way into a speech by New Brunswicker George Eulas Foster before the Parliament of Canada in 1896.

28. Gerald Alexanderson, "Euler and Königsberg's Bridges: A Historical View," *Bulletin of the American Mathematical Society* 43 (July 2006): 567.

29. Michael Church, "Immanuel Kant and the Emergence of Modern Geography," in *Reading Kant's Geography*, ed. Stuart Elden and Eduardo Mendieta (Albany: State University of New York Press, 2011); and Robert Louden, "'The Play of Nature': Human Beings in Kant's *Geography*," in *Reading Kant's Geography*, ed. Stuart Elden and Eduardo Mendieta (Albany: State University of New York Press, 2011).

30. See, for example, Kant's "lecture announcement" for the summer semester of 1757: "Entwurf und Ankündigung eines Collegii der physischen Geographie nebst dem Anhange einer kurzen Betrachtung über die Frage: Ob die Westwinde in unsern Gegenden darum feucht seien, weil sie über ein großes Meer streichen" (Königsberg: Johann Friedrich Driest, 1757). Werner Stark and Reinhard Brandt prepared the physical geography notes for publication in the Academy edition.

31. "We have now not merely explored the territory of pure understanding, and carefully surveyed every part of it, but have also measured its extent, and assigned to everything in it its rightful place." Immanuel Kant, *Critique of Pure Reason*, trans. F. Max Müller (London: Macmillan, 1881), chap. 3, "Of the Ground of the Division of All Objects into Phenomena and Noumena," A235/B294.

32. The island is "the land of truth [*das Land der Wahrheit*]—enchanting name!—surrounded by a wide and stormy ocean, the native home of illusion, where many a fog bank and many a swiftly melting iceberg [*manches bald wegschmelzende Eis*] give the deceptive appearance of farther shores, deluding the adventurous seafarer ever anew with empty hopes, and engaging him in enterprises which he can never abandon and yet is unable to carry to completion." Immanuel Kant, *Critique of Pure Reason*, trans. Norman Kemp Smith (New York: St. Martin's, 1929). On differences among the various English-language translations of this famous passage, see Chapter 14.

33. "The region of illusion" is the translation of *Sitz des Scheins*, in Immanuel Kant, *Critique of Pure Reason*, trans. John Miller Dow Meiklejohn (London: Bell & Daldy, 1871). See Chapter 6.

34. A *noumenon* is "a posited object that is known, if at all, without the senses."

35. Friedrich Nietzsche, *The Gay Science*, ed. Bernard Williams, trans. Josefine Nauckhoff and Adrian Del Caro (Cambridge: Cambridge University Press, 2001), sec. 343.

36. See Richard Rorty, *Objectivity, Relativism, and Truth* (Cambridge: Cambridge University Press, 1991), 14, 216.

37. Jean-François Lyotard, *Phrases in Dispute*, trans. Georges Van Den Abbeele (1983; Minneapolis: University of Minnesota Press, 1988), 27–28; see also "Each genre of discourse would be like an island . . ." (130–131). Relevant here is "The Archipelago," chap. 2 in Lyotard's *Kantian Critique of History*, trans. Georges Van Den Abbeele (Stanford, CA: Stanford University Press, 2009).

38. Quoted by Pappus of Alexandria, *Synagoge* [Collection], bk. 7.

39. The day that Kant received his copy of Jean-Jacques Rousseau's *Émile*, he amazed his neighbors by missing his otherwise entirely regular afternoon walk. Or so goes the story.

40. Jacques Derrida, *The Beast and the Sovereign: Part II*, trans. Geoffrey Bennington (Chicago: University of Chicago Press, 2011), chap. 1, sess. 1, p. 11, December 2002; see also Gilles Deleuze, "Causes and Reasons of Desert Islands," in *Desert Islands and Other Texts 1953–74*, ed. David Lapoujade, trans. Mike Taormina (Los Angeles: Semiotext(e), 2004).

41. According to the *OED*, "the simple *ie* . . . derived from *ahwa* 'water' . . . with sense 'of or pertaining to water,' 'watery,' 'watered.' . . . A cognate compound frequent in OE was *éaland*, lit. 'water-land.' . . . In 15th c. the first part of the word began to be associated with [the same-sounding] . . . *ile, yle* (of Fr. origin), and sometimes . . . written *ile-land*; and when *ile* was spelt *isle, iland* erroneously followed it as *isle-land, island*; the latter spelling [island] became established as the current form before 1700."

42. Gillian Beer briefly notes these two meanings of *island*. See Beer, "The Island and the Aeroplane: The Case of Virginia Woolf," in *Nation and Narration*, ed. Homi Bhabha (London: Routledge, 1990), 271.

43. The *OED* confirms thus: "Island is . . . a compound of O[ld] E[nglish] *íeg, íg*, O[ld] N[orwegian] *ey* (Norw[egian] *öy*), O[ld] Fris[ian] *ey* 'isle' + LAND. The simple *íeg*[. . .] derived from *ahwa* 'water' (O[ld] S[wedish] and O[ld] H[igh] G[erman] *aha*, O[ld] Frisian and O[ld] N[orwegian] *á*, OE *éa*), with sense 'of or pertaining to water,' 'watery.'"

44. Joshua Sylvester, in his translation of the French poet Guillaume de Saluste Du Bartas's *Devine Works and Weeks* (1605), writes about "the spungie Globe of th' . . . Earth," 2.1.382.

45. This is the clay, the material from which, says the Bible, the first man, Adam (meaning "clay" in Hebrew) was made.

46. Hegel vacationed on the island of Rügen in 1819; see the review/translation of Karl Rosenkranz's *Hegel's Leben*, *National Quarterly Review* 18 (December 1869–March 1869): 131. On the island of Rügen and the development of German national ideology, see Chapter 13.

47. Plato, *Republic*, 524; *Theaetetus*, 185; and *Hippias Major*, 300.

48. Heraclitus, *Fragments*, ed. Hermann Diels and Walther Kranz, frag. 36.

49. *OED*, s.v. "island," etymology.

50. Another name for Eysysla is Saaremaa (isle's land).

51. Kamau Brathwaite, "Caribbean Culture: Two Paradigms," in *Missile and Capsule*, ed. Jürgen Martini (Bremen: University of Bremen Press, 1983), 42.

52. Nathaniel Mackey, "An Interview with Edward Kamau Brathwaite," *Hambone* 9 (1991): 44.

53. Kamau Brathwaite, "New Gods of the Middle Passage," *Caribbean Quarterly* 46, no. 3–4 (2000): 12–58.

54. Brathwaite, "Caribbean Culture," 49.

55. Kamau Brathwaite, "Rex Nettleford and the Renaissance of Caribbean Culture," *Caribbean Quarterly* 43, no. 1–2 (1997): 34–69.

56. Elizabeth M. DeLoughrey, introduction to *Routes and Roots: Navigating Caribbean and Pacific Island Literatures* (Honolulu: University of Hawaii Press, 2007).

57. Paul A. Griffith, *Afro-Caribbean Poetry and Ritual* (London: Palgrave Macmillan, 2010).

58. "Editorial Review" of Griffith's *Afro-Caribbean Poetry*, by the poet and anthologist R. S. Gwynn.

59. Silvio Torres-Saillant, "The Trials of Authenticity in Kamau Brathwaite," *World Literature Today* 68, no. 4 (1994): 704.

60. On the Renaissance genre of the *isolario* (island book), see Georgios Tolias, "Isolarii: Fifteenth Century to Seventeenth Century," in *Cartography in the European Renaissance*, ed. David Woodward (Chicago: University of Chicago Press, 2007), 3:1, 263–284. Older Greek prototypes include the island sections of Dionysius the Traveler's geographical poem "The Surveye of the World; or, Situation of the Earth," trans. Thomas Twyne (London, 1572). For an overview, see Frank Lestringant, *Le livre des îles: Atlas et récits insulaires de la Genèse à Jules Verne* (Geneva: Librairie Droz, 2002).

61. See Owe Ronström, "Island Words, Island Worlds: The Origins and Meanings of Words for 'Islands' in North-West Europe," *Island Studies Journal* 4, no. 4 (2009): 163–182.

62. Eric Partridge, *Origins: A Short Etymological Dictionary of Modern English* (London: Routledge, 1961), s.v. "-nese." An ancient Greek word for "duck" is *nēssa/nētta*. Stanley Mayer Burstein, in his work on Agatharchides—a writer from the peninsular and/or island-city of Knidos and author of the incomplete *On the Erythraean Sea* (second century BC)—points to a telling confusion of *nēssa* with *nēsos*. See Burstein, ed. and trans., *On the Erythraean Sea* (London: Hakluyt Society, 1989), 147–148.

63. The apparently Greek-language term *chersonese* means "peninsula." It is a Roman invention: Romans, not Greeks, stitched together its two Greek constituent parts (*chersos* means "dry land"). See Pascal Payen, *Les îles nomades: Conquérir et résister dans "l'Enquête" d'Hérodote* (Paris: Éditions de l'École des Hautes Études en Sciences Sociales, 1997).

64. See Herodotus, *Inquiries*, 7.22–24; see also Chapter 5.

65. Cristoforo Buondelmonti, *Liber Insularum Archipelagi*, Biblioteca Nazionale Marciana, Venice, lat. X.215.

66. *Nēsos* as such arrives in English as a suffix added to toponyms of archipelagoes: *Indonesia, Micronesia, Polynesia, Melanesia,* and *Austronesia.* Thus John Callander writes, in his English-language translation of Charles De Brosses's *History of Navigations* (1756), that "we call the third division *polynesia*, being composed of all those islands, which are found dispersed in the vast Pacific Ocean." John Callander, trans., *Terra Australis Cognita: or,*

Voyages to the Terra Australis, or Southern Hemisphere, during the Sixteenth, Seventeenth, and Eighteenth Centuries, 3 vols. (Edinburgh: A. Donaldson, 1766–1768), 1:49. For the Roman *Heptanesia,* see Chapter 5.

67. On other apparent ambiguities in the meaning of the term *nēsos,* see Giorgos Tolias, *Ta nesologia: E monaxia kai e suntrofia ton nesion* [The Isolarii: The Solitude and the Companion of Islands] (Athens: Oikos, 2002).

68. Christian Depraetere, "The Challenge of Nissology: A Global Outlook on the World Archipelago," *Island Studies Journal* 3, no. 1 (2008): 3–36.

69. Herman Melville, *Redburn,* chap. 23. On floating islands, see Chapter 4.

70. Herman Melville, *Mardi,* chap. 57.

71. For Gilles Deleuze's endorsement of the archipelago notion, see his "Bartleby; or the Formula," in *Essays Critical and Clinical* (Minneapolis: University of Minnesota Press, 1997), 86.

72. Gaius Sallustius Crispus, *Jughurtine War* (first century BC).

73. *OED,* s.v. "archipelago," etymology: "No such word [as *archipelago*] occurs in ancient or mediaeval Greek; Ἀρχιπέλαγος in mod[ern] Greek Dict[ionaries], is introduced from [W]estern languages. *Arcipelago* occurs in a Treaty of 30th June 1268, between the Venetians and the emperor Michael Palaeologus: 'Item, quod pertinet ad insulas de Arcipelago'; it is used also by Villani c1345."

74. Alonso's work is housed in the Bibliotheca Nacional de España, Madrid, Res. Ms. 38, esp. f.18v.

75. Umberto Eco, *The Island of the Day Before,* trans. William Weaver (New York: Harcourt, 1995), 130.

76. See Chapter 7.

77. See Chapter 11. The tradition of assembling artful *isolarii* reached a philological peak with Marco Boschini's pleasantly decorated *L'arcipelago: Con tutte le isole* (Venice: Francesco Nicolini, 1658). See Mitchell Frank Merling, "Marco Boschini's 'La carta del navegar pitoresco': Art Theory and Virtuoso Culture in Seventeenth-Century Venice," PhD diss., Brown University, 1992.

78. See Chapter 5.

79. For the term *madol,* see Kenneth L. Rehg and Damian G. Sohl, *Ponapean-English Dictionary,* PALI Language Texts: Micronesia (Honolulu: University of Hawaii Press, 1979), 134, 239.

80. *Islandology* pursues the general subjects of geography and the cultural topology of islandness in general, partly by way of naming particular islands as examples, here Nan Madol and Venice, and partly by way of extended case studies of particular places, including Hormuz, Zealand, and Venice. The names of these places and likewise the etymologies of those names are always contested (see Chapter 2).

81. Among these are Venise-en-Québec, Venice (Florida), and Venezuela.

82. Herman Melville, *Moby-Dick* (New York: Harper & Brothers, 1851), 278.

83. "O islands" (איים = *'iyiym*). This word means "islands" according to variants of the Vulgate, the Septuagint, the Chaldee, the Syriac, and the Arabic translations. The same word is also used to denote "maritime countries" and/or "any lands or coasts far remote, or beyond sea." See Psalm 72:10; Isaiah 24:15, 41:5, 42:4, 42:10, 42:12, and 49:1; Jeremiah 25:22; and Daniel 11:18.

84. There is the English town of Eye: marshland prone to flooding surrounded its first settlement. Eye is located in East Anglia at the River Dove. See Clive Paine, with contributions by Jan Perry, *The History of Eye* (Eye, UK: Benyon de Beauvoir, 1993). In addition, there is Ireland's Eye (the island near Dublin) and very many others. Consider Dursey (island of *dur*: *dwr*, "water"), Dalkey, Lambay on the Irish coast, Anglesey, Orkney, Eday, Sanday, Bressay, Housay, Neay, Oxney (Isle of Oxen), Stokesay, Sheppey, Colonsay, Oronsay, Bardsey, Lundy, Guernsey, Jersey, Alderney, Menai, and Thorney.

85. Compare *Québec's Eye*, which is the other name for René-Levasseur Island in annular Lake Manicouagan, the second-largest lake island in the world.

86. See Genesis 10:5.

87. See Charles D. Wright, "'Insulae Gentium': Biblical Influence on Old English Poetic Vocabulary," in *Magister Regis: Studies in Honor of Robert Earl Kaske*, ed. Arthur Groos (New York: Fordham University Press, 1986), 9–21; cited in Winfried Rudolf, "The Spiritual Islescape of the Anglo-Saxons," in *The Sea and Englishness in the Middle Ages: Maritime Narratives, Identity and Culture*, ed. Sebastian I. Sobecki (Cambridge: D. S. Brewer, 2011), 31. See also Alfred Hiatt, "'From Hulle to Cartage': Maps, England, and the Sea," in *The Sea and Englishness in the Middle Ages: Maritime Narratives, Identity and Culture*, ed. Sebastian I. Sobecki (Cambridge: D. S. Brewer, 2011).

88. In this respect Illyria is a counterpart to the island of Sicily, which some readers call the home of the shipwrecked twin heroes of *Twelfth Night*.

89. Greg Dening, *Islands and Beaches: Discourse on a Silent Land, Marquesas 1774–1880* (Honolulu: University of Hawaii Press, 1980); cited in Rod Edmond and Vanessa Smith, "Editors' Introduction," in *Islands in History and Representation*, ed. Rod Edmond and Vanessa Smith (London: Routledge, 2003), 3.

90. Greg Dening, *Beach Crossings: Voyaging across Times, Cultures, and Self* (Philadelphia: University of Pennsylvania Press, 2004).

91. Yi-fu Tuan, *Topophilia: A Study of Environmental Perception, Attitudes, and Values* (London: Columbia University Press, 1998), 118; cited in John R. Gillis, "Taking History Offshore: Atlantic Islands in the European Minds, 1400–1800," in *Islands in History and Representation*, ed. Rod Edmond and Vanessa Smith (London: Routledge, 2003), 19.

92. Emmanuel-Louis-Eugène de Martonne, "Regions of Interior Basin Drainage," *Geographical Review* 17 (1927): 397.

93. Endorheic deltas include the delta of the Sacramento and San Joaquin rivers in California, the Inner Niger Delta in Mali, and the Okavanga Delta in Botswana.

94. So goes the "Black Sea Deluge" theory. See William B. F. Ryan, "Status of the Black Sea Flood Hypothesis," in *The Black Sea Flood Question: Changes in Coastline, Climate and Human Settlement*, ed. Valentina Yanko-Hombach, Alan S. Gilbert, Nicolae Panin, and Pavel M. Dolukhanov (Dordrecht: Springer, 2007).

95. Muhammad Abu'l-Qasim Ibn Hawqal, "Map of the Caspian Region," in Ibn Hawqal, *Configuration of the Earth* [*Safarnamah-'i Ibn Hawqal: Iran dar "Surat al-ard"*] (Tehran: Mu'assasah-'i Intisharat-i Amir Kabir, 1986).

96. Muhammad Abu'l-Qasim Ibn Hawqal, "Map of the World," in Ibn Hawqal, *Configuration of the Earth*.

97. Muhammad Abu'l-Qasim Ibn Hawqal, *The Oriental Geography of Ebn Haukal, Arabian Traveller of the Tenth Century*, trans. William Ouseley (London: Oriental Press, 1800), 184.

98. See *The Book of Curiosities* (Bodleian Library, MS Arab, c.90), bk. 2, chap. 2, "On the Depiction of the Earth" [*surat al-ard*]; and chap. 6, "On the Depiction of the Seas and Islands and Havens." For information about this manuscript, see Jeremy Johns and Emilie Savage-Smith, "*The Book of Curiosities*: A Newly Discovered Series of Islamic Maps," *Imago Mundi* 55 (2003): 7–24.

99. John Smith, *A Sea Grammar* (London: I. Hauiland), 45. This is the first edition with this title.

100. On Jamestown Island's insular status, see Chapter 15. For the term *water-locked*, see the coastal poet James Hurdis, *The Favorite Village: A Poem* (Bishopstone, UK: printed at the author's own press, 1800), 81: "Forlorn and water-lock'd stands the lone mill." Hurdis refers to the now-derelict nearby hamlet of Tide Mills in Sussex, England.

101. Herodotus, *Histories*, trans. Aubrey de Selincourt, 4.42.

102. See W. F. G. Lacroix, *Africa in Antiquity: A Linguistic and Toponymic Analysis of Ptolemy's Map of Africa* (Saarbrücken: Verlag für Entwicklungspolitik, 1998), esp. app. 3.

103. Peter Clayton, *Chronicle of the Pharaohs* (London: Thames & Hudson, 1994), 195.

104. See Aristotle, *Meteorology*, 1.14.

105. "He undertook to conquer the inhabited earth." Diodorus Siculus, *Library of History*, trans. Charles Henry Oldfather, Loeb Classical Library 279 (Cambridge, MA: Harvard University Press), 1.53.7. (Sesoösis, to whom Diodorus refers, is the same person as Sesostris.)

106. Necho "was the first who attempted the channel leading to the Erythraian [Red] Sea." Herodotus, *Histories*, 2.158.

107. Ptolemy II's canal included a navigable lock with sluice gates to keep salt water and fresh water separate. See Carol A. Redmount, "The Wadi Tumilat and the Canal of the Pharaohs," *Journal of Near Eastern Studies* 54, no. 2 (1995): 127–135.

108. "Ptolemy II had the most powerful navy in the Mediterranean for a few decades." Christelle Fischer-Bovet, "Army and Society in Ptolemaic Egypt," PhD diss., Stanford University, 2008, 140. On Ptolemy II's Red Sea ports—and Herodotus's reports of them—also see Fischer-Bovet, "Army and Society," 41.

CHAPTER 2

1. For a discussion of "*mainland* versus *island*" in ancient Greece, see Christy Constantakopoulou, *The Dance of the Islands: Insularity, Networks, The Athenian Empire, and the Aegean World* (Oxford: Oxford University Press, 2007), 112ff.

2. *OED*, s.v. "terra firma."

3. Ephraim Chambers, *Cyclopaedia* (1727–1741).

4. *The Collected Short Stories [of] D. H. Lawrence* (London: Heinemann, 1974), 671. This fable was published posthumously and cited in Elizabeth McMahon, "The Gilded Cage: From Utopia to Monad in Australia's Island Imaginary," in *Islands in History and Representation*, ed. Rod Edmond and Vanessa Smith (London: Routledge, 2003), 200.

5. Johann Gottfried von Herder, *On the Spirit of Hebrew Poetry: An Instruction for Lovers of the Same and the Oldest History of the Human Spirit*, trans. James Marsh (Burlington, VT: Edward Smith, 1833), 1:263.

6. *Island*, like *Mainland*, often names a particular island. Examples include the tiny disputed island in the Strait of Gibraltar, less than a kilometer from Morocco, which Spaniards often call *Isla de Perejil* and which Moroccans call *Leila*.

7. Walter Scott, gloss to *Lady of the Lake*, in *The Poetical Works of Sir Walter Scott*, ed. J. Rogie Robertson with an introduction by Walter Scott (1810; London: Oxford University Press, 1909), 198. The Danish poem is from Anders Sørensen Vedel, *Et hundrede udvalde danske viser* (1591).

8. John Venn, *Symbolic Logic* (London: Macmillan, 1881), xxiv–xxvn.

9. Stephen A. Royle, *A Geography of Islands: Small Island Insularity* (London: Routledge, 2001); and David Weale, "Islandness," *Island Journal* 8 (1991): 81–82.

10. Hermann Diels and Walther Kranz, eds., *Die Fragmente der Vorsokratiker* (Berlin: Weidmannsche Buchhandlung, 1922), 1:172, citing Heraclitus B94. For the passage from Plutarch's *On Exile*, see Plutarch, *Moralia*, trans. Phillip H. De Lacy and Benedict Einarson, Loeb Classical Library 405 (Cambridge, MA: Harvard University Press, 1959), 7:549.

11. Freeman Dyson, *Disturbing the Universe* (New York: Harper & Row, 1979), 245.

12. The illustration on the cover of *Islandology* shows plate 32 in Thomas Wright's *An Original Theory or New Hypothesis of the Universe* (London, 1750): "The universe filled with many galactic systems like our own, each a star-filled shell surrounding its own providential eye."

13. Immanuel Kant, appendix to *Allgemeine Naturgeschichte und Theorie des Himmels*, in *Universal History and Theory of the Heavens*, trans. Ian Johnston, illus. Ian Crowe (1755; Arlington, VA: Richer Resources Publications, 2009).

14. Alexander von Humboldt uses the term *Weltinsel* in *Kosmos: Entwurf einer physischen Weltbeschreibung* (Stuttgart: J. G. Gotta'schen Verlag, 1847), 1.93: "Unter den vielen selbstleuchtenden ihren Ort verändernden Sonnen . . . welche unsre Weltinsel bilden." The term has also been attributed to Sir William Herschel. Richard A. Proctor writes in *Other Suns Than Ours* (1887), 1.1, "Our 'island universe,' as Humboldt poetically called the stellar system" (ibid., 11), but he attributes the idea to Herschel: "The results which [Herschel] published in 1817 and 1818 justify the belief that . . . large numbers of the nebulae must be regarded as external galaxies. This grand conception fascinated . . . some who, like Humboldt, had understood and appreciated the work of the great observer. The idea of 'island universes' strewn throughout the ocean of space impressed the world."

15. See Alfred Hiatt, "'From Hulle to Cartage': Maps, England, and the Sea," in *The Sea and Englishness in the Middle Ages: Maritime Narratives, Identity and Culture*, ed. Sebastian I. Sobecki (Cambridge: D. S. Brewer, 2011).

16. MS Bodley 340, fol. 39r.

17. Macrobius Ambrosius Theodosius, *Commentary on Somnium Scipionis*, ed. James Willis (1963; Stuttgart: B. G. Teubner, 1994), 66.

18. Boethius, *Theological Tractates. The Consolation of Philosophy*, trans. H. F. Stewart, E. K. Rand, and S. J. Tester, Loeb Classical Library 74 (Cambridge, MA: Harvard University Press, 1973), 2.7.

19. Richard Anthony Proctor, *Other Worlds Than Ours* (London: Longmans, 1870), 2.36.

20. In Wolfgang Schirmacher, ed., *German Socialist Philosophy: Ludwig Feuerbach, Karl Marx, Friedrich Engels* (New York: Continuum, 1997), 223.

21. Raymond F. Jones, *This Island Earth* (Chicago: Shasta Publications, 1952). The cover illustration is by Robert Johnson.

22. Quoted by Pappus of Alexandria, *Synagoge* [Collection], bk. 7.

23. V. M. Hillyer and Mary S. W. Jones, *A Child's Geography of the World* (New York: Appleton-Century Crofts, 1951).

24. Thomas Nashe, *The Complete Works*, ed. Alexander B. Grosart (London: printed for private circulation, 1885), 6:88.

25. This differs from "above mean sea level" (AMSL), which is the elevation (on the ground) or altitude (in the air) of any object in relationship to the average sea level.

26. All terms are listed in the *OED*.

27. For example, Bar Island, Maine.

28. For example, Ministers Island, New Brunswick; Hull Island, Massachusetts; Lindisfarne Island, United Kingdom; and Mont Saint-Michel, France.

29. P. G. Winslow, *The Counsellor Heart* (New York: Collins Crime Club, 1980), 221.

30. The "Marsh Arabs," also known as the *Ma'dān*, inhabit the Tigris-Euphrates marshlands in the south and east of Iraq and those along the Iranian border. They developed a unique culture centered on the marshes' natural resources. Many Marsh Arabs were displaced when the wetlands were drained during and after the uprisings in Iraq in the early 1990s. See Gavin Young, with photographs by Nik Wheeler, *Return to the Marshes: Life with the Marsh Arabs of Iraq* (London: Collins, 1977).

31. Fox Point is where Thomas Riedelsheimer filmed the documentary *Rivers and Tides* (2001), which features site-specific artist Andy Goldsworthy making sculptures of naturally found stone and wood that disappear into the rising and falling waters and thus presumably fulfill themselves in the waters and participate with them.

32. Ptolemy, *Geografia*, 4-6-34.

33. "There was an island out there that sometimes broke the horizon. I've never seen it myself. The old men used to speak about it. But always when they rowed toward the island, it melted like a dream. They called it 'The Island of the Women.'" George Mackay Brown, "The Island of the Women," in *The Island of the Women and Other Stories* (London: J. Murray, 1998), 12; discussed by Simon Hall, *The History of Orkney Literature* (Edinburgh: John Donald, 2010), 136.

34. Relevant specifically to the notion of an island of women is Batya Weinbaum's comparative study *Islands of Women and Amazons: Representations and Realities* (Austin: University of Texas Press, 1999).

35. Cited in Markman Ellis, "'The Cane-Land Isles': Commerce and Empire in Late Eighteenth-Century Georgic and Pastoral Poetry," in *Islands in History and Representation*, ed. Rod Edmond and Vanessa Smith (London: Routledge, 2003), 47.

36. Charles Darwin, *Earthworms* (London: John Murray, 1881), 313.

37. Charles Darwin, *The Structure and Distribution of Coral Reefs. Being the first part of the geology of the voyage of the Beagle, under the command of Capt. Fitzroy, R.N. during the years 1832 to 1836* (London: Smith, Elder & Co., 1842).

38. Quoted in Desmond King-Hele, "The Furtive Evolutionist," *New Scientist*, April 12, 2003, 48–49.

39. See, for example, in Erasmus Darwin's poetical *Economy of Vegetation* (1791), his gloss on the phrase "On ice built isles," 1.529, and on the phrase "Raised her primeval islands," 2.36.

40. Charles Darwin, "On Certain Areas of Elevation and Subsidence in the Pacific and Indian Oceans, as Deduced from the Study of Coral Formations," *Proceedings of the Geological Society of London* 2 (1837): 552–554.

41. Charles Darwin, *Geological Observations on the Volcanic Islands Visited during the Voyage of H.M.S. Beagle* (London: Smith, Elder & Co., 1844).

42. It is also difficult to know the realistic biological limits of human beings. Here, too, island hypotheses can be useful for the imagination. Robert Paltock's *Life and Adventures of Peter Wilkins* (1751) is an imaginary voyage inspired by the writings of Daniel Defoe and Jonathan Swift, centering on the "what if" discovery of a race of winged people on an isolated island.

43. E. Alfred and A. C. Seward, *The New Flora of the Volcanic Island of Krakatau* (1908; Cambridge: Cambridge University Press, 2009). See also Ian Thornton, *Krakatau: The Destruction and Reassembly of an Island Ecosystem* (Cambridge, MA: Harvard University Press, 1997).

44. *Nature* 10 (September 3, 1874): 353; cited in Gillian Beer, "Island Bounds," in *Islands in History and Representation*, ed. Rod Edmond and Vanessa Smith (London: Routledge, 2003), 32.

45. Darwin, "On the Absence of Volcanoes in the Areas of Subsidence, and on Their Frequent Presence in the Areas of Elevation," in *Coral Reefs*.

CHAPTER 3

1. Édouard Glissant, *Poetics of Relation* [*Poétique de la relation*], trans. Betsy Wing (Ann Arbor: University of Michigan Press, 1997). There is a second epigraph: Derek Walcott's "Sea Is History."

2. Kamau Brathwaite, "Caribbean Man in Space and Time," *Savacou* 11–12 (1975): 1.

3. In John Donne, *Devotions upon Emergent Occasions* (1624).

4. Compare the ironic gist of the lyrics of the Simon and Garfunkel song "I Am a Rock" (1964): "I am an island."

5. The Hebrew *adam*, which names human beings' oldest ancestor (according to the Bible), means "clay." "Ashes to ashes" is the phrase from *The Order for the Burial of the Dead*, according to the Church of England; it is spoken "while the earth shall be cast upon the body by some standing by." See *The Book of Common Prayer* (1552), chap. 18.

6. A *land-crab* is "a person who is not a fisherman." Memorial University Folklore and Language Archive, Manuscript Collection: B. Trask, Summerville/Ellison, 1968 (M 68-24). See *Dictionary of Newfoundland English*, s.v. "land."

7. "All Land Is One Land under the Sea, . . ." *Island Journal* 10 (1993): 1–2.

8. Whaleback Island is actually part of a jagged ledge, also known as Whaleback, that is completely underwater at high tide and is a continuation of the southern portion of Gerrish Island in Maine.

9. A poet warns against this mere camouflage: "Ye are mad, ye have taken / A slumbering Kraken / For firm land of the Past." James Russell Lowell, "Ode to France" (1848), 30.

10. "The Whales, the Seas Leviathan . . . like so many floating Ilands concomitating us." Thomas Herbert, *Some Yeares Travels into Africa and Asia the Great*, rev. ed. (1638), 13.

11. The ship *Abraham Crijnssen* was stationed in the Dutch East Indies when World War II began. After the Battle of the Java Sea in 1942, it was covered with tree branches. It crossed the Japanese naval lines camouflaged as a tropical island.

12. *Memoir of the Life and Character of the Right Hon. Edmund Burke, with Specimens of his Poetry and Letters and an Estimate of his Genius and Talents, compared with those of his*

Great Contemporaries, ed. James Prior (London: Baldwyn, Cradock, & Joy, 1825), 2:471. See Samuel Arthur Bent, *Short Sayings of Great Men: With Historical and Explanatory Notes* (Boston: James R. Osgood, 1882), 85.

13. "Etymology (Supplied by a Late Consumptive Usher to a Grammar School)," in Herman Melville, *Moby-Dick*.

14. Thence Carl Schmitt incorporates the quotation without attribution to Melville. See Schmitt, *Land und Meer: Eine weltgeschichtliche Betrachtung* (Leipzig: Reclam Verlag, 1942); translated by Simona Draghici in 1954 as *Land and Sea*, rev. Greg Johnson (Washington, DC: Plutarch Press, 1997), chap. 17.

15. Eric Partridge, *Origins: A Short Etymological Dictionary of Modern English*, s.v. "-nesia."

16. Job 41:33.

17. The influential Rabbi Johanan said, "Once we went in a ship and saw a fish which put his head out of the water. He had horns upon which was written: 'I am one of the meanest creatures that inhabit the sea. I am three hundred miles in length, and enter this day into the jaws of the Leviathan.'" *Baba Batra* 74a.

18. The myth of "the island that lives like a whale" links *Moby-Dick* to Melville's earlier writings. These include *Omoo: A Narrative of Adventures in the South Seas* (1847) and *Typee: A Peep at Polynesian Life during a Four Months' Residence in a Valley of the Marquesas* (1846).

19. Erik Pontoppidan, *Natural History of Norway* (1752–1753; trans. 1755), 2.7, sec. 11.211.

20. *Exeter Book*, Exeter (UK) Cathedral Library, fol. 96b–97b.

21. *Popeye the Sailor Meets Sinbad the Sailor* (Max Fleischer Studios, 1936).

22. *Physiologus*, Latina Versio B, XXIV: Aspis Chelone.

23. *De proprietatibus rerum* (13th century), bk. 13.

24. British Library, Arundel MS 292 (13th century).

25. Konungs skuggsjá.

26. *Voyage of Saint Brendan the Navigator*, ed. John O'Meara (Chester Springs, PA: Dufour Editions, 1991).

27. Schmitt, *Land and Sea*, chap. 17.

28. Herman Melville, *White-Jacket*, chap. 46. For *ships* as "floating islands" in Melville's work, see also Chapter 1. The reference here is to the Sulu Archipelago in southwestern Philippines.

29. There are communities of fishing people who live almost their entire lives in boats at sea, almost never touching shore, seven days a week, twenty-four hours a day, year-round. Anthropo-geographers sometimes cite the Moro Bajan of the southern Philippines as such "sea gypsies." See Ellen Churchill Semple, *Influences of Geographic Environment on the Basis of Ratzel's System of Anthropo-Geography* (New York: Henry Holt, 1911), 318, citing *Census of the Philippine Islands*, vol. 1 (Washington, DC: Government Printing Office, 1905). Or at least there are tales and legends of the same. Likewise, there are those people, who are also fisherman and who may seek a modicum of protection from potential enemies, who live always in pile dwellings. As long ago as Herodotus, historians have been fascinated by such people; see Herodotus, *Inquiries*, 5.16. Island dwellers often have something in common with that sort of group.

30. Thomas Moore, "Fire-Worshippers," in *Lalla Rookh* (1817).

31. Melville, preface to *Omoo* (1847).

32. Published as one of Jules Verne's *Voyages Extraordinaires*.

33. Verne is reported as saying, "The action [of *L'île à hélice*] will take place on a floating island created by the ingenuity of man, a kind of Great Eastern magnified 10,000 times,
and containing . . . the whole of what in this case may be truly called a moving population."
Marie A. Belloc, "Jules Verne at Home," *Strand Magazine*, February 1895, 213.

34. So wrote Verne's publisher Pierre-Jules Hetzel.

35. Francis Bacon, *Sylva Sylvarum* (London: William Lee, 1627), sec. 790 (margin).

36. This last term is etymologically Welsh; see *OED*, s.v. "tussock."

37. The reed isles of Uros on pre-Incan Lake Titicaca float at the border between Bolivia and Peru. The "floating island" of Umbagog Lake is the 860-acre island near Harper's
Meadow, where the Umbagog empties into the Androscoggin River. The long-problematic
politico-geographic status of this lake, to which I refer, is one subject of a novel by the Vermont secretary of state and abolitionist Daniel Pierce Thompson: *The Trappers of Umbagog:
A Tale of Border Life* (Boston: J. P. Jewett, 1857). The novel concerns both the international
border between Canada and the United States and the interstate border between Maine and
New Hampshire. Thompson was already well known for his *Green Mountain Boys* (1839),
which concerns the land-grant dispute between Vermont and New York. Thompson's great-
great-nephew was U.S. Supreme Court justice William O. Douglas.

38. For example, there are floating islands on Lake Upemba (Democratic Republic of
Congo), including muddy Mitala Island.

39. The great 1815 volcano at Tambora on Sumbawa Island (part of the Lesser Sunda
Islands in Indonesia) produced pumice rafts 3.1 miles across. See Richard B. Stothers, "The
Great Tambora Eruption in 1815 and Its Aftermath," *Science* 224, no. 4654 (1984): 1191–1198.

40. Some of these rafts floated all the way to the general maritime region of Calcutta.
See Clive Oppenheimer, "Climatic, Environmental and Human Consequences of the Largest Known Historic Eruption: Tambora Volcano (Indonesia) 1815," *Progress in Physical Geography* 27, no. 2 (2003): 230–259.

41. "New Island and Pumice Raft, Tonga," NASA Earth Observatory photo with commentary, November 2006; and "New Island and Pumice Raft, Tonga," NASA Earth Observatory photo with commentary, August 2006.

42. So we read in an issue of *National Geographic* (June 1942), 715/1: "The 'sand bar'
turned out to be a thick layer of floating pumice, a reminder of the volcanic eruption of a
few months previous."

43. "[They] swim upon water." Frederick Collier Bakewell, *Geology . . . or, Former
Worlds* (London: National Illustrated Library, 1854), 80.

44. "In the 1970s, a US citizen was killed by another US citizen on an ice island floating in the Arctic Ocean where they were both working as members of a research team.
The incident took place while the ice island was floating in the Canadian Arctic Sector. A
US investigation team was sent to bring the offender back to the United States. The plane
carrying him first landed in the state of Virginia, where the offender was charged with
murder.—The issue that is raised with this case is whether the US, or Canada, or both
national entities, have jurisdiction over this crime. Canada did not want to interfere with
the course of justice in a case that concerned two US nationals. Nor did it want its lack
of involvement in this specific case to have any repercussions on its claim over the specific

territory. The complexity of the issue consists in the fact that the territorial sovereignty claimed by Canada forms the basis of exclusive jurisdiction." Giorgos Lagoudakis, "The Escamilla Case: Floating Territories or Handmade Ships?" *Floater Magazine* 1 (Fall 2008); see also Daniel Wilkes, "Law for Special Environments: Ice Islands and Questions Raised by the T-3 Case," *Polar Record* 16 (1972): 23–27.

45. "There is a fletyng Island. . . . Sume saied it was a shred of the bankes of . . . Paradise." William Bullein, *Dialogue against the Feuer Pestilence* ([1564], 1578; Early English Text Society, 1888).

46. Lucretius, *On the Nature of Things*, trans. W. H. D. Rouse (London: Heinemann, 1924), 5.261ff.; cf. 6.608ff.

47. Henry David Thoreau, "Where I Lived, and What I Lived For," in *Walden* (Boston: Ticknor & Fields, 1854). A fuller passage is the epigraph to *Islandology*.

48. See the Postamble.

CHAPTER 4

1. See Morgan Peter Kavanagh, *Myths Traced to Their Primary Source through Language* (London: T. C. Newby, 1856), 1:5. For *insula*, see Thomas Kerchever Arnold, *Latin Word-Building, with an Etymological Vocabulary* (London: Rivingtons, 1855), no. 1109.

2. Nancy LeClaire and George Cardinal, *Alberta Elders' Cree Dictionary*, ed. Earle Waugh (Edmonton: University of Alberta Press, 1998).

3. Also known as *The Wreck of Hope*.

4. *Wanderer über dem Nebelmeer*.

5. "When first the mass separates from the land-berg or glacier. . . ." Elisha Kent Kane, *The United States Grinnel Expedition in Search of Sir John Franklin: A Personal Narrative* (Philadelphia: Childs & Peterson, 1856), 420.

6. Pobeda was named "Termination Land" by Charles Wilkes because it blocked his passage (in 1840).

7. Important here is the density of water, which freezes at 32°F (for the fresh waters of glaciers and icebergs) and at lower temperatures for the varying other kinds of ice in the polar regions (averaging around 28.8°F for most salt waters).

8. *Island pan* means "a flat piece of floating ice." *Dictionary of Newfoundland English*, ed. G. M. Story, W. J. Kirwin, and J. D. A. Widdowson, 2nd ed., with supplement and foreword by Rex Murphy (Toronto: University of Toronto Press, 1990), s.v. "island."

9. Geoffrey K. Pullum summarizes thus the relevant part of Boaz's "Introduction": "Just as English uses derived terms for a variety of forms of water (liquid, lake, river, brook, rain, dew, wave, foam) that might be formed by derivational morphology from a single root meaning 'water' in some other language, so Eskimo uses the apparently distinct roots *aput* 'snow on the ground,' *gana* 'falling snow,' *piqsirpoq* 'drifting snow,' and *qimuqsuq* 'a snow drift.'" See Pullum, *The Great Eskimo Vocabulary Hoax* (Chicago: University of Chicago Press, 1991), 162.

10. See, for example, Christian Wilhelm Schult-Lorentzen, *Dictionary of the West Greenlandic Eskimo Language* (Copenhagen: C. A. Reitzel, 1927).

11. The smaller floes have formed an important part of "Eskimo culture." For the Eskimos and their depiction in popular culture, the ice floes are a crucial part of life and death. The movie *The Savage Innocents* (1959), directed by Nicholas Ray—as well as Hans Ruesch's novel *Top of the World* (1950) on which the movie was based—includes two women (Powtee

and Asiak) who are put out on the ice floes. Powtee is put out on floating aqua firma; Asiak walks across the sea ice to drown herself, floating along first on a small ice floe. Robert Flaherty's "documentary" *Nanook of the North* (1922) and the Inuit "historical" movie *Before Tomorrow* (2008)—directed by Marie-Hélène Cousineau and Madeline Ivalu, based on the novel by the Danish writer Jørn Riel and set near Puvirnituq (in the Nunavik region of northernmost Quebec)—put forward incidents where people are variably stranded on islands by movements of ice. There is also the case of *The Viking* (1931), directed by Varick Frissell and George Melford, in which two men "hunt each other" on the ice floes. *The Viking* is, to some extent, an expansion of Frissell's relatively short documentary *The Swilin' [Sealing] Racket*, also known as *The Great Arctic Seal Hunt* (1928), which likewise shows sealers working on the ice floes.

12. Philemon Holland thus uses the term *land-passage* instead of *isthmus* in his 1601 English-language translation of Pliny the Elder's *Natural History*. Philemon Holland, trans., *The Historie of the World, Commonly Called, The Naturall Historie of C. Plinius Secundus* (1601), 1:78: "Another land passage or Isthmus there is of like streightness . . . and of equall breadth with that of Corinth." Even so, Matthew Hale employs the word *land-passage* in his 1676 *Primitive Origination of Mankind*: "There is no Land-passage [or bridge] from this Elder World unto that of America," 2.7.190. Robert Johnson likewise uses the term *land strait* in his *Travellers Breviat* (1601), an English-language translation of part of Giovanni Botero's 1591 *Relationi universali*: "Peruana is . . . enuironed on al sides with the sea, saue wheras the forsaid Land-streight doth ioyn the same to Mexicana."

13. So Richard Montagu writes in his *Appello Caesarem* (1625), 2.5.158.

14. Charles Merivale, *A History of the Romans under the Empire* (1865), vol. 7, 60.260.

15. On Newfoundland, a *landwash* is "the sea shore between high and low tide marks, washed by the sea." *Dictionary of Newfoundland English*, s.v. "landwash."

16. "These were never referred to as icebergs. We always used the term Island of Ice." Greta Hussey, *Our Life on Lear's Room, Labrador*, ed. Susan Shiner (St. John's: Robinson-Blackmore, 1981).

17. What results is a *jökulhlaup* (Icelandic for "glacier run [or burst]") and consequently often also an accompanying *lahar* (Javanese for "mudflow of volcanic debris").

18. The CIA's *World Factbook* says Canada has 202,000 kilometers of coastline and the world has 356,000 kilometers. https://www.cia.gov/library/publications/the-world-factbook/fields/2060.html. The *World Resources Handbook* has it that the world has 1,634,000 kilometers of coastline and that Canada has 265,000 kilometers.

19. Some geographers say that Finland's "Archipelago Sea" has more; see Altti Holmroos, *Salaisuuksien saaristo* (Kustavi: Ulkosaariyhdistys, 2003).

20. See Chapter 6.

21. See Chapter 9.

22. "Prime Minister Stephen Harper Announces New Arctic Offshore Patrol Ships," Reuters, July 9, 2007.

23. Pierre Berton, *The Arctic Grail: The Quest for the North West Passage and the North Pole, 1818–1909* (New York: Viking, 1988).

24. Glynder Williams, *Voyages of Delusion: The Quest for the Northwest Passage* (New Haven, CT: Yale University Press, 2003).

25. Writes Captain James Cook: "Some curious and interesting experiments are want-ing to know what effect cold has on Sea Water in some of the following instances: does it freeze or does it not, if it does, what degree of cold is necessary and what becomes of the salt brine." See H. F. P. Herdman, "Some Notes on Sea Ice Observed by Captain James Cook, R.N., during His Circumnavigation of Antarctica, 1772–75," *Journal of Glaciology* 3, no. 26 (1959): 534–541, esp. 535. See also Johann Reinhold Forster, *Observations Made during a Voy-age round the World* (London: G. Robinson, 1778).

26. *OED*, s.v. "meer," 2 (for "lake") and 1 (for "sea"). Ellesmere Island was named by the British to honor Francis Egerton, First Earl of Ellesmere. The toponym in the Inuit language is *Umingmak Nuna* (land of muskox), on which see Lyle Dick, *Muskox Land: Ellesmere Island in the Age of Contact* (Calgary: University of Calgary Press, 2001).

27. For the description of this song as a national hymn, see Benoît L'Herbier, *La chan-son québécoise* (Montreal: L'Édition de l'Homme, 1974).

28. On the west side of the imagined Northwest Passage, Francisco de Ulloa sailed along Baja California; he concluded incorrectly that the Gulf of California was the south-ern part of a passage going to the Gulf of Saint Lawrence (1539). The search for the "Strait of Anian" (another toponym for the Northwest Passage—probably recalling the Chinese province noted by Marco Polo)—began around the same time. Sir Francis Drake sought it. Juan de Fuca claimed (without proof) that he sailed it all the way across and back.

29. On the east side, Jacques Cartier believed that the Saint Lawrence River was the way to China. Stymied at the river rapids near Montreal, he called the place *China* [La-chine]. Martin Frobisher and Humphrey Gilbert explored the northerly regions in the sixteenth century; John Davis and Henry Hudson did the same in the next century. René-Robert Cavelier, Sieur de La Salle, sought the passage via the Great Lakes.

30. Knud Rasmussen, *Across Arctic America: Narrative of the Fifth Thule Expedition* (New York: G. P. Putnam's Sons, 1927).

31. Phillip Vannini, "Recontinentalizing Canada: Arctic Ice's Liquid Modernity and the Imagining of a Canadian Archipelago," *Island Studies Journal* 4, no. 2 (2009): 121–138.

32. On April 9, 2006, Canada's Joint Task Force North declared that the Canadian military establishment would no longer refer to the region as the "Northwest Passage" but as the "Canadian Internal Waters."

33. Peter Høeg, *Smilla's Sense of Snow*, trans. Tiina Nunnally (New York: Farrar, Straus & Giroux, 1993), 418.

34. See Ted S. Clarke and Keith Echelmeyer, "Seismic-Reflection Evidence for a Deep Subglacial Trough," *Journal of Glaciology* 43 (1996): 141.

35. Greenland: 836,000 square miles; Bentley Subglacial Trench: around 759,000 square miles.

36. The English-language translation of Høeg's novel by Tiina Nunnally for the Amer-ican edition of the novel is *Smilla's Sense of Snow*. Høeg, who knows English, prefers the British English-language translation by F. Felicity (pseud.), *Miss Smilla's Feelings for Snow*. (The movie is mostly spoken in English.)

37. Prem Poddar and Cheralyn Mealor, "Danish Imperial Fantasies: Peter Høeg's *Smilla's Feeling for Snow*," in *Translating Nations*, ed. Prem Poddar, The Dolphin 30 (Aarhus: Aarhus University Press, 2000), 161–202.

38. Gardar was the place to which King Christian IV of Denmark sent expeditions in 1605–1607. On the *ting* as national assembly, see Chapter 12.

39. Smilla's birthplace, Siorapaluk, lies in the Qaanaaq area. It is, by some measures, the northernmost "permanently inhabited settlement" in the world.

40. Høeg, *Smilla*, 316.

41. "But there remained something in the back of my mind about Bergman's limiting and depleting himself by confining himself to the island of his mind, and venturing less and less to the world outside. Sweden was a small country in a big cosmos, but even Sweden itself remained a remote entity in Bergman's later films. Time seemed to stand still for him, and even recede." Andrew Sarris, "Ingmar Bergman: The Island of His Mind," *New York Observer*, July 31, 2007. See Marie Nyreröd's 2004 television documentary *Bergman Island*.

42. The Swedish word *ö* means "island." For the pun that conjoins meanings of *eye* with those of *island*, see such island place toponyms as *Ireland's Eye*, which names both the island at the entrance to Smith Sound in Trinity Bay, Newfoundland, and the island in County Dublin, Ireland; see also Chapter 1.

43. Ingmar Bergman, *Fårö Document*, Bergman Center, http://bergmancenter.se/en/about-bergmancenter/news-archive/faro-document-released-on-dvd-for-the-first-time/.

44. Ingmar Bergman, *The Magic Lantern: An Autobiography*, trans. Joan Tate (New York: Penguin, 1989), 208.

45. See Erik Hedling, "The Welfare State Depicted: Post-Utopian Landscapes in Ingmar Bergman's Films," in *Ingmar Bergman Revisited: Performance, Cinema and the Arts*, ed. Maaret Koskinen (London: Wallflower Press, 2008), 180–185.

46. *Shame* was partly shot in Visby on Gotland Island, also the shooting location of *The Touch* (1971).

47. Gotland was also the location for Andrei Tarkovsky's Bergman-like movie *The Sacrifice* (1986), whose particular location was mostly the Narsholmen Peninsula of Gotland Island.

48. See Mel Gussow, "A *Hamlet* Stamped with a Bergman Seal," *New York Times*, June 10, 1988.

49. Raphael Shargel, *Ingmar Bergman: Interviews* (Jackson: University Press of Mississippi, 2007), 48. Bergman never completed *The People Eaters*.

50. For Strindberg's islandology, see his psychological novel *By the Open Sea* (1890) and *The Life of the Men of the Skerries* (1888), *The People of Hemsö* (1887), and *In the Outer Skerries* (1890).

51. For the role of this painting at the end of Strindberg's piece, see Strindberg, *The Chamber Plays*, ed. Evert Sprinchorn (New York: Dutton, 1962), 152.

52. *Toten-Insel* was published posthumously in 1918. On this work, see Carl Dahlstrom, *Strindberg's Dramatic Expressionism* (New York: Benjamin Blom, 1965), 209.

53. A reproduction hung in his parents' Stockholm home; see Egil Törnqvist, *Bergman och Strindberg: Spöksonaten* (Stockholm: Prisma, 1973), 226.

54. See Chapter 14.

55. See Ingmar Bergman, *The Magic Lantern* (Chicago: University of Chicago Press, 2007). Compare Karsten Larson, "The Birth of Evil: Genesis According to Bergman," in *The Christian Century*, June 7–14, 1978, 615–619.

56. Shakespeare, *Julius Caesar*: "And therefore think him as a serpent's egg / Which hatch'd, would, as his kind grow mischievous, / And kill him in the shell" (act 2, sc. 1).

57. The thirteenth-century *Gutalagen* (Gotlandic law book) partly promulgates the laws of the *ting*, on which see Chapter 12.

58. On the *ham-* and *hamn-*, see Chapter 10. Laertes, whose "necessaries are inbark'd" (*Hamlet*) takes his leave at the Elsinore dock (*Hamn*), where "the wind sits in the shoulder of the sail." He thus makes his exit from Denmark in the way that Hamlet is both compelled not to do ("go not to Wittenberg") and to do ("The bark is ready, and the wind at help"). Eventually Laertes returns to Denmark. It is as if he were a migratory bird, "as a woodcock to my own springe."

59. Britt Halqvist, *Hamlet* (Stockholm: Ordfront, 1986).

60. James Fisher, *Rockall* (London: Geoffrey Bles), 12–13.

61. Rockall is still claimed by these nation-states.

62. Todd Field, "Bergman, Ever the Provoker," *Los Angeles Times*, August 1, 2007.

63. Even in the twentieth century, the Orkney islanders had a pronounced dialect: a mixture of insular Scots and Norse lexis and pronunciation. Hugh Marwick, introduction to *The Orkney Norn* (London: Oxford University Press, 1929).

64. Lasse Bergstrom, "Bergman's Best Intentions," in *Ingmar Bergman: Interviews*, ed. Raphael Shargel (Jackson: University Press of Mississippi, 2007), 176.

65. On the role of insular landscape in *Through a Glass Darkly*, see Robert Daudelin, "Entre la mer et le coin aux fraises: La nature chez Ingmar Bergman," *Revue 24 Images* 144 (2009): 20–22.

66. "Fårö . . . became both a real and a symbolic place to Ingmar Bergman." Birgitta Steene, *Ingmar Bergman: A Reference Guide* (Amsterdam: Amsterdam University Press, 2005), 40.

67. Michael Billington, "Walking a Tightrope to Great Acting," *The Guardian*, March 23, 2000. Gielgud approached other filmmakers about making a movie of *The Tempest*, among them Alain Resnais, Akira Kurosawa, and Orson Welles. See Douglas Brode, *Shakespeare in the Movies: From the Silent Era to Today* (New York: Berkley Boulevard Books, 2001), 228–229.

68. Steene, *Ingmar Bergman*, 418.

69. Danielle Pergamnet, "The Enchanted Island That Bergman Called Home," *New York Times*, October 7, 2007. The innkeeper was Thomas Soderlund.

CHAPTER 5

1. The list of cities that entirely occupy islands—such that their "city limits" match their "natural coastlines"—is long. Examples include Lindau (in Lake Constance, whose shores border Switzerland and Austria as well as Germany); Santa Cruz del Islote (Colombia); Isola dei Pescatori (in Lake Maggiore, Italy); Mexcaltitán (Mexico); Trogir (Croatia); Nesebar (Bulgaria); Flores (Guatemala); and Malé (Maldives). Larger cities that occupy islands include Abu Dhabi, Mumbai, and Hong Kong. There is also Venice, the "Floating City," whose estuarine embayment includes the mouths of the rivers Po and Piave as well as those of the Sile River and Riviera del Brenta. In this same group one might include Singapore, Conakry (on Tombo Island), and Recife (on the islands of Recife, Santo Antônio, and Boa Vista).

2. As in the volumes jointly titled *International Straits of the World*, ed. Gerard J. Mangone (The Hague: Sijthoff & Nordhoff, 1978). See also Gunnar Alexandersson, *The Baltic Straits* (The Hague: Nijhoff, 1982).

3. *OED*, s.vv. "hals," "hawse." Examples of other straits, artificial and natural, include the Panama Canal, the Russian Arctic Straits, the Korean Straits, the Red Sea, the Gulf of Aden, the Baltic Straits, and the Torres Strait.

4. Jacques Cartier believed that the island of Montreal was the "thwart" to the Northwest Passage.

5. Examples include the Riau Islands in the Strait of Malacca and the Princes' Islands in the Sea of Marmara (near the Bosporus). Turkish writer Sait Faik, who lived on one of these, had for one pen name *Adalı* (island dweller).

6. Examples include the near-island Gibraltar at the strait of the same name, which is disputed by Spain and the United Kingdom. Miyoun Island (formerly Perim Island) at the Bab-el-Mandeb Strait in the Red Sea is disputed by several states, including Yemen. The Yijiangshan Islands in the Taiwan Strait are disputed by China and Taiwan. Three islands—Picton, Lennox, and Nueva—in the Beagle Channel (south of Tierra del Fuego) are intermittently disputed by Chile and Argentina.

7. Examples include Tierra del Fuego in the Strait of Magellan.

8. Benedetto Bordone, *Libro . . . de tutte l'isole del mondo* (Venice: Nicollo Zoppino, 1528).

9. This delta is often confused with the Inner (endorheic) Niger Delta.

10. The intensely urban "Golden Triangle" of the Yangtze lies at the heart of the Chinese region traditionally called "Jiangnan," home to more than 105 million people as of 2010, of which 80 percent live in cities.

11. Ptolemy, *Geography*, 7.195.

12. Henry Yule and Arthur Coke Burnell, *Hobson-Jobson: A Glossary of Colloquial Anglo-Indian Words and Phrases*, ed. William Crooke (London: John Murray, 1903).

13. See Frank Lestringant, "Fictions de l'espace brésilen à la Renaissance: L'example de Guanabara," in *Arts et légendes d'espaces: Figures du voyage et rhétorique du monde*, ed. Christian Jacob and Frank Lestringant (Paris: Presses de l'École Normal Supérieure, 1981), 205–256.

14. Ellen Churchill Semple, *Influences of Geographic Environment on the Basis of Ratzel's System of Anthropo-Geography* (New York: Henry Holt, 1911), 427.

15. They include, at different times in their histories, such islands as Aegina, Cyprus, Rhodes, Crete, Malta, Corfu, Sicily, and Sardinia, as well as the Channel Islands, Corsica, and Taiwan. Popular literature often takes up this theme. The British television miniseries *Island at War* (2004) concerns the experience of residents of the Channel Islands during the Nazi invasion.

16. For this term, see *OED*, s.v. "islanded."

17. Thomas More, *Utopia*, bk. 2.

18. On how Mount Athos is usually conceived in any case as an island in much later Christian European thinking and practice, see Veronica della Dora, "Mapping a Holy Quasi-Island: Mount Athos in Early Renaissance *Isolarii*," *Imago Mundi* 60, no. 2 (2008): 139–165.

19. *OED*, s.v. "disinsulation." The toponym *Tyre*, meaning "rock" (Phoenician *sur*), refers to the formerly offshore islet.

20. See the illustration in Johann Bernhard Fischer von Erlach, *Entwurf einer historischen Architektur* (Leipzig, 1721); see also Veronica della Dora, *Imagining Mount Athos: Visions of a Holy Place, from Homer to World War II* (Charlottesville: University of Virginia Press, 2011).

21. See Marcus Vitruvius, *De architectura*, bk. 2.

22. Plutarch, *Life of Alexander*.

23. Strabo, *Geographica*, 7.6.792. For a map of ancient Alexandria showing the island, see Mostafa El-Abbadi, *The Life and Fate of the Ancient Library of Alexandria* (Mayenne, France: Imprimerie Floch, 1990).

24. Philo of Alexandria, *The Letter of Aristeas*, trans. H. St. J. Thackeray (London: Macmillan, 1904), 301; Philo of Alexandria, *On the Life of Moses*, 2:35–44; and *[Aulus Avilius] Flaccus*, 27.110.

25. Philo, *Moses*, 2:41–43.

26. Luciano Bosio, *Le origini di Venezia* (Novara: Istituto Geografico de Agostini, 1986).

27. More, *Utopia*, bk. 2.

28. It is the second edition (1534) of Bordone's work that employs the term *isolario*. The book was republished in 1560–1570.

29. George Tolias, "The Politics of the Isolario: Maritime Cosmography and Overseas Expansion during the Renaissance," *Historical Review* 9 (2012): 27–52.

30. Baedeker, *Venice*, 198.

31. The name of the bridge is a reference to the nearby mint building.

32. Thomas Coryate, *Crudities: Hastily Gobbled Up in Five Month's Travels* (1611), sig. O6.

33. See Marc Shell, "The Wether and the Ewe: Verbal Usury in *The Merchant of Venice*," in *Money, Language, and Thought* (Berkeley: University of California Press, 1982).

34. On this map, see Svat Soucek, *Piri Reis and Turkish Mapmaking after Columbus: The Khalili Portolan Atlas* (London: Nour Foundation in association with Azmimuth and Oxford University Press, 1996).

35. We will turn our attention to these stilt homes in Chapter 15.

36. See Betty Fussell, "Exploring Twin Cities by Canal Boat," *New York Times*, March 13, 1988.

37. At the time in which the novel is set, the Yellow River still passed through the area. Dongping Lake—in west Shandong Province—is now all that is left of the great ancient marshes and Eight Hundreds Li Liangshan Lake.

38. For the title translations, see *Maxine Hong Kingston: A Critical Companion*, ed. E. D. Huntley (Westport, CT: Greenwood Press, 2001), 172.

39. Shi Nai'an and/or Guanzhong Luo, *The Scattered Flock: Part Five of The Marshes of Mount Liang*, trans. John Dent-Young and Alex Dent-Young (Hong Kong: Chinese University of Hong Kong, 2002), 44.

40. Muḥammad ibn Aḥmad Shams al-Dīn al-Muqaddasī, *Ahsan al taqasim fi Ma'rifat al Aqalim* [The Best Divisions in the Knowledge of the Regions] (ca. 900).

41. The earliest known tide mill in Europe was Little Island in County Cork, which drew energy between the millpond and the sea. See Colin Rynne, "Milling in the Seventh Century—Europe's Earliest Tide Mills," *Archaeology Ireland* 6, no. 2 (1992): 22–24.

42. I refer here to the Nendrum Monastery on Mahee Island in Ireland. See Thomas C. McErlean, Caroline Earwood, Dermot Moore, and Eileen Murphy, "The Sequence of Early Christian Period Horizontal Tide Mills at Nendrum Monastery: An Interim Statement," *Historical Archaeology* 41, no. 3 (2007): 63–75.

43. In November 1994, Iraq formally accepted the UN-demarcated border with Kuwait that had been spelled out in Security Council Resolutions 687 (1991), 773 (1992), and 833 (1993), which formally ended an earlier claim to Bubiyan Island.

44. See Rudi Mathee, "Between Arabs, Turks and Iranians: The Town of Basra, 1600–1700," *Bulletin of the School of Oriental and African Studies, University of London* 69, no. 1 (2006): 66.

45. Piri Reis, *Kitab-i bahriye* [Book of Navigation] (Ankara, 1935), 65–66; see Dejanirah Couto and Rui Manuel Loureiro, eds., *Revisiting Hormuz: Portuguese Interactions in the Persian Gulf Region in the Early Modern Period* (Wiesbaden: Harrassowitz, 2008), esp. Svat Soucek, trans., "The Portuguese and the Turks in the Persian Gulf," 44.

46. The modern Iranian province of Hormozgán also includes islands such as Kish and Qeshm.

47. Christina P. Harris, "The Persian Gulf Submarine Telegraph of 1864," *Geographical Journal* 135, no. 2 (1969): 169–190.

48. In popular belief, the derivation is the Persian god *Hormoz*. Some scholars, though, link the name with the (local) Persian word *Hur-mogh* (date palm).

49. A. W. Stiffe, "The Island of Hormuz (Ormuz)," *Geographical Magazine* 1 (April 1874): 14.

50. Albert Gray, *The Voyage of François Pyrard of Laval to the East Indies, the Maldives, the Moluccas and Brazil* (London: Hakluyt Society, 1888), AC.6172/63, 238–245.

51. The "yolk" of an egg is surrounded by albumen and shell in the way that an island like Hormuz is surrounded by water or a planet like Earth is surrounded by atmosphere; the island or planet is the "center," innermost part," "core," or "best part." The terms in quotation marks are "figurative meanings" of *yolk* (according to the *OED*).

52. The island in *The Tempest*, if it is anywhere real, is located strangely not only somewhere in the "brave new world" being discovered in Shakespeare's day. (It was Bermuda, say some literalists, who stress the "source" in Strachey's eyewitness report of a shipwreck there in 1609–1610.) It is located also somewhere in the middle of the Earth, or the Mediterranean (say the more metaphorical).

53. Peter B. Rowland, "Essays on Hormuz (History)" and "Images of Hormuz," http://www.dataxinfo.com/.

54. The language(s) were quite different on either side of the Gulf even as late as the twentieth century. See Bertram Thomas, "The Kumzari Dialect of the Shihuh Tribe (Musandam), Arabia, and a Vocabulary," *Journal of the Royal Asiatic Society of Great Britain and Ireland* 62, no. 4 (1930): 785–854.

55. There is some question here as to whether Marco Polo is accurate as to the precise name.

56. *Travels of Marco Polo*, the complete Yule-Cordier edition, including the unabridged third edition (1903) of Henry Yule's annotated translation, as revised by Henri Cordier, together with Cordier's later volume of notes and addenda, 2 vols. (New York: Dover, 1920), 661n.

57. Valeria Fiorani Piacentini, "Salghur Shah, Malik of Hormúz, and His Embargo of Iranian Harbours," in *Revisiting Hormuz: Portuguese Interactions in the Persian Gulf Region in the Early Modern Period*, ed. Dejanirah Couto and Rui Manuel Loureiro (Wiesbaden: Harrassowitz, 2008), 8.

58. John Harris, *Complete Collection of Voyages and Travels* (London, 1705). See also *The Voyage of Master Joseph Salbancke through India, Persia, part of Turkie, the Persian Gulfe and Arabia* (1609).

59. What we know about the vegetation today bears out this view. See Günther Künkel, *The Vegetation of Hormoz, Qeshm and Neighbouring Islands (Southern Persian Gulf Area)* (Vaduz: J. Cramer, 1977), 392.

60. Jacob d'Ancona, *The City of Light: The Hidden Journal of the Man Who Entered China Four Years before Marco Polo*, trans. David Selbourne (New York: Citadel Press, 1997), 80.

61. W. J. Fischel, "The Region of the Persian Gulf and Its Jewish Settlements in Islamic Time," in *Alexander Marx Jubilee Volume on the Occasion of His Seventieth Birthday* (New York: Jewish Theological Seminary of America, 1950), 203–230. The earliest image of Hormuz that I have seen is that supplied on the central panels of the *Catalan Atlas* (1375) by the Jewish cartographers Abraham and Jehuda Cresques, residents of the island, then a city-state of Majorca. Mss. Esp. 30, Bibliothèque Nationale, Paris.

62. Richard Henry Major, ed., *India in the Fifteenth Century, Being a Collection of Narratives of Voyages to India, in the Century Preceeding the Portuguese Discovery of the Cape of Good Hope from Latin, Persian, Russian and Italian Sources, now first translated into English* (London: Hakluyt Society, 1857; New Delhi: Asian Educational Services, 1992), 1:5–7.

63. John W. Draper, "Milton's Ormus," *Modern Language Review* 20, no. 3 (1925): 323.

64. John Milton, *Paradise Lost*, 2.1–5.

65. Samuel Purchas, *Purchas His Pilgrimes* (London: William Stansby for Henry Fetherstone, 1625), chap. 9, 1787. Compare Peter Padfield, *Tide of Empires: Decisive Naval Campaigns in the Rise of the West* (London: Routledge, 1979), 65.

66. Many Chinese artifacts have been found on Hormuz. See Peter Morgan, "New Thoughts on Old Hormuz: Chinese Ceramics in the Hormuz Region in the Thirteenth and Fourteenth Centuries," *Iran: Journal of the British Institute of Persian Studies* 29 (1991): 67–83.

67. For a translation, see "The Travels of Athanasius Nikitin," in *India in the Fifteenth Century*, ed. Richard H. Major, trans. Mikhail M. Wielhorsky (London: Hakluyt Society, 1857), ser. 1, vol. 2.

68. Patricia Risso, *Oman and Muscat: An Early Modern History* (London: Croom Helm, 1986), 10.

69. See Shell, "Wether and Ewe."

70. Georg Schurhammer, "Die Trinitätspredigt Mag. Gaspars in der Synagoge von Ormuz 1549," *Archivum Historicum Societatis Iesu* 1 (1933), 279–309.

71. In 1549, the Jesuit missionary Caspar Baertz, under the direction of Francis Xavier, preached at one of the larger Hormuz synagogues. When he failed to convert the congregation, relations became very bitter.

72. *The Book of Duarte Barbosa: An Account of the Countries Bordering on the Indian Ocean and Their Inhabitants* (ca. 1518), ed. and trans. Mansel Longworth Dames (New Delhi: Asian Educational Services, 1989), chap. 42.

73. Henry James Coleridge, *The Life and Letters of St. Francis Xavier (1506–1556)* (New Delhi: Asian Educational Services, 1997), 104–105.

74. C. R. Boxer, *The Portuguese Seaborne Empire 1415–1825* (London: Pelican Books, 1973), 324.

75. That is the view of Stephen Neil, *A History of Christianity in India: The Beginnings to 1707* (Cambridge: Cambridge University Press, 1984), 549.

76. Relevant here is the history of the Jews of Sohar in Oman.

77. The modern town of Khasab, built by the Portuguese in the seventeenth century on the Musandam Peninsula (with its many fjords and, until recently, its relative inaccessibility by land), has several creoles and dialects, among them Kumzari (spoken by the Shihu tribe). It seems to me that the complex linguistic heritage of the place is a remnant of what once existed at Hormuz.

78. Niccolao Manucci, *Storia do Mogor; or, Mogul India*, trans. William Irvine (London: John Murray, 1907), 1:57.

79. The proposal, "Hormuz Corridor: Building a Cross-Border Region between Iran and the United Arab Emirates," suggests how this might happen, if only . . . See Ali Parsa and Ramin Keivani, "The Hormuz Corridor: Building a Cross-Border Region between Iran and the United Arab Emirates" in *Global Networks: Linked Cities*, ed. Saskia Sassen (London: Routledge, 2002).

80. Najla Moussa, "Bridge Connecting Egypt, Saudi Arabia Considered," *Daily News Egypt*, March 2, 2006.

81. Thus *Jejudo* often refers to "island" and *Jeju-do* to the island's "government."

82. Deryck Scarr, *The History of the Pacific Islands: Kingdoms of the Reefs* (South Melbourne: Macmillan, 1990).

83. See William Elliot Griffis's pejorative (generally "anti-Korean") *Corea: The Hermit Nation* (New York: Scribner, 1882).

84. For this Nagasaki-based Japanese translator's interpretation of the Dutch edition of the German Engelbert Kaempfer's influential *History of Japan* (1729, 1733), see Geoffrey C. Gunn, *First Globalization: The Eurasian Exchange, 1500–1800* (Lanham, MD: Rowman & Littlefield, 2003), 151.

85. See David Mervart, "A Closed Country in the Open Seas: Engelbert Kaempfer's Japanese Solution for European Modernity's Predicament," *History of European Ideas* 35, no. 3 (2009): 321–329.

86. *Journael van de ongeluckige voyagie van 't jacht de Sperwer van Batavia gedestineert na Tayowan in 't jaar 1653, en van daar op Japan; hoe 't selve jacht door storm op 't Quel-paarts eylant is ghestrant, hoe de maats van daar naar 't Coninckrijck Coeree sijn vervoert . . .* (Rotterdam: Johannes Stichter, 1668).

87. Stéphane Bois, *Connaissance par les îles: Relations coréenes de la Pérousse à Zuber*, illus. Henri Zuber (Paju-si, Gyeonggi, South Korea: Jaimimage, 2009).

88. Compare the Japanese *tsuri*.

89. First aired on the MBC Network, August–September 2009.

90. Another such movie involving free diving on Jeju is *My Mother, the Mermaid* (2004).

91. Jean-François de Galaup La Pérouse, *Atlas du voyage de La Pérouse* (Paris: Imprimerie de la République, an V [= 1797]); see plate 45.

92. Bois, *Connaissance par les îles*, 37.

93. Claire Le Chatelier, "L'église Saint Joseph et le Père Louis Helot," *Le Souvenir Français* 34 (November 2009): 6.

94. Bois, *Connaissance par les îles*, 38.

95. Ibid., 39.

96. Robert Neff, "An Expedition to Korea to Rescue the Crew of the *Narwal* in April 1851," *Transactions of the Royal Asiatic Society—Korea Branch* 83 (2008): 27–73.

97. Wrote Roze: "[T]he destruction of one of the avenues of Seoul, and the considerable losses suffered by the Korean government should render it more cautious in the future." Pierre Gustave-Roze is quoted likewise in Bois, *Connaissance par les îles*, 106.

98. Kim Hyung-yoon, "Ganghwa Island: A Prism for Viewing Korean History," *Koreana: A Quarterly on Korea Art and Culture* (2012).

99. Jean-Marie Thiébaud, *La présence française en Corée de la fin du XVIIIème siècle à nos jours* (Paris: Harmattan, 2005), 22.

100. Lee Kyong-hee, "Joseon Royal Books Return Home after 145 Years in France," *Koreana: A Quarterly on Korea Art and Culture.*

101. Jean Henri Zuber, "Une expedition en Corée," *Le Tour du Monde: Nouveau Journal des Voyages* 25 (1873): 401, 414. See also Alain Génetiot, "Henri Zuber, un étonnant témoin de l'expédition de l'amiral Roze," *Culture Coréenne* 78 (Spring–Summer 2009): 10–13.

102. The Mongols invaded Korea in 1234 with the result that the Koryo dynasty moved its headquarters to Ganghwa Island and presumably printed books there.

103. See John Merrill, *Korea: The Peninsular Origins of the War* (Newark: University of Delaware Press, 1989).

104. The Peace Dam, which has no reservoir, has since been built only in order to prevent purposeful flooding by the North Koreans. The same happened in 2009. See Choe Sang-Hun, "North Korea Opens Dam Flow, Sweeping Away 6 in the South," *New York Times*, September 6, 2009.

105. China as well has been preparing its new territory inside North Korea—the Rajin-Sonbong Economic Special Zone—not only to ship coal south to Shanghai via its ice-free ports but also to enter the Arctic fray more easily through La Pérouse Strait—dividing the southern part of the Russian island of Sakhalin (Karafuto) from the northern part of the Japanese island of Hokkaido, and connecting the Sea of Japan on the west with the Sea of Okhotsk on the east—where Japan's territorial waters extend only three miles out, perhaps to allow U.S. nuclear-armed ships to enter without violating "Japanese law."

106. *Pyoryu* means "to float or drift," and *gi* means "a record." A record of one's experience having washed up on some foreign shore is a subgenre when it comes to Korea and Korea-related literature, as for Hendrick Hamel's *pyoryugi*.

107. Thus the American Korean author Emanuel Pastreich translates the title of his Korean-language book *Insaeng eun sokudo anira banghyang ida: Habodeu baksa eui hanguk*

pyoryugi as "Life Is a Matter of Direction, Not of Speed: Records of a Robinson Crusoe in Korea" (London: Nomad Books, 2011). On the translation of Hamel's book, see *Natsbŏn Chosŏn ttang-esŏ ponaen 13-nyŏn 20-ir-ŏi kirok: Hamel p'yoryugi* (Seoul: Sŏhae Munjip, 2003).

108. The usual urban notion, that rural island life is idyllic or utopian, is quickly dispelled in modern Korean cinema by such works as Chul-soo Jang's 2010 "box office hit" *Bedevilled*, where the retreat to a remote island (Mudo) reveals the horrid workings of isolated life.

109. See Erich Schonfeld, "Cyworld Ready to Attack MySpace," *CNN Money*, July 27, 2006.

110. See Marc Shell, *Polio and Its Aftermath: The Paralysis of Culture* (Cambridge, MA: Harvard University Press, 2005), chap. 6.

111. There is an island of the same name in Korea's North Jeolla Province.

112. Javier Gaete, "Seoul Floating Islands / Haeahn Architecture + H Architecture," *ArchDaily*, July 12, 2012.

113. First aired on the Korean Broadcasting System's KBS-2TV television network.

114. Margaret Blunden, "Geopolitics and the Northern Sea Route," *International Affairs* 88, no. 1 (2012): 119.

115. In 2011, the Korean Gas Corporation of South Korea bought 20 percent ownership of the Umiak SDL 131 gas field.

116. Rosemary Neering, *Continental Dash: The Russian-American Telegraph* (Ganges, BC: Horsdal & Schubart, 2000).

117. William Gilpin, *The Cosmopolitan Railway: Compacting and Fusing Together All the World's Continents* (San Francisco: The History Company, 1890). Gilpin first proposed the idea in *The Central Gold Region: The Grain, Pastoral and Gold Regions of North America. With Some New Views of Its Physical Geography; and Observations on the Pacific Railroad* (Philadelphia: Sower, Barnes & Co., 1860).

118. Kevin Starr, *Endangered Dreams: The Great Depression in California* (New York: Oxford University Press, 1996), 330. See also Richard B. Cathcart, Alexander A. Bolonkin, and Radu D. Rugescu, "The Bering Strait Seawater Deflector (BSSD): Arctic Tundra Preservation Using an Immersed, Scalable and Removable Fiberglass Curtain," in *Macro-Engineering Seawater in Unique Environments: Arid Lowlands and Water Bodies Rehabilitation*, ed. Viorel Badescu and Richard B. Cathcart (Berlin: Springer Verlag, 2011), 758.

119. See James A. Good, *The Ohio Hegelians* (Bristol, UK: Thoemmes Press, 2004).

120. See Roelof Schuiling, Viorel Badescu, Richard Cathcart, and Piet van Overveld, "The Hormuz Strait Dam Macroproject," in *Macro-Engineering Seawater in Unique Environments: Arid Lowlands and Water Bodies Rehabilitation*, ed. Viorel Badescu and Richard B. Cathcart (Berlin: Springer Verlag, 2011), 149–166.

121. "The Island of Ganghwa must have been selected by the government of Seoul as the military avenue of Korea." Pierre-Gustave Roze, quoted in *Han-pul kwan'gye charyo, 1846–1887* [Original Documents Relating to Korea-French Relations], ed. Andreas Choe (Seoul: Han'guk Kyohoesa Yŏn'guso, 1986), 326.

122. Pernilla Ouis, "'And an Island Never Cries': Cultural and Societal Perspectives on the Mega Development of Islands in the United Arab Emirates," in *Macro-Engineering Seawater in Unique Environments: Arid Lowlands and Water Bodies Rehabilitation*, ed. Viorel Badescu and Richard B. Cathcart (Berlin: Springer Verlag, 2011), 59–67.

123. In September 2009, the *Times of London* reported that work on "The World" had been suspended because of a global financial crisis.

124. See "'The World' Is Sinking," *The Age*, January 28, 2011; and Claire Bates, "Is It the End of the World? NASA Picture Suggests Dubai Globe Is Sinking Back into the Sea," *Daily Mail*, February 2, 2010.

125. For Athens, see Chapter 8; for the British *ting*, see Chapter 12.

CHAPTER 6

1. The name changed officially from Watling to San Salvador in 1925. At that time, the Bahamas Parliament decided that the place we now call Cat Island (but then called San Salvador) was not the island that Columbus first discovered.

2. In this same way, *terra firma* becomes the toponym *Terra Firma*, which names various locations in South America (as, for example, indicated on Herman Moll's 1701 map) and certain areas around the Venetian Lagoon. Similarly, *land's end* becomes *Lands End*—the place on Canada's Prince Patrick Island. So, too, *island* becomes *Island*—a translation of the Danish *Zealand*—as *empty* becomes *Empty*—the translation of *Rub' al Khali* in the Arabian Peninsula and of *Tura*, the disputed island just north of the Kingdom of Morocco.

3. Among the works consulted is the old-school *Merriam Webster's Geographical Dictionary*, with the subtitle *A Dictionary of Names of Places, with Geographical and Historical Information and Pronunciations* (Springfield, MA: C. & C. Merriam, 1960), and the new online *Getty Thesaurus of Geographic Names* (www.getty.edu). There were gazetteers already in Hellenist Greece and in first-century China; see Robert C. White, "Early Geographical Dictionaries," *Geographical Review* 58, no. 4 (1968): 652–659.

4. Quoted by Pappus of Alexandria, *Synagoge* [Collection], bk. 7, in *Greek Mathematical Works: Aristarchus to Pappus*, trans. Ivor Thomas, Loeb Classical Library 362 (Cambridge, MA: Harvard University Press, 1941), 2:35. Compare "Give me where to stand, and I will move the earth," as translated in John Bartlett, *Familiar Quotations*, ed. Emily Morison Beck, 14th ed. (Boston: Little, Brown, 1968), 105.

5. Muḥammad Shafīq, *Arabic-Berber Dictionary* [Arabic and Tamazight] (Rabat: Akādīmīyat al-Mamlakah al-Maghribīyah, 1990), 1:346.

6. "Nous avons rejeté l'agression armée du gouvernement espagnol contre l'îlot de Toura qui a toujours fait partie intégrante du territoire national." "Discours de S. M. le Roi Mohammed VI à l'occasion du troisième anniversaire de l'accession du Souverain au Trône de ses glorieux ancêtres," Tangier, July 30, 2002, http://www.maroc.ma/NR/exeres/83B9DF1C-E33F-4523-A797-6BA1E523EC9F.

7. On the island of Hispaniola, the way one pronounces *Perejil* can be fatal. That is what happened in 1937 at the shallow Massacre River, which forms part of the northern border between Haiti and the Dominican Republic. On the island of Ireland, one's use of the nationalist "Derry" instead of the unionist "Londonderry" likewise can get one killed. (*Derry*, from the Irish *doire*, refers to the oak grove [*doire*] island in the Foyle River's bogs and waters.)

8. Relevant here is the methodologically interesting work of Richard Coates on the names of the Channel Islands; Eilert Ekwall's useful *Concise Oxford Dictionary of English Place-Names* (1936; new editions: 1940, 1947, 1951, 1960); and Margaret Gelling and Ann Cole's *Landscape of Place-Names* (Donington, UK: Shaun Tyas, 2000).

9. "Island fragments of broken empires are found everywhere. They figure conspicuously in that scattered location indicative of declining power. Little St. Pierre and Miquelon [just off the coast of Newfoundland] are the last geographical evidence of France's former dominion in Canada. The English Bermudes and Bahames point back to the time when Great Britain held the long-drawn opposite coast." Ellen Churchill Semple, *Influences of Geographic Environment on the Basis of Ratzel's System of Anthropo-Geography* (New York: Henry Holt, 1911), 430.

10. Kenneth Jackson, *Language and History in Early Britain: A Chronological Survey of the Brittonic Languages, First to Twelfth Century A.D.* (Edinburgh: Edinburgh University Press, 1953), 220–223, summarized in H. R. Loyn, *Anglo-Saxon England and the Norman Conquest*, 2nd ed. (Harlow, UK: Longman, 1991), 7–9. See also Mattias Jacobsson, *Wells, Meres, and Pools: Hydronymic Terms in the Anglo-Saxon Landscape* (Uppsala: Ubsaliensis S. Academiae, 1997).

11. The name *Avon* is cognate with the Welsh *afon*, both related to the old British *abona* (river).

12. Among other English river names of Celtic origin are Wear, Don, Avon, Axe, Exe, Esk, Usk, and Wiske.

13. Consider Welsh *ynys*, Irish *innis*, Old Irish *inis*, Cornish *enys*, and Breton *enez*.

14. Alexander MacBain, *An Etymological Dictionary of the Gaelic Language* (Stirling, UK: E. Mackay, 1911), s.v. "innis." Compare John Strachan, *Keltische Etymologien* (1893).

15. *Geiriadur Prifysgol Cymru* (Cardiff: Gwasg Prifysgol Cymru, 1950–2002), 4:3819.

16. John Carey, "The Location of the Otherworld in Irish Tradition," *Église* 19 (1982): 36–43; and Patrick Sims-Williams, "Some Celtic Otherworldly Terms," *Celtic Language, Celtic Literature: A Festschrift for Eric P. Hamp*, ed. A. T. E. Matonis and Daniel F. Melia (Van Nuys, CA: Ford & Baillie, 1990), 57–81.

17. Herman Melville, *Moby-Dick*, chap. 89.

18. For Justinian on the law of the sea, see Percy Thomas Fenn Jr., "Justinian and the Freedom of the Sea," *American Journal of International Law* 19, no. 4 (October 1925): 716–727.

19. Justinian, "Of the Different Kinds of Things," *Pandects*, bk. 2, title 1, 22, also considers variations, as when a river island rises in such a way that it belongs in common to the landowners on either bank in proportion to the extent of their riparian interest.

20. *Dictionary of Newfoundland English*, s.v. "shore-fast."

21. Goethe, *Faust Part One*.

22. Immanuel Kant, "Metaphysik der Sitten," in *Kant's Gesammelte Schriften*, Königliche Preussiche Akademie der Wissenschaften (Berlin: G. Reimer, 1902), 6:262.

23. Hershel Parker, *Herman Melville: A Biography* (Baltimore: Johns Hopkins University Press, 1996), 1:185.

24. Or, as some biographers have it, from Fairhaven, Massachusetts.

25. Melville refers to the Nore mutiny at Spithead as an "episode in the [British] Island's grand naval story." *Billy Budd, Sailor*, ed. Harrison Hayford and Merton M. Sealts Jr. (Chicago: University of Chicago Press, 1962), chap. 3. On the Nore mutiny and British naval history, see Chapter 15.

26. Among these are Antigua and Barbuda, Bahamas, Bahrain, Barbados, Cape Verde, Comoros, Cuba, Cyprus, Dominica, Fiji, Grenada, Iceland, Jamaica, Japan, Kiribati, Mad-

agascar, and Maldives. There are also Malta, Marshall Islands, Mauritius, Federated States of Micronesia, Nauru, New Zealand, Palau, Philippines, Saint Kitts and Nevis, Saint Lucia, Saint Vincent and the Grenadines, Samoa, São Tomé and Príncipe, Seychelles, Singapore, Solomon Islands, Sri Lanka, Taiwan, Tonga, Trinidad and Tobago, Tuvalu, and Vanuatu. The smallest such state is the Republic of Nauru, a guano-phosphate rock (eight square miles) that was once a colony of the German Empire. The largest is Australia (insofar as it *is* an island).

27. Warwick Fox, *A Theory of General Ethics: Human Relationships, Nature, and the Built Environment* (Cambridge, MA: MIT Press, 2007).

28. Semple, *Influences of Geographic Environment*.

29. See Woodruff D. Smith, "Friedrich Ratzel and the Origins of Lebensraum," *German Studies Review* 3, no. 1 (1980): 51–68.

30. The Principality of Sealand is a micronation located off the coast of Suffolk, England. Since 1967, it has been occupied by former major Paddy Roy Bates (also known as His Royal Highness Prince Roy of Sealand); his associates and family claim that it is an independent sovereign state.

31. The Republic of Minerva, in the Minerva Reefs (Pacific Ocean), was a 1972 attempt at creating a sovereign micronation on reclaimed land on an artificial island. The architect was Michael Oliver. These days Minerva has been "reclaimed by the sea."

32. The Republic of Rose Island was a micronation on a man-made platform in the Adriatic Sea, 11 kilometers (7 miles) off the coast of Rimini, Italy. Giorgio Rosa funded the construction of a 400-square-meter platform. The artificial island declared independence on June 24, 1968, under its Esperanto name *Respubliko de la Insulo de la Rozoj*. Eventually, carabinieri and tax inspectors landed on the "Isola delle Rose" and assumed control.

33. They include Alcatraz (United States), Black Water (India; in Hindi, Kālā Pānī), Coiba (Panama), Devil's Island (French Guiana), Gorgona Island (Colombia), Île d'If (France), İmralı Island (Turkey), and Prison Island (Zanzibar). There is also Johnson's Island (United States), María Madre (Mexico), Poulo Condor (Vietnam), Rikers Island (United States), Robben Island (South Africa), Roosevelt Island (United States), Saint Helena (United Kingdom), and Solovki (Russia). Australia, an island-continent founded by the United Kingdom as a prison colony, has itself experimented with the use of islands as detention camps (Nauru) as well as with the use of prison ships.

34. One is Franklin J. Schaffner's *Papillon* (1973), based on Henri Charrière's novel of the same name (1969), about the penal colony on Devil's Island, part of offshore French Guiana. A cult television series is *The Prisoner* (1967–1968). Its setting is Portmeirion, in partly peninsular Penrhyndeudraeth, in the boldly tidal estuary of the River Dwyryd in Wales. However, *The Prisoner* always seems to take place on an actual island whose name and location, as in *The Tempest*, are never reliably disclosed. The Kafkaesque predicament of the series is that the hero can escape neither the prison that is the island nor the prison that is the mind. In 2004, *TV Guide* rated *The Prisoner* its seventh most important "cult show" ever. The episode "Many Happy Returns" has a hint that the whole is meant to take place on an island near Portugal or Spain; "The Chimes of Big Ben" suggests a Baltic island in the vicinity of Lithuania and Poland; and a character in "Many Happy Returns" proposes that the "Village" is on an island.

35. Thomas Hobbes, *Leviathan*, chap. 28, "Of Punishments and Rewards."

36. See Mark Weber, "The 1945 Sinkings of the Cap Arcona and the Thielbek: Allied Attacks Killed Thousands of Concentration Camp Inmates," *Journal of Historical Review* 19, no. 4 (July–August 2000): 2–3.

37. See Theresa Arnold-Scriber, *Ship Island, Mississippi: Rosters and History of the Civil War Prison* (Jefferson, NC: McFarland, 2008); see also Chapter 11.

38. See Rod Edmond, "Islands of Disease: Colonialism and Leprosy," in *Islands in History and Representation*, ed. Rod Edmond and Vanessa Smith (London: Routledge, 2003).

39. Ibid., 133.

40. Spinalonga (the island leper colony) is the subject of Werner Herzog's movie *Last Words* (1968).

41. By the year 2006, nineteen memoranda of understanding with twenty-eight maps had been signed between the two countries pertaining to the survey and demarcation of the border covering a distance of 1,822.3 kilometers (1,132 miles) of the 2,019.5-kilometer (1,254-mile) border.

42. "International Boundary Study: Greece-Turkey Boundary," The Geographer, Office of Research in Economics and Science, Bureau of Intelligence and Research, U.S. Department of State, Washington, DC, November 23, 1964.

43. During World War II, U.S. "island hopping" started from the Tarawa Atoll in the Gilbert Islands and then proceeded to Kwajalein and Enewetak in the Marshall Islands to the islands of Saipan, Guam, and Tinian in the Marianas, to Peleliu in the Palau Islands, and then to Leyte in the eastern Philippines. (The Battle of Leyte Gulf was the largest in naval history anywhere and at any time.) It proceeded to Samar and Luzon, to the large Mindanao, and then on to Iwo Jima, and finally to Okinawa. Flying from Tinian, the B-29 *Enola Gay* dropped the first atom bomb on Hiroshima. What an islands epic that is!

44. See the discussion of indigenous literary representations of precolonial voyaging in the Cook islander Thomas R. A. H. Davis's novel *Vaka: Saga of a Polynesian Canoe* (Auckland: Polynesian Press, Samoa House, 1992).

45. See Elizabeth M. DeLoughrey, "Vessels of the Pacific: An Ocean in the Blood," in *Routes and Roots: Navigating Caribbean and Pacific Island Literatures* (Honolulu: University of Hawaii Press, 2007), 96–160, for a short history of international military forces in the Pacific Ocean area. In the 1930s, the U.S. government sent colonists to some of the Line Islands (including Jarvis) and Phoenix Islands (including Howland and Baker) in the central Pacific Ocean overtly in order to lay claim to them for commercial aviation and covertly for long-term military purposes. That is the gist of Noelle Kahanu's documentary *Under a Jarvis Moon* (2010). In the late 1930s, the British Empire made its last attempt at colonization in the same area: the Phoenix Islands Settlement Scheme on Nikumaroro, Manra Island, and Orona Atoll. "The Colonization of the Phoenix Islands" appears as a chapter in H. E. Maude's fine book, *Of Islands and Men: Studies in Pacific History* (Melbourne: Oxford University Press, 1968). As it is no longer readily available in print, we are excerpting this section here, with all due apologies to Professor Maude.

46. The border island Mbanie in Corsico Bay is disputed by Equatorial Guinea and Gabon. The lake islands in Mbamba Bay and Lake Nyasa are disputed by Tanzania and Malawi. Lete Island and Tondi Kwara Barou are disputed by Benin and Niger. The river island Rukwanzi is disputed by the Democratic Republic of Congo and Uganda. The Sedudu

River islands are disputed by Namibia and Botswana. The Ntem River islands are disputed by Cameroon and Equatorial Guinea.

47. China is in dispute with Taiwan and Vietnam over the Paracel Islands, with Japan over the Senkaku Islands, with Russia over Zhenbao Island (as well as islands at the confluence of the Amur and Ussuri rivers and one island in the Argun River), and with Vietnam over the Spratly Islands. There is also the dispute between China and South Korea over (the usually submerged) Socotra Rock. Japan disputes Liancourt Rocks with South Korea and Sakhalin Island (as well as several islands in the Kuril and Habomai archipelagoes) with Russia. Russia disputes the river islands Ukatny, Zhestky, and Malozhemchuzny with Kazakhstan, and quarrels over Kosa Tuzla Island with Ukraine. Kazakhstan disputes the river island (now a peninsula) Vozrozhdeniya with Uzbekistan. In addition to its various border disagreements over Borneo, Indonesia disputes Sipadan and Ligitan with Malaysia, and Pedra Branca and Middle Rocks (in the Strait of Malacca) with Singapore (which dispute was recently settled at the International Court of Justice in The Hague). Yeonpyeong Island in the Yellow Sea, at the border between North and South Korea, is disputed by the two Koreas. Scarborough Shoal (Huangyan Island), located between the Macclesfield Bank and the Philippines Luzon Island in the South China Sea, is disputed by China and the Philippines.

48. Disputed islands include the river island Ankoko (Venezuela versus Colombia), Los Monjes Archipelago (Venezuela versus Colombia), Bird Island (Venezuela versus Dominica), and the river islands Rio Quarai and Arroio Invernada (Brazil versus Uruguay).

49. Nicaragua and Colombia dispute the Archipelago of San Andrés, Providencia and Santa Catalina. Quita Sueño Bank is claimed by the United States as well as Colombia and Nicaragua. The United States claims Serranilla and Bajo Nuevo as unincorporated American territories; Colombia, Honduras, and Nicaragua have similar claims. Conejo Island is disputed by Honduras and El Salvador. Haiti claims Navassa Island, which is administered nonetheless by the United States.

50. The Minerva Reefs are disputed by Tonga and Fiji. The Cocos (Keeling) Islands are disputed by Australia and the Netherlands. Koh Ta Kiev, Koh Thmey, Koh Ses, and Koh Tonsay (as well as the Northern Pirates) are disputed by Cambodia and Vietnam. Minicoy Island is disputed by India and the Maldives, and Kachatheevu Island is disputed by India and Sri Lanka. Bangladesh disputes the Naf River islands with Myanmar.

51. The United Kingdom disputes the Falkland Islands, South Georgia Island, and South Sandwich Islands with Argentina. Britain disagrees about the ownership of Rockall Island with Denmark. (The Faroe Islands, Ireland, and Iceland also have claims on Rockall.) Spain and the United Kingdom both claim Gibraltar (the island rock connected to mainland Europe by a low tombolo).

52. France disputes Matthew and Hunter islands (as well as Rocky Tale) with Vanuatu, and it disputes Europa Island and the Glorioso Islands with Madagascar. Bulgaria disputes several islands in the Danube with Romania. Ukraine and Romania spar over Cernofca. Croatia and Serbia quarrel over the islands of Šarengrad and Vukovar. Spain disputes the islands of Peñón de Alhucemas and Peñón de Vélez de la Gomera, as well as the Chafarinas Islands—including Perejil—with Morocco. The Spanish *Isla de Perejil* means "Parsley Island." Its "original" name, *Tura*, means "empty" (in Berber). Some Moroccan media refer to it instead as *Leila* (ليلى), which is a local pronunciation of "La isla" (Spanish). In Moroccan historical references, it is only known as "Tura."

53. The United States disputes Seal Island and North Rock (in the Grand Manan Archipelago) with Canada; Wake Island, with the Marshall Islands; Cordova Island (in the Rio Grande), with Mexico; Navassa Island, with Haiti; and Swains Island, with Tokelau / New Zealand.

54. Already, Lohachara Island and New Moore Island have disappeared under the sea, and Ghoramara is half-submerged.

55. The water level in Lake Chad, on the border between Nigeria and Chad, has now fallen so low that the islands that these two states shared in the Bogomerom Archipelago have become part of the mainland. Falling water levels in the Neusiedler See, on the border between Austria and Hungary, have now exposed the island flats there.

56. Courtney Hunt's movie *The Frozen River* (2008) concerns the smuggling of "illegal immigrants" on the "reservation." For the sovereignty claim, see Mort Ransen's documentary *You Are on Indian Land* (1968), produced by the National Film Board of Canada.

57. Doug Struck, "Russia's Deep-Sea Flag-Planting at North Pole Strikes a Chill," *Washington Post Foreign Service*, August 7, 2007.

58. Steps were taken in 2010 toward resolution.

59. George Armstrong Nares first navigated through the Nares Strait in 1875. He got to the Lincoln Sea only to discover that it was not the ice-free "open polar sea"—the way to the Northwest Passage—that so many people had thought it would be.

60. Alexander Island, the largest island in Antarctica, is claimed by the United Kingdom, Chile, and Argentina; South Orkney, South Georgia, and the Sandwich Islands are claimed by the United Kingdom and Argentina; the South Shetland Islands are claimed by the United Kingdom, Argentina, and Chile.

61. On the various laws of the sea as they pertain to islands, see Chapter 13.

CHAPTER 7

1. R. H. MacArthur and E. O. Wilson, *The Theory of Island Biogeography* (Princeton, NJ: Princeton University Press, 1967). See also David Quammen, *The Song of the Dodo: Island Biogeography in an Age of Extinctions* (London: Pimlico, 1996).

2. That is how Edwin Muir puts it, as quoted by Simon Hall in *The History of Orkney Literature* (Edinburgh: John Donald Publishers, 2010), 7. See Muir, *Collected Poems: 1921–1958*, ed. J. C. Hall and Willa Muir (London: Faber & Faber, 1960). "The Sufficient Place" is also the title of the unpublished autobiography of Orkney Islands anthologizer Ernest Marwick, as cited by Hall, *History of Orkney Literature*, 130. "Utmost corners of the world" is what William Fowler calls those islands in his sonnet about them, as cited by Hall, *History of Orkney Literature*, 8.

3. Aldous Huxley, *Island*, 41. Richard Rorty, though he writes about *Brave New World*, does not discuss, or even mention, *Island*.

4. Herman Melville, *Moby-Dick*, chap. 12.

5. Robert J. Flaherty, "How I Filmed 'Nanook of the North': Adventures with the Eskimos to Get Pictures of Their Home Life and Their Battles with Nature to Get Food. The Walrus Fight," *World's Work* 44 (September 1922): 553–560; and Robert J. Flaherty, "Life among the Eskimos: The Difficulties and Hardships of the Arctic. How Motion Pictures Were Secured of Nanook of the North and His Hardy and Generous People," *World's Work* 44 (October 1922): 632–640.

6. Pathé Pictures promotional poster (1922).

7. One need know only that the "Owl of Minerva" is resting there for a while, or could rest there, as the German idealist Georg Wilhelm Friedrich Hegel might put it in the preface to *Hegel's Philosophy of Right*, trans. T. M. Knox (Oxford: Oxford University Press, 1967).

8. Quoted by Pappus of Alexandria, *Synagoge* [Collection], bk. 7.

9. Henry David Thoreau, "Where I Lived, and What I Lived For," in *Walden* (Boston: Ticknor & Fields, 1854).

10. For Darwin and subsidence and uplift, see Chapter 2; and Wagner, Chapter 14; and T. S. Eliot, Chapter 15; and terra infirma, the Postamble.

11. *OED*, s.v. "speciation."

12. Charles Darwin, Notebook E #135, in *Notebooks 1836–1844: Geology, Transmutation of Species, Metaphysical Inquiries*, ed. Paul H. Barrett, Peter J. Gautry, Sandra Herbert, David Kohn, and Sydney Smith (Ithaca, NY: Cornell University Press, 1987), 138.

13. In "Utopian tongue," Latin, and English. Another translation is:

> My king and conqueror Utopos by name,
> A prince of much renown and immortal fame,
> Hath made me an isle that erst no island was,
> Full fraught with worldly wealth, with pleasure and solace.
> I one of all other without philosophy
> Have shaped for man a philosophical city.
> As mine I am nothing dangerous to impart,
> So better to receive I am ready with all my heart.

14. MacArthur and Wilson, *Theory of Island Biogeography*; and Quammen, *Song of the Dodo*.

15. Some islands are so "isolated" that no one has ever been able to set foot on them. Among these are Peter I Island in Antarctica (Norway). See Judith Schalansky, *Atlas of Remote Islands: Fifty Islands I Have Not Visited and Never Will*, trans. Christine Lo (London: Particular Books, 2010), s.v. "Peter I Island."

16. See Rebecca Maura Lemov, *World as Laboratory: Experiments with Mice, Mazes and Men* (New York: Hill & Wang, 2005).

17. See Vanessa Smith, "Pitcairn's 'Guilty Stock': The Island as Breeding Ground," in *Islands in History and Representation*, ed. Rod Edmond and Vanessa Smith (London: Routledge, 2003).

18. Mary Russell Mitford, *Christina*; cited in Smith, "Pitcairn's 'Guilty Stock,'" 127.

19. Schalansky, *Atlas of Remote Islands*.

20. Consider the case of the four-island archipelago of Socotra in Yemen, which is continental (i.e., nonvolcanic) while being also one of the islands that is most isolated from any mainland. A third of its plant life—some 307 species—are found nowhere else on Earth. (Environmentalists still know very little about its fog oasis in the Haggier Mountains.) Socotra's language is Soqotri, which differs significantly from the dialectal Arabic that islanders also speak; it is related to such discrete Arabian Peninsula languages as Mehri and Bathari.

21. Relevant here is the still mostly (and unfairly) discredited German geographer Friedrich Ratzel. Ratzel wrote about *Lebensraum* (living space) and *Realpolitik* in the wake of Charles Darwin and the German zoologist Ernst Heinrich Haeckel.

22. The title in Italian is *Travolti da un insolito destino nell'azzurro mare d'agosto.*

23. Antonioni returned to Lisca Bianca in 1983 to make the movie *Ritorno a Lisca Bianca* (1983), a ten-minute short.

24. In mainland Europe, imitations of *Robinson Crusoe* abound, among them Johann Gottfried Schnabel's *Die Insel Felsenberg* (1731; republished in abridged version by Ludwig Tieck in 1828), about which Arno Schmidt writes: "It is attested that around and after 1750 the library of a commoner consisted of at least two volumes: the Bible and the *Insel Felsenberg.*"

25. The first known novels to be set on a deserted island were *Philosophus Autodidactus,* written by Ibn Tufail (1105–1185), and *Theologus Autodidactus,* written by Ibn al-Nafis (1213–1288). The protagonists are feral children living in seclusion on a deserted island until they eventually encounter castaways from the outside world who are stranded on the island.

26. In this same vein, some language scientists are often struck in unexpected ways, as with the language of Rapa Iti (the Austral Islands in French Polynesia). Some reports have it that Marc Liblin "learned" this language entirely "on his own"—from dreams in the Vosges region of mainland France.

27. Christine Marion Fraser, *Rhanna at War* (London: Blond & Briggs, 1980), 3; cited in Manfred Malzahn, "Aspects of Identity: The Contemporary Scottish Novel (1978–1981) as National Self-Expression," PhD diss., University of Wuppertal, Wuppertal, Germany, 1983, 42.

28. The work is reported by Diodorus Siculus and Eusebius of Caesarea.

29. Felix Jacoby, *Fragmente der griechischen Historiker,* 115, F7.

30. See James S. Romm, *The Edges of the Earth in Ancient Thought: Geography, Exploration, and Fiction* (Princeton, NJ: Princeton University Press, 1992).

31. Diskin Clay and Andrea Purvis, preface to *Plato's Atlantis, Euhemeros of Messene's Panchaia; Iambulos' Island of the Sun; Sir Francis Bacon's New Atlantis* (Newburyport, MA: R. Pullins, 1999).

32. Heinz Günther Nesselrath, "Theopomps Meropis und Platon: Nachahmung und Parodie," *Göttinger Forum für Altertumswissenschaft* 1 (1998): 1–8.

33. "Le mieux est l'ennemi du bien." Voltaire, *La Bégueule: Conte moral* (1752).

34. *Qui tacit consentire videtur.* On this maxim, see *Latin for Lawyers* (New York: Lawbook Exchange, 1992), 230.

35. "On December 2, 1942, in an abandoned squash court at the University of Chicago, [Enrico] Fermi consummated the first man-made [atomic] chain reaction. It was called the 'K-Factor'; though among the scientists it was known as the 'Great God K.'" Mark Albertson, "Road to Hiroshima Began in Hitler's Germany," *Hour Newspaper* [Norwalk, CT], August 5, 1995.

36. See Norbert Aping, *The Final Film of Laurel and Hardy: A Study of the Chaotic Making and Marketing of* Atoll K (Jefferson, NC: McFarland, 2008), 16.

37. Compare Ezekiel 3:15: "I came to the exiles who lived at Tel Aviv." Nahum Skolow was the translator.

38. The same statement is recalled in the epilogue. Theodor Herzl, *Altneuland, Roman* (Leipzig: Hermann Seemann Nachfolger, 1902); *Tel Aviv, Sippur* (Hebrew), trans. Nahum Sokolov (Warsaw: Hatzefirah, 1902); *Altneuland* (Yiddish), trans. Isadore Eliashov (Warsaw: Hatzefirah, 1902); and "OldNewLand," *The Maccabean,* 1902.

39. Mordecai Manuel Noah tried to found a Jewish homeland on Grand Island. See his *Discourse on the Restoration of the Jews* (1824); and Selig Adler and Thomas E. Connolly, *From Ararat to Suburbia: The History of the Jewish Community of Buffalo* (Philadelphia: Jewish Publication Society of America, 1960). From the realm of literature: Ben Katchor's graphic novel *The Jew of New York* (New York: Pantheon, 1999). See *Slattery Report: The Problem of Alaskan Development* (Washington, DC: U.S. Department of the Interior, 1939) for the presentation of the plan to establish a Jewish refuge on Baranof Island (Sitka Island) in Alaska. From the realm of literature, see Michael Chabon's "alternate history" novel, *The Yiddish Policeman's Union: A Novel* (New York: Harper Perennial, 2007).

40. More includes the following poem:

> Me Utopie cleped Antiquity,
> Void of haunt and herborough,
> Now am I like to Plato's city,
> Whose fame flieth the world thorough;
> Yea, like, or rather more likely Plato's plat to excel and pass.
> For what Plato's pen hath platted briefly
> In naked words, as in a glass,
> The same have I performed fully,
> With laws, with men, and treasure fitly.
> Wherefore not Utopie, but rather rightly
> My name is Eutopie: a place of felicity.

41. Veronica della Dora, "Mapping a Holy Quasi-Island: Mount Athos in Early Renaissance *Isolarii,*" *Imago Mundi* 60, no. 2 (2008): 153.

42. Ernst Bloch, *The Principle of Hope,* trans. Neville Plaice, Stephen Plaice, and Paul Knight (Cambridge, MA: MIT Press, 1986), 2:517.

43. Plato, *Timaeus,* trans. Benjamin Jowett.

44. On Ambrosius Holbein's *Memento Mori* map for the third edition (Johann Froben) of *Utopia,* see Malcolm Bishop's essay in *British Dental Journal* 199 (2005): 107–112, which focuses mainly on the "ship of teeth" aspect of the skull here represented.

45. See Laurent Gervereau, "Symbolic Collapse: Utopia Challenged by Its Representations," in *Utopia: The Search for the Ideal Society in the Western World,* ed. Roland Schaer, Gregory Claeys, and Lyman Tower Sargent (New York: New York Public Library, 2000), 358.

46. For an exposition of the link between the cartographic artwork of Holbein and Martellus's *Insularium Illustratum,* see Evangelos Livieratos and Maria Parzale, "Ou topos é topos Orous?" in *Orous Athó gés thalasés perimetron khartón metamorpheis,* ed. Evangelos Livieratos (Thessalonica: Ethniké Khartothéke, 2003), 246.

47. See Giancarlo Motta, Antonia Pizzigoni, and Carlo Ravagnati, *L'architettura delle acque e della terra* (Milan: F. Angeli, 2006), esp. 85; and S. Gély, "L'île comme concrétisation ambivalent de l'utopie, de Platon à Thomas More," in *Les îles, du mythe à la réalité,* ed. Monique Pelletier (Paris: Éditions du CTHS, 2002).

48. The play was first printed in 1844; it was re-edited from Harleian MS 7368 in the British Museum.

49. Samuel Butler's *Erewhon* (1872) uses a similar plot as a vehicle for Swiftian social satire. Another early example is Simon Tyssot de Patot's *Voyages et Aventures de Jacques*

Masse (1710), which includes prehistoric fauna and flora. There is also Robert Paltock's *Life and Adventures of Peter Wilkins* (1751), an imaginary voyage where a man named Peter Wilkins discovers a race of winged people on an isolated island (as in Edgar Rice Burroughs's *Caspak*).

50. Jonathan Swift, *Prose Works*, ed. Herbert Davis (1736; Oxford: Basil Blackwell, 1959), 13:123.

51. On Swift's fictional Lilliput Island, the inhabitants are six inches tall. On the peninsula of Brobdingnag, the inhabitants are sixty feet tall. Laputa is a flying island of strange scientists. On Luggnagg, the people live forever in decrepit bodies. The king of Laputa controls the "mainland" by threatening to cover rebel regions with the island's shadow (blocking sunlight and rain), throwing rocks at rebellious surface cities (aerial bombardment), or lowering the island onto the cities below in order to crush them. On the island of the Houyhnhnms is a race of intelligent horses, but the place is also home to the depraved Yahoos. When Gulliver returns home to England, all people there remind him of the Yahoos.

52. Burton J. Fishman, "Defoe, Herman Moll, and the Geography of South America," *Huntington Library Quarterly* 36, no. 3 (1973): 227–238.

53. Arnold Bennett, *The Bright Island* (London: Chatto & Windus, 1924).

54. *Las sergas de Esplandián* was published in Seville, Spain. For a translation of a portion of the original story, see Edward Everett Hale, "The Queen of California," *Atlantic Monthly*, March 1864, 266.

55. Dennis Reinhartz, *The Cartographer and the Literati: Herman Moll and His Intellectual Circle* (Lewiston, NY: E. Mellen, 1997).

56. The map by Moll showing California as an island is sometimes also referred to as *To the Right Honourable John Lord Sommers*. See Glen McLaughlin, with Nancy H. Mayo, *The Mapping of California as an Island: An Illustrated Checklist* (Saratoga, CA: California Map Society, 1995).

CHAPTER 8

1. "*Euripus* has . . . become a general name for all streights where the water is in great motion or agitation." Ephraim Chambers, *Cyclopaedia* (1751).

2. Diodorus of Sicily reports that the Boeotians agreed to help the Euboeans in the construction of a bridge over the strait because "it was to their special advantage that Euboea should be an island to everybody else but a part of the mainland to themselves." Christy Constantakopoulou, *The Dance of the Islands: Insularity, Networks, the Athenian Empire, and the Aegean World* (Oxford: Oxford University Press, 2007), 15.

3. "The sea-battle [at Euripus] which followed was much like the fighting at Thermopylae; for the Persians were resolved to overwhelm the Greeks and force their way through the Euripus, while the Greeks, blocking the narrows, were fighting to preserve their allies in Euboea." Diodorus Siculus, *Library of History*, trans. Charles Henry Oldfather, Loeb Classical Library 375 (Cambridge, MA: Harvard University Press), bk. 11.13.

4. "The moaning straits," in Sophocles, *Antigone*, 1265.

5. Herodotus, *Inquiries*, 8.16.

6. James Morwood, "A Note on the Euripus in Euripides' *Iphigenia at Aulis*," *Classical Quarterly (New Series)* 51, no. 2 (2001): 607–608.

7. Pindar, *Pythian*, ed. Diane Arnson Svarlien, bk. P, poem 11.

8. According to some alternate versions of the story, Iphigenia was spirited off to the Greek colony of Tauris. This was, some say, an island in the Black Sea linked to the mainland by the narrow Isthmus of Perekop (meaning "transit").

9. Euripides, *The Plays of Euripides*, trans. E. P. Coleridge (London: George Bell & Sons, 1891), vol. 2.

10. Simonides himself was from the island of Amorgos in the Cyclades.

11. See *The Contest of Homer and Hesiod* [Certamen Homeri et Hesiodi], 12.

12. Hesiod, *Works and Days*, 650–659.

13. Richard Francis Burton, *Two Trips to Gorilla Land and the Cataracts of the Congo* (London: S. Low, Marston, Low, & Searle, 1876), 2:15.

14. One etymology has it that *Hellespont* (originally) meant "narrow Pontus" or "entrance to Pontus." *Pontus* referred to the region around the Black Sea and hence also to the sea itself. See George Rippey Stewart, *Names on the Globe* (New York: Oxford University Press, 1975).

15. The same oddity exists for the similarly brackish and meromictic Øresund, which connects the interior Baltic Sea to the North Sea and hence the Atlantic Ocean by way of Elsinore. The Øresund's brackish surface layer runs northward into the North Sea; its more saline subsurface layer runs southward into the Baltic. On the Øresund, see Chapters 9 and 10.

16. Hermann Diels and Walther Kranz, eds., *Fragmente der Vorsokratiker*, 5th ed. (Berlin: Weidmannsche Buchhandlung, 1934), 22B60. See Marc Shell, *The Economy of Literature* (Baltimore: Johns Hopkins University Press, 1978), chap. 1.

17. Herodotus, *Histories*, 7.22–24.

18. Traces of the old canal, once thought to be mythical, have been found at the site. See B. S. J. Isserlin, R. E. Jones, S. Papamarinopoulos, and J. Uren, "The Canal of Xerxes on the Mount Athos Peninsula," *Annual of the British School at Athens* 89 (1994): 277–284.

19. Constantakopoulou, *Dance of the Islands*, 26, 219.

20. Irwin L. Merker, "The Ptolemaic Officials and the League of the Islanders," *Historia* 19, no. 2 (1970): 141–160. See also Christy Constantakopoulou, "Identity and Resistance: The Islanders' League, the Aegean Islands and the Hellenistic Kings," *Mediterranean Historical Review* 27, no. 1 (2012): 51–72; and Pierre Roussel, "La confédération de Nésiotes," *Bulletin de Correspondance Hellénique* 35 (1911), 441–455.

21. "These leagues, known as Amphictyonies, were not political alliances." George Rawlinson, *A Manual of Ancient History* (Oxford: Clarendon Press, 1869), 122.

22. See Luisa Breglia, "The Amphictyony of Calaureia," *Ancient World* 36 (2005) [Studies in Honor of John M. Fossey]: 18–33; and Ernst Curtius, "Der Seebund von Kalaureia," *Hermes* 10 (1876): 385–392. See also Strabo, *Geography*, 8.6.14.

23. Paola Ceccarelli, "Sans thalassocratie, pas de démocratie: La rapport entre thalassocratie et démocratie à Athènes dans la discussion du Vᵉ et IVᵉ siècle," *Historia* 42 (1993): 444–470.

24. Constantakopoulou, *Dance of the Islands*, 91.

25. Rudolf Kassel and Colin Austin, *Poetae Comici Graeci* (Berlin: de Gruyter, 1983), F402–F414.

26. Kassel and Austin, *Poetae*, F19–F26; cited in Constantakopoulou, *Dance of the Islands*, 79.

27. On Aretades, see Plutarch, *Parallel Lives*, 11, 27.

28. The genre of *nēsiōtika* was well established in antiquity. See Paola Ceccarelli, "Nesiotika," *Annali della Scuola Normale Superiore di Pisa* 19, no. 3 (1989): 903–935.

29. Constantakopoulou, *Dance of the Islands*, 174.

30. See Raoul Baladié, *Le Péloponnèse de Strabon: Étude de géographie historique* (Paris: Société d'Édition "Les Belles Lettres," 1989).

31. Émile Yerahmiel Kolodny, *La population des îles de la Grèce: Essai de géographie insulaire en Méditerranée orientale*, 3 vols. (Aix-en-Provence: Edisud, 1974).

32. "The How [presumably the Hoo Peninsula in southeast England], whiche is not an Ilande . . . but (if I may giue such peeces a new name) a bylande, bycause we may passe thyther from the maine Isle, by an Isthmus." William Harrison, *An Historicall Description of the Islande of Britayne* (London, 1577), 1.8, f. 12/1, in Raphael Holinshed, *Chronicles of England, Scotlande, and Irelande*, 2 vols. (1577).

33. "[The river] Tamer [Tamar] . . . leaveth Cornwall, as it were a Peninsula, or Byland." Tristram Risdon, *The Chorographical Survey of . . . Devon* ([1630], 1714), 2:302.

34. While some scholars have suggested that *hoo* derives from Old Norse *haugr* (burial mound)—one among them W. G. Arnot in *The Place Names of the Deben Valley Parishes* (Ipswich, UK: Adlard, 1946), 70—others argue for Old English *hóh* or Middle English *howe* (projecting ridge of land)—for example, Allen Mawer in *The Chief Elements Used in English Place-Names* (Cambridge: English Place-Name Society, 1930), 1:2. See also Judith Glover, *The Place Names of Kent* (London: B. T. Batsford, 1976) for a discussion of the Old English *ho*, meaning "spur of land."

35. Homer, *Odyssey*, bks. 11.128 and 23.

36. Pirkei Avot 1:1.

37. Borca, *Terra Mari Cincta: Insularità e cultura romana* (Rome: Carocci, 2000), 92.

38. See François Doumenge, "Les îles et micro-états insulaires," *Hérodote* 37–38 (1985): 297–327.

39. On Aretades, see Plutarch, *Parallel Lives*, 11, 27.

40. Herodotus, *Inquiries*, 1.174.

41. Pindar, *Second Olympian*, 70–77.

42. "[E]n premier lieu, l'action de fortifier un isthme et celle de creuser sont mises sur un même plan, ce qui marque une évolution vers l'abstraction dans le concept d'île." Paola Ceccarelli, "De la Sardaigne à Naxos: Le rôle des îles dans les *Histoires d'Herodote*," in *Impression d'îles*, ed. F. Létoublon (Toulouse: Presses Universitaires du Mirail, 1996), 43–44.

43. See also Frederic James, "Of Islands and Trenches: Naturalization and the Production of Utopian Discourse," *Diacritics* 7 (1977): 2–12.

44. Three centuries after Periander, in 307 BC, Demetrius Poliorcetes made up his mind to cut a naval passage through the isthmus. He actually began excavations before he was convinced not to continue by Egyptian engineers, who predicted that the different sea levels between the Corinthian and Saronic gulfs would cause Aegina and nearby islands to be inundated.

45. D. K. Pettegrew, "The Diolkos and the Emporion: How a Land Bridge Framed the Commercial Economy of Roman Corinth," in *Corinth in Contrast: Studies in Inequality*, ed. S. J. Friesen, S. James, and D. N. Schowalter (Leiden: Brill, 2002). See also

D. K. Pettegrew, *Corinth on the Isthmus: Studies of the End of an Ancient Landscape*, PhD diss., Ohio State University, 2006.

46. On the railway at Corinth, see Michael Jonathan Taunton Lewis, "Railways in the Greek and Roman World," in *Early Railways. A Selection of Papers from the First International Early Railways Conference*, ed. Andy Guy and Jim Rees (London: Newcomen Society, 2001), 8–19. On the *diolkos* being better called a "trackway," see Walter Werner, "The Largest Ship Trackway in Ancient Times: The Diolkos of the Isthmus of Corinth, Greece, and Early Attempts to Build a Canal," *International Journal of Nautical Archaeology* 26 (1997): 98–119.

47. Thucydides, *History of the Peloponnesian War*, 1.93.

48. Ibid.

49. Herodotus, *Inquiries*, 8.61.

50. Constantakopoulou, *Dance of the Islands*, 141, referring to Herodotus, *Inquiries*, 8.61.

51. Ibid., 151.

52. Marcel Piérart, "If Athens Were an Island," *Studia Europaea Gnesnensia* 4 (2011): 135–151.

53. In the *Quarterly Review* (April 1905), 352. Murray here reviews Victor Bérard's book *Les Phéniciens et l'Odyssée* (Paris: Armand Colin, 1902–1903).

54. John R. Hale, *Lords of the Sea: The Epic Story of the Athenian Navy and the Birth of Democracy* (New York: Viking, 2009), 104.

55. Constantakopoulou, *Dance of the Islands*, 143; see *Cimon*, 13.16.

56. Donald Kagan, *The Peloponnesian War* (New York: Viking, 2003).

57. Hale, *Lords of the Sea*, 105.

58. David H. Conwell, *Connecting a City to the Sea: The History of the Athenian Long Walls* (Leiden: Brill, 2008).

59. Old Oligarch (Pseudo-Xenophon), *Constitution of the Athenians*, in *Xenophon in Seven Volumes*, trans. Edgar Cardew Marchant (Cambridge, MA: Harvard University Press, 1968), 2.16.

60. Hale, *Lords of the Sea*, 146.

61. Thucydides, *History of the Peloponnesian War*, 1.143, quoting Pericles speaking to the Athenian citizens.

62. Old Oligarch, *Constitution of the Athenians*, 2.15 (translation modified).

63. Ibid., 2.16.

64. Francis Hartog, *The Mirror of Herodotus: An Essay on the Representation of "the Other"* (Berkeley: University of California Press, 1988), 202ff.; and Joseph Nicholas Jansen, "After Empire: Xenophon's 'Poroi' and the Reorientation of Athens' Political Economy," PhD diss., University of Texas, 2007, 232n.

65. Josette Elayi and A. G. Elayi, *Le monnayage de la cité phénicienne de Sidon à l'époque perse (Ve–IVe siècles avant J.-C.)*, 2 vols. (Paris: Gabalda, 2004).

66. *A Guide to the Principal Coins of the Greeks*, 2nd ed. (London: British Museum, 1932), plates 7 and 51.

67. In the Pergamon Museum, Berlin.

68. See, for example, the silver legionary denarius of Mark Antony minted in Patrae in 32 BC; listed in David R. Sear, *Roman Coins and Their Values*, 4th ed. (London: Seaby, 1988), BW reference 004 043 147.

69. Herodotus, *Inquiries*, 2.97.1.

70. Ibid., 9.51.1–2.

71. Whether or not geology and hydrology "underpin" philosophy was an important question in the eighteenth century. During the Enlightenment, Neptunists like Abraham Werner sparred with Plutonists like James Hutton. The Euripian flux seemed to much bother Aristotle.

72. "What kind of rheological substance is a turbulent fluid?" Gerald V. Middleton and Peter R. Wilcock, *Mechanics in the Earth and Environmental Sciences* (Cambridge: Cambridge University Press, 1994), 393.

73. Plato, *Phaedo*, 90c.

74. The aphorism is from Simplicius of Cilicia, as cited in Jonathan Barnes, *The Presocratic Philosophers*, rev. ed. (London: Routledge / Taylor & Francis, 1982), 65. See the similar reference in Simplicius's commentary on Aristotle's *Physics*.

75. Aristotle, *Meteorologica*, 2.8.

76. Aristotle, *Nicomachean Ethics*, trans. W. D. Ross, 9.6:
Now such unanimity is found among good men; for they are unanimous both in themselves and with one another, being, so to say, of one mind (for the wishes of such men are constant and not at the mercy of opposing currents like a strait of the sea [literally, Euripus]), and they wish for what is just and what is advantageous, and these are the objects of their common endeavour as well. But bad men cannot be unanimous except to a small extent, any more than they can be friends, since they aim at getting more than their share of advantages, while in labour and public service they fall short of their share; and each man wishing for advantage to himself criticizes his neighbour and stands in his way; for if people do not watch it carefully the common weal is soon destroyed. The result is that they are in a state of faction, putting compulsion on each other but unwilling themselves to do what is just.

77. Anton-Hermann Chroust, "The Myth of Aristotle's Suicide," *Modern Schoolman* 44 (1967): 177–178.

78. Thomas Browne, *Pseudodoxia Epidemica* (1646), chap. 13. Browne continues:
Wherein, because we perceive men have but an imperfect knowledge, some conceiving Euripus to be a River, others not knowing where or in what part to place it; we first advertise, it generally signifieth any strait, fret, or channel of the Sea, running betweene two shoars, as Julius Pollux hath defined it; as we read of Euripus Hellespontiacus, Pyrrhaeus, and this whereof we treat, Euripus Euboicus or Chalcidicus, that is, a narrow passage of Sea dividing Attica, and the Island of Euboea, now called Golfo de Negroponte, from the name of the Island and chief City thereof; famous in the wars of Antiochus, and taken from the Venetians by Mahomet the Great.

79. Collected in Ingemar Düring, *Aristotle in the Ancient Biographical Tradition*, Studia Graeca et Latina Gothoburgensia 5 (Stockholm: Almqvist & Wiksell, 1957), sec. 48.

80. Omitted here are these words:
. . . (for if that were the case, the Strait of Sicily would not be changing its current twice a day, as Eratosthenes says it does, but the strait of Chalcis seven times a day, while the strait at Byzantium makes no change at all but continues to have its outflow only from the Pontus into the Propontis, and, as Hipparchus reports, even stands still sometimes). . . .

81. Strabo, *Geography*, 1.3.12. The first critical edition of his *Geographica* was published in 1587.

82. Hven was part of Denmark until May 6, 1658, when Sweden sent troops to seize the island. It has been Swedish ever since.

83. Quoted in Charles Hutton, *A Philosophical and Mathematical Dictionary* (London, 1815), 2:515.

CHAPTER 9

1. Quoted by Pappus of Alexandria, *Synagoge* [Collection], bk. 7.

2. Claudian [Claudius Claudianus], *Carmina Minora*, trans. M. Platnauer, Loeb Classical Library 136 (Cambridge, MA: Harvard University Press, 1922), 279–281. G. Wilhelm Leibniz quotes these lines in *Theodicy* (1710), ed. Austin M. Farrer, trans. E. M. Huggard (London: Routledge & Kegan Paul, 1951), 216.

3. Shakespeare, sonnet 55, lines 1–2.

4. *OED*, s.v. "distracted," adj., 1.

5. In Genesis 1:6, the King James Bible has it this way: "And God said, Let there be a firmament in the midst of the waters, and let it divide the waters from the waters." The translators seem to prefer the Latinate English-language term *division* (*duyde*) to the Norse English-language term *sound* (or *sunder*). The Hebrew term is מבדיל.

6. http://en.wikipedia.org/wiki/File:Denmark_location_sjalland.svg. "In Norse mythology, the island was created by the goddess Gefjun after she tricked Gylfi, the king of Sweden, as told in the story of Gylfaginning. She removed a piece of land and transported it to Denmark, and it became the island of Zealand." Gefjun is the "mother" of Norway, Sweden, and Denmark in *Gefion, a Poem in Four Cantos* by Eleonora Charlotta d'Albedyhll (1770–1835). A fountain depicting Gefjun driving her oxen sons to pull her plough, sculpted by Anders Bundgaard in 1908, stands in Copenhagen. In Snorri Sturluson's thirteenth-century *Ynglinga Saga* (the first part of his *Heimskringla*), trans. Samuel Laing, sec. 5, one reads the following:

> Gefion from Gylve drove away,
> To add new land to Denmark's sway—
> Blythe Gefion ploughing in the smoke
> That steamed up from her oxen-yoke:
> Four heads, eight forehead stars had they,
> Bright gleaming, as she ploughed away;
> Dragging new lands from the deep main
> To join them to the sweet isle's plain.

The Scandinavian epic *Heimskringla* [World Circle] thus reports that "the earth's circle which the human race inhabits is torn across into many bights [coastline indentations], so that great seas run into the land from the out-ocean." The Danish translation of this work was made in 1660.

7. William Caxton, England's first printer, uses the term *distract/district* in his influential 1480 translation of Ovid's *Metamorphoses*, writing that "[the winds] remysed us in to the cruel dystraytis of Eolus" (14.6). See *OED*, s.vv. "distrait," 1a; "destrayt."

8. James Wilson Bright, MLA Presidential Address (1902), xli–xlxii.

9. *OED*, s.v. "continent," n., 2.4.b.

10. *OED*, s.v. "continent," n., 2.4.a.

11. Grímur Jónsson Thorkelín, ed., *De Danorum Rebus Gestis Secul. III & IV: Poëma Danicum Dialecto Anglo-Saxonica* (1815).

12. In his "Greetings to the Reader," with which he prefaced his 1815 edition of *Beowulf*, we read: "I came home [from England] with great success and rich reward, and with me a poem that had been absent for more than a thousand years returned to its country of origin." Quoted in Robert Bjork, "Grímur Jónsson Thorkelín's Preface to the First Edition of *Beowulf*, 1815," *Scandinavian Studies* 68 (1996): 291.

13. For further discussion, see Bjork, "Thorkelin's Preface," 293; see also T. A. Shippey and Andreas Haarder, eds., *Beowulf: The Critical Heritage* (London: Routledge, 1998), 10–11.

14. See Chapter 10.

15. "Beowulf's Contest with Breca," lines 505–580. The phrase *on . . . ymb sund* can be understood to refer to swimming or rowing. In *Beowulf*, *sund* and its variants can refer to the sea or ocean strait (e.g., lines 213 and 223), swimming (e.g., line 1436), being safe or sound (e.g., lines 1628 and 1998), a commotion of water (e.g., line 1450), being asunder (e.g., line 2422), and a ship (line 1906).

16. Already in 1562, John Shute made the etymologically based alignment of *district* (region) with *distrait* (narrow passage of water). "If this distraite of ye land were cut through, Peloponesso shold be an isle." Shute, *Two Very Notable Commentaries: "The Turcks and Empire of the House of Ottomanno" and "The Warres of the Turcke against George Scanderbeg, Prince of Epiro,"* trans. Andrea Cambini (London, 1562), 7b. See also *OED*, s.v. "distrait," b. Compare the Franco-American urban toponym *Detroit*.

17. Saxo's translation was lost by fire in 1728.

18. Christiern Pedersen, *Richt Vay to the Kingdome of Heuine*, trans. John Gau (Malmö: Hochstraten, 1533). At the time of publication, Malmö was located in then-Danish Scania.

19. When Anne of Denmark married King James VI in Elsinore (1589), Vedel was there and impressed Queen Sophie of Denmark so much that she supported the later publication of Vedel's collection of Danish folk songs, *Hundredvisebogen* (1591), which is still an important source-base of understanding the history of Danish literature.

20. Scholars often note the likeness of his name to Shakespeare's "Rosencrantz."

21. In that decade two authors produced relevant works. One was the Dutchman John Isaac Pontanus, whose family was from Haarlem, with its famous dike, and who had studied with Tycho Brahe on Hven Island. His *Rerum Danicarum Historia* (Amsterdam, 1631) contains a section titled "Chorographica Descriptio" about the water topography of Denmark, including its islands, along the lines that Vedel earlier projected. The second author was the Dutchman Johannes Meursius, who took a professorship on Zealand Island in Denmark in 1625 and published *Historia Danica* (Copenhagen, 1630).

22. Edward Gordon Craig—actor, director, set designer, and son of the actress Ellen Terry and the architect Edward Godwin—was also an influential author on the theater and, at the invitation of Constantin Stanislavski, the director of the famous 1911 production of *Hamlet* at the Moscow Theatre. He writes: "*Hamlet* has not the nature of a stage representation. *Hamlet* and the other plays of Shakespeare have so vast and so complete a form when read, that they can but lose heavily when presented to us after having undergone stage treatment. That they were acted in Shakespeare's day proves nothing." Craig, "First Dialogue" [in which a stage director and a playgoer are conversing], *On the Art of the Theater*; quoted in C. D. Innes, *Edward Gordon Craig: A Vision of Theatre* (Amsterdam: Overseas

Publication Association, 1998), 246. The plays have adapted well to such twentieth-century cinematographic presentation as we discuss in *Islandology*.

23. Saxo Grammaticus, *Gesta Danorum*, bks. 3 and 4.

24. Ibid., bk. 5.

25. Saxo Grammaticus, *The Nine Books of the Danish History of Saxo Grammaticus*, ed. Rasmus B. Anderson and J. W. Buel, trans. Oliver Elton (New York: Norroena Society, 1905), opening passage. I refer also to the translation by the Count de Falbe in Chapter 10. See also Geoffrey Bullough, ed., *Narrative and Dramatic Sources* (London: Routledge & Kegan Paul, 1973); Hilda Ellis Davidson, ed., Peter Fisher, trans., *Saxo Grammaticus: The History of the Danes, Books I–IX* (Cambridge: D. S. Brewer, 1979); and William F. Hansen, *Saxo Grammaticus and the Life of Hamlet: A Translation, History, and Commentary* (Lincoln: University of Nebraska Press, 1983). Also useful here is Eric Christiansen, *Saxo Grammaticus: Danorum Regum Heroumque Historia, Books X–XVI*, BAR International Series 84 (Oxford: British Archaeological Reports, 1980–1981).

26. Ibid.

27. "How long will a man lie i' the earth ere he rot?" wonders Hamlet. On this role of island settings in Norse literature, see Kristel Zilmer, "The Powers and Purposes of an Insular Setting: On Some Motifs in Old-Norse Literature," in *Isolated Islands in Medieval Nature, Culture and Mind*, ed. Gerhard Jaritz and Torstein Jørgensen (Budapest: Central European University Press, 2011).

28. On the notion of "rim of the world," especially in medieval times, see Felicitas Schmeider, "Paradise Islands in the East and West: Tradition and Meaning in Some Cartographical Places on the Medieval Rim of the World," and Eldar Heide, "Holy Islands and the Otherworld: Places beyond Water," both in Jaritz and Jørgensen, *Isolated Islands*.

29. Belleforest's English translator (London, 1608) spells *iland* as *island*. For the significance, in terms of these spellings, for understanding the tension, conveyed here by the history of the English language, between "land as water" and "land versus water," see the *OED* quotation in note 43 of Chapter 1.

30. Saxo Grammaticus, *Nine Books*, opening passage.

31. That would be the Schlei Estuary, which has Hedeby at its head. This town developed as a trading center thanks to the fact that there was a short portage there to the Treene River, which flows into the Eider with its North Sea estuary. One was thus able to avoid the Jutland pirates and shallows by taking the Hedeby route.

32. Laurence Marcellus Larson, *Canute the Great: 995 (circ)–1035 and the Rise of Danish Imperialism during the Viking Age* (New York: Putnam, 1912), 257; and "North Sea Empire," http://en.wikipedia.org/wiki/North_Sea_Empire.

33. This Thorney Island (near Bosham) should be distinguished from the other Thorney Island (in the Thames River).

34. Diana Greenway, ed. and trans., *Henry Archdeacon of Huntingdon: Historia Anglorum (History of the English People)*, Oxford Medieval Texts (Oxford: Oxford University Press, 1996), 366–369.

35. When mere kings set their hands to this sort of thing, they either failed outright or employed cheap trickery in order to seem to succeed. King Maelgwn Gwynedd, in the sixth century, convened a group of kings at the estuary of the River Dyfi to see which of them was able to stay on his throne the longest when the tide came up; Maelgwn won by

employing a tricky "flotation device" made of bird feathers. John Rhys, *Celtic Britain*, 2nd ed. (London: Society for Promoting Christian Knowledge, 1884), 125. There is also the tale about Essarg, also known as Tuirbe Tramár, who stops the tide by throwing an ax at it. "Tuirbe's strand [in Brittany], whence was it named? Not hard to say. / Tuirbe Trágmar, father of the Gobbán Saer, 'tis he that / owned it. 'Tis from that heritage he used to hurl a cast of his / axe, from *Tulach in Bela* [the Hill of the Axe] in the face / of the flood-tide, so that he forbade the sea, and it would not come over the axe." Whitley Stokes, ed., "The Prose Tales in the Rennes *Dindshenchas*," *Revue Celtique* 16 (1895): 31–83.

36. "Island of Thorns." The Thames River *eyot* called "Thorney" is not the same as the Thorney Island near Bosham.

37. Illustrated by Arnold Bennett (London: Chatto & Windus, 1918), 22.

38. Canute the Great should be distinguished from Saint Canute IV, king of Denmark from 1080 to 1086, who considered that he was the rightful king of England and that the Norman William the Conqueror, who successfully invaded Britain in 1066, was a usurper.

39. *Kalmar* names the town and strait where the Kalmar Union participants— Denmark, Sweden, Finland, and Norway, as well as Iceland, Greenland, the Faroe Islands, the Shetland Islands, and the Orkney Islands—signed the agreement. The relatively un-friendly and independentist Sweden had dropped out earlier in the sixteenth century. The hostility of Sweden to Denmark's control of the Øresund grew to open warfare during the Kalmar War of 1611. This involved Sweden's unwillingness to pay the Sound Dues for its use of the Øresund. That same year was the last year for Shakespeare the dramatist: he completed *The Tempest*, and that was that.

40. The tale is told (in Latin) by Johann von Posilge (ca. 1340–1405) of Pomerania.

41. The Polish province "Royal Prussia" was part of the Kingdom of Poland beginning in 1466. From 1569 to 1772, it was part of the Lithuanian-Polish Commonwealth.

42. Johann von Posilge reports in *Chronik des Landes Preussen* that, around 1400, [a] poor sick man came to the country and stayed near the village of Grudziadz [in "Poland"; in Shakespeare's day, this town was in the Polish province called "Royal Prussia," a region of the Kingdom of Poland beginning in 1466 and then, from 1569 to 1772, part of the Lithuanian-Polish Commonwealth]. A group of merchants from Denmark asked him if he was not well-known in Denmark since he looked very much like the late King Olaf. The merchants left to find another [person] who had seen the king and returned with him. When the newcomer saw the [person whom they took for Olaf], [the newcomer] cried out, "My lord king!" Many people especially in Norway didn't believe that Olaf had died. They thought Queen Margaret had poi-soned young Olaf to get him out of the way, so she could rule. According to the rumors, young Olaf hid himself and escaped. The news reached a merchant, Tyme von der Nelow, who took the man to Gdansk. The high born of the town welcomed Olaf as the rightful King of Denmark and Norway and gave him fine clothes and presents.

43. The Danish National Council released a detailed explanation of the real Olaf's death in 1387 in order to contradict the rumor.

44. See Marc Shell, *Children of the Earth: Literature, Politics and Nationhood* (Oxford: Oxford University Press, 1993), chap. 5.

45. Eric had close ties with Poland: his father—Wartislaw VII, Duke of Pomerania— had pledged vassalage to Wartislaw Jagiello, the king of Poland.

46. In *Henry V*, Shakespeare focuses on Hal's marriage with Catherine of Valois.

47. King James married Anne first by proxy in 1589 and then in person in 1590.

48. Ethel Carleton Williams, *Anne of Denmark* (London: Longman, 1970), 10.

49. She attended Shakespeare's *Love's Labour's Lost* and organized masques at court with Ben Jonson.

50. The title as it appears in *Hamlet*, Second Quarto (1604).

51. Many lake islands on the island of Britain are called that, and so too are some of its offshore islands. Many such islands are called that in the Lake District. See Tom Holman, *Lake District Miscellany* (London: Frances Lincoln, 2007), 66.

52. Paul Henri Mallet, *Manners, Customs, Religion and Laws, Maritime Expeditions and Discoveries, Language and Literature of the Ancient Scandinavians (Danes, Swedes, Norwegians, and Icelanders). With Incidental Notices Respecting our Saxon Ancestors. With a translation of the Prose Edda from the original Old Norse Text; and notes critical and explanatory by I. A. Blackwell, Esq. To which is added an Abstract of the Eyrbyggja Saga by Sir Walter Scott*, trans. Thomas Percy (London, 1847).

53. British-English toponymy includes many examples. In Scotland, there is *Sweyn Holm*. There are dozens of examples in the Shetland, Orkney, and Faroe islands. Sweden has its *Stockholm*. In the Channel Islands, one has the (Norman) term *hou*. In Denmark, the main example is *Bornholm*. In the Severn Channel, offshore from King Lear's Dunraven Castle, there is *Steep Holm*.

54. *OED*, s.v. "holmgang," etymology.

55. See David Carnegie, "Selden's Duello as a Source for Webster's 'The Devil's Law-Case,'" *Notes and Queries* 244, no. 2 (1999): 260–262.

56. *OED*, s.v. "holmgang." See also R. S. Radford, "Going to the Island: A Legal and Economic Analysis of the Medieval Icelandic Duel," *Southern California Law Review* 62 (1989): 615–644.

57. The swordsman Miyamoto Musashi, author of *The Book of Five Rings* (ca. 1645), met Sasaki Kojiro on Ganryu and defeated him there. This *holmgang* is the subject of much film and literature, such as Hiroshi Inagaki's film *Samurai III: Duel at Ganryu Island* (1956), and Eiji Yoshikawa's historical novel *Musashi* (1935). There is still a quarrel about whether Musashi may have cheated.

58. Compare Koushun Takami's novel *Battle Royale* (1999; San Francisco: VIZ Media, 2003), now adapted as a *manga* series and feature film. The story concerns students who are abducted to an island, roughly modeled on Ogijima in the Inland Sea. They must fight each other to the finish until one alone stands victor.

59. See *Egil's Saga*, chap 56, trans. Gwyn Jones, 138ff.

60. Adapted from Frederick York Powell's introduction to Saxo Grammaticus, *Nine Books*, 1:30–31.

61. Belleforest, *Tragic Histories*.

62. My translation of "Horwendillus, triennio tyrannide gesta, per summam rerum gloriam piraticae incubuerat."

63. Saxo, *Nine Books*, bk. 3.

64. Ibid.

65. Ibid.

66. Compare how Old Hamlet broke off his "parle" with the Poles. A *parle* is "a meeting between enemies or opposing parties to discuss the terms of an armistice" (*OED*, s.v. "parle," 1).

67. As in Saxo Grammaticus; see Marc Shell, *The End of Kinship: "Measure for Measure," Incest, and the Ideal of Universal Siblinghood* (Stanford, CA: Stanford University Press, 1988). Saxo writes: "Inter quos forte quidam Amlethi collacteus aderat, a cuius animo nondum sociae educationis respectus exciderat." For the familiarity with Saxo's counterpart to Ophelia: "Pari igitur studio petitum ac promissum est silentium: maximam enim Amletho puellae familiaritatem vetus educationis societas conciliabat, quod uterque eosdem infantiae procuratores habuerit."

68. For the relevant meaning of *ham*, see Chapter 10.

69. Bornholm had been in Danish hands for centuries until, from 1525 to 1575, the Danes pawned it to Lübeck.

70. In 1559, King Frederick II of Denmark bought Oesel. He gave it to his brother Magnus. Magnus then landed on the island with an army in 1560. The entire island became a Danish possession in 1573. It became Swedish territory in the next century.

71. In 1346, Denmark sold Oesel to the Livonian Order.

72. That was thanks to an agreement whereby the Danes ceded both Wiek Island and Dagö Island to Poland. In the 1580s, much of "Estonia," including Wiek, was part of the Polish-Lithuanian Commonwealth; later on, it was conquered by Sweden.

73. These days, Poland and Germany share Wolin, condominium style.

74. The Swedish historian Lauritz Weibull dismissed Jomsborg and Vineta as legendary.

75. Thomas William Shore, *Origin of the Anglo-Saxon Race: A Study of the Settlement of England and the Tribal Origin of the Old English People* (London: E. Stock, 1906).

76. Thomas Downing Kendrick, *A History of the Vikings* (New York: Courier Dover Publications, 2004), 180. According to Saxo Grammaticus, the Curonians and Estonians fought in the Battle of Bravalla against the Danes and the Wends of Pomerania. On the view that the island is real but submerged, see Adolf Hofmeister, *Der Kampf um dies Ostsee vom 9. bis 12. Jahrhundert* (1931; Darmstadt: Wissenschaftliche Buchgesellschaft, 1960).

77. T. M. Evans, "The Ancient Britons and the Lake Dwelling at Ulrome in Holderness," in *The Hull Quarterly and East Riding Portfolio*, ed. W. G. B. Page (Hull: Brown & Sons, 1884), 29m.

78. On the Danish origin of the toponym *Holderness*, see David Hey, *A History of Yorkshire* (Lancaster, PA: Carnegie Publishing, 2004). See also Richard Stead, *Holderness and the Holdernessians: A Few Notes on the History, Topography, Dialect, Manners, and Customs of the District* (Hull, UK: William Hunt, 1878).

79. See also Edward Thomas Stevens, *Flint Chips: A Guide to Pre-Historic Archeology, as Illustrated by the Collection in the Blackmore Museum* (Salisbury, UK: Bell & Daldy, 1870), 409.

80. *OED*, s.v. "flat," n3, 5a.

81. *OED*, s.v. "flat," n3, 6.

82. *OED*, s.v. "flat," n3, 5b. Shakespeare's *Tempest* reads, "All the infections that the Sunne suckes vp / From Bogs, Fens, Flats" (act 2, sc. 2).

83. Walter MacFarlane, *Geographic Collections Relating to Scotland*, ed. A. Mitchell (1724; Edinburgh: Scottish History Society, 1906–1908), 1:8.

84. *OED*, s.v. "moat."

85. The *OED* lists Scottish, Irish-English, Anglo-Norman, French, post-classical Latin, and early Italian. See also George Thomas Clark, *Mediaeval Military Architecture of England* (London: Wyman & Sons, 1884), 1.2.27, who writes, "Many of these mounds, under the name of motes (*motae*) retained their timber defences to the twelfth and thirteenth centuries."

86. The *OED* remarks that "the name *dic* was given to either the excavation or the bank, and evolved to both the words "dike"/"dyke" and "ditch."

87. Shakespeare, *Richard II*, act 2, sc. 1; my emphasis.

88. *OED*, s.v. "dig," vi, 1.

89. *OED*, s.v. "dig," vi, 2.

90. Ezekiel 4:2.

91. For the term *geoglyph*, see Mick Aston, *Interpreting the Landscape from the Air* (Stroud, UK: Tempus, 2002).

92. On the confusion in the *Anglo-Saxon Chronicle* of the names *Offa* and *Ossa*, see Steve Blake and Scott Lloyd, *The Keys to Avalon: The True Location of Arthur's Kingdom Revealed* (Shaftesbury, UK: Element, 2000).

93. Eutropius suggests that "Offa had his most recent war in Britain, and to fortify the conquered provinces with all security, he built a wall for 133 miles from sea to sea" (*Historiae Romanae Breviarium*), but he was writing in the fourth century, and it is generally thought that he was referring to some other construction.

94. Paolo Squatriti, "Digging Ditches in Early Medieval Europe," *Past and Present* 176, no. 1 (2002): 11–65.

95. An *arpent* is a unit of land area, usually around four-fifths of an acre.

96. This is the only occurrence of the word *sled* in Shakespeare's currently (extant) works. Some interpreters say that *sleaded*, as *sled* appears in some versions of the play, might mean "loaded with lead" and that *Polacks* might mean "pole-axe."

97. "[It] is brittle and bad for skating, 'shell-ice' as it is called." *United Service Magazine* 139 (1875): 42; *OED*, s.v. "shell," c7.

98. *OED*, s.v. "cat," c2a.

99. John A. Matthews and Keith R. Briffa, "The 'Little Ice Age': Re-Evaluation of an Evolving Concept," *Geografiska Annaler: Series A, Physical Geography* 87 (2005): 17–36.

100. The historical novelist Edward Rutherfurd's *London: The Novel* (London: Century, 1997) makes this point.

101. Quoted by Ian Currie, *Frost, Freezes and Fairs: Chronicles of the Frozen Thames and Harsh Winters in Britain from 1000 AD* (Coulsdon, UK: Frosted Earth, 1996).

102. *The Friend: A Religious and Literary Journal* 36, no. 11 (1862): 82.

103. Maurice Rickards, *Encyclopedia of Ephemera: A Guide to the Fragmentary Documents of Everyday Life for the Collector, Curator, and Historian*, edited and completed by Michael Twyman, Sally de Beaumont, and Amoret Tanner (New York: Routledge, 2000), 154; quoting the following handbill of February 4, 1814: Bankside, Southwark, London. Letterpress, printed at "Crown & Constitution, near Thames Street Stairs."

104. Compare *Poland*.

105. The waters off both Polish Pomerania (southern Baltic) and Polish Livonia (northern Baltic) often freeze firmly at the surface. The sea here is neither *solid* earth (*terra*

firma) nor liquid *water* (*aqua liquida*); it is solid water (*aqua firma*). In present decades, the Baltic Sea is firmly, or solidly, ice covered for about 45 percent of its surface area at the maximum annually.

106. Aeschylus, *The Persians*, trans. Robert Potter, 492–507.

107. For the first scene of *Alexander Nevsky*, 1:102:16; for the second, 1:23:14.

108. See Catherine A. M. Clarke, "Edges and Otherworlds: Tidal Spaces in Early Medieval Britain," in *The Sea and Englishness in the Middle Ages: Maritime Narratives, Identity and Culture*, ed. Sebastian I. Sobecki (Cambridge: D. S. Brewer, 2011), 87.

109. Bodleian Library, MS Junius 11.

110. 1594 edition, sig. Cᵛ.

111. Edgar Allan Poe, "A Descent into the Maelström" (1841).

112. The Elizabethans knew the maelstrom: "There is between the said Rost Islands, and Lofoote, a whirle poole, called Malestrand, which . . . maketh such a terrible noise, that it shaketh the rings in the doores of the inhabitants houses of the said Islands, ten miles of [f]." Anthony Jenkinson, quoted in Richard Hakluyt, *The Principal Navigations, Voiages, Traffiques and Discoueries of the English Nation* (1589–1600), 2:334.

113. Israel Gollancz, *Hamlet in Iceland: Being the Icelandic Romantic Ambales Saga* (London: David Nutt, 1898).

114. Giorgio de Santillana and Hertha von Dechend, *Hamlet's Mill: An Essay on Myth and the Frame of Time* (Boston: Gambit, 1969).

115. Immanuel Kant, *The Critique of Judgment*, trans. James Creed Meredith (Oxford: Clarendon Press, 1952); Meredith translates Kant's term *Felsen* as the English-language "rocks." Guyer and Matthews render it as "cliffs"; Immanuel Kant, *Critique of the Power of Judgment*, ed. Paul Guyer, trans. Paul Guyer and Eric Matthews (Cambridge: Cambridge University Press, 2001), 145.

116. See Chapter 13.

117. Geoff Brown, "Sister of the Stage: British Film and British Theatre," in *All Our Yesterdays: 90 Years of British Cinema*, ed. Charles Barr (London: British Film Institute, 1986), 1–29. The interior scenes were shot at the Hepworth Studio in Walton-on-Thames. Johnston Forbes-Robertson plays Hamlet. For a clip, see http://www.youtube.com/watch?v=bN5eHQhMjg0&feature=player_detailpage.

118. See Simon Williams, *Shakespeare on the German Stage* (Cambridge: Cambridge University Press, 1990), 1:75.

CHAPTER 10

1. See Marc Shell, *Children of the Earth: Literature, Politics and Nationhood* (New York: Oxford University Press, 1995), chap. 5.

2. Lucius Junius Brutus also has a part in Shakespeare's *Rape of Lucrece* and *Coriolanus*.

3. James Joyce picks up on these identities (or likenesses) when Stephen Dedalus says, "Hamlet, the black prince, is Hamnet Shakespeare." Joyce, *Ulysses* (Paris: Shakespeare & Co., 1922).

4. See Marc Shell, *Polio and Its Aftermath: The Paralysis of Culture* (Cambridge, MA: Harvard University Press, 2006); and Marc Shell, *Stutter* (Cambridge, MA: Harvard University Press, 2005).

5. Pt. 1, sec. 3.

6. John Minsheu, Ἡγεμὼν εἰς τὰς γλῶσσας: *Ductor in Linguas* [Guide into Tongues] (1617).

7. *OED*.

8. *Hem* is "from the same root as *ham* n2, and North Ger[man] *hamm* enclosure; the radical sense being 'border.'" *OED*, s.vv. "ham," n2; "hem," etymology.

9. For a nineteenth-century view of the opposition between mainland and island Denmark, see Illustration 51.

10. A related question is this: Saxo set part of the story of Hamlet on the natural island of Britain, but Shakespeare replaces that with Hamlet's report of an offstage scuffle that takes place on artificial floating islands, or ships. Why?

11. See also Hamlet's remark to Rosencrantz and Guildenstern, "Gentlemen, you are welcome to Elsinore," and to the Players, ". . . you are welcome to Elsinore."

12. "What is your affair in Elsinore?" "What make you at Elsinore?"

13. Count de Falbe, trans., "The Hamlet Saga," *The Nineteenth Century: A Monthly Review* 12 (July–December 1882): 957.

14. See Chapter 9.

15. The Norn language was still spoken on the Orkneys in 1800. See Glanville Price, *The Languages of Britain* (London: Edward Arnold, 1984), esp. 203–206; and Michael Barnes, *The Norn Language of Orkney and Shetland* (Lerwick, UK: Shetland Times, 1998).

16. The flagship of the Spanish Armada ran aground on islands that were former "colonies" of the Danish Empire, such as Frjóey (Fair Isle). Shakespeare rarely discusses the Armada outright. It was a touchy subject, especially among recusant Catholics. *Hamlet*, though, is one text where the Armada has a role. Its reference to the hobbling Spanish "bilboes" points to the Spanish instrument of torture used to hobble mutineers. (One can see one of these, collected by the British in the wake of the Armada, in the Tower of London.) By way of comparison to the British defeat of the Spanish Armada, in *Othello*, whose last four acts are set on the island of Cyprus, we read "News, lads! our wars are done. / The desperate tempest hath so bang'd the Turks, / That their designment halts" (act 2, sc. 1). One might say that the Prince of Arragon in Shakespeare's marine play *The Merchant of Venice* recalls King Philip II of Spain (and Aragon), who was Queen Mary's unpopular Catholic husband (1554–1558). After Mary's death, Philip became a spurned suitor of Princess Elizabeth.

17. Scania was part of Denmark until 1658, when it became part of Sweden. Hven, the island in the Øresund, became part of Sweden two years later. The "sundering" of Scandinavia at Elsinore is the subject of Norse myths in which the goddess Gefjun creates Zealand. See Chapter 9.

18. Other straits that make the connection within Denmark are the Great Belt (which flows between Samsø and Langeland islands) and the Little Belt (which flows between Funen Island and the Jutland Peninsula). For the etymology, see "Balteus" in the encyclopedic *Nordisk Familjebok*, 801–802.

19. As it appears in Virgil's *Aeneid* translated by the Scotsman Gavin Douglas (1513), whose work Shakespeare knew. Alastair Fowler, "Two Notes on *Hamlet*," in *New Essays on Hamlet*, ed. John Manning and Mark Thornton Burnett (New York: AMS Press, 1994). The term is also used in the Scotsman John Bellenden's translation of Hector Boece's history of Scotland (1540).

20. There is the passage in the novel *Pudd'nhead Wilson* (1894) where Mark Twain compares a change of moral viewpoint to the explosion on the volcanic island of Krakatoa. "A gigantic eruption, like that of Krakatoa a few years ago, with the accompanying earthquakes, tidal waves, and clouds of volcanic dust, changes the face of the surrounding landscape beyond recognition, bringing down the high lands, elevating the low, making fair lakes where deserts had been, and deserts where green prairies had smiled before. The tremendous catastrophe which had befallen Tom had changed his moral landscape in much the same way. Some of his low places he found lifted to ideals, some of his ideas had sunk to the valleys, and lay there with the sackcloth and ashes of pumice stone and sulphur on their ruined heads." Twain, *Pudd'nhead Wilson* (Hartford, CT: American Publishing, 1894), 122–123.

21. An example of the conjunction of strait geography and moral life is the passage in the Roman statesman Marcus Tullius Cicero's *Oration for L[ucinus Licinius] Morena* (63 BC), where Cicero compares the commotion at the narrows of waters (at the Euripus Strait) to a political commotion (in Rome): "For what sea [*fretum*], what Euripus do you think exists, which is liable to such commotions, to such great and various agitations of waves, as the storms and tides by which the comitia [the official assembly of Roman citizens] are influenced?" M. Tullius Cicero, *The Orations of Marcus Tullius Cicero*, trans. C. D. Yonge (London: Henry G. Bohn, 1856), 2:17.35.

22. Claudius tries (unsuccessfully) to pray—"My words fly up, my thoughts remain below," says he, noting sadly, "Words without thoughts never to heaven go" (act 3, sc. 3). The king's would-prayers are comparable to the *orysownes* about which we read in *Legends of the Saints* (1380): "He knelyt done, & to God mad[e] his orysowne." John Barbour, *Legends of the Saints*, ed. W. M. Metcalfe, 13, 18, 23, 25, 35, 37, 3 vols. (1887–1895; Edinburgh: Scottish Text Society, 1896), 432.

23. The English *ore* is cognate with Old Icelandic *óss* and means both "an edge or bank" (*OED*, s.v. "ore," n4) and a "shore or coast." Compare Latin *ora*.

24. *OED*, s.v. "sound," n1.

25. Consider *Ore sound* (the mouth of a river or estuary). The English *ore* means "an edge or bank" (*OED*, s.v. "ore," n4) like that where Ophelia dies. *Ore* means "shore or coast" in such toponyms for places in England as *Oare* and *Windsor* in Buckinghamshire and *Pershore* in Worcestershire. Compare English *ore*, meaning "mouth[,] including that of a river or sound" (*OED*, s.v. "ore," n3).

26. This work is one of the three parts of Berlioz's *Tristia* (Op. 18).

27. The Romantic period's fascination with Ophelia began with Harriet Smithson's performance of *Hamlet* in English in 1827—in Paris! Hector Berlioz saw that performance and fell in love with Smithson at first sight. They married in 1833, but the marriage dissolved within seven years. In view of Ophelia's "nun"-like status, it is worth noting that Millais married Effie, the "virgin wife" of John Ruskin, in 1856.

28. The English *oration* is often spelled *oriso(u)n* and *orison*. See *Dictionary of the Scots Language*, s.v. "oriso(un)." The grammarian John Vaus spells out *orations* as *orisonis* when he refers to the "orisonis of [the Roman orator] Cicero" (1531). Vaus, *Rudimenta Puerorum in Artem Grammaticam* (Paris, 1531), iii. John Bellenden likewise writes in his translation of the Roman Livy's *History of Rome* (1533), how "in the myddis of [an] orisoun ["oration"]

enterit Romulus & Remus with ane band of armit men." *Livy's History of Rome: The First Five Books*, trans. John Bellenden (1533), 20/26.

29. John Boroughs, *Sovereignty of the British Seas* (1651), 383.

30. The comedic *Philotus* (1603) says that "zour orisoun sir sounds with sic skil / In Cupids court as ze had bene vpbrocht." *Ane Verie Excellent and Delectabill Treatise Intitulit Philotus* (Edinburgh, 1603), lxv. Nowadays, there are Danish musical groups with names like Øre-Sound. For a posting from one of these groups, see www.ore-sound/dk.

31. Some would-be scientific etymologists have it that the Øresund got its name from the "fact" that the sound has the shape of a human ear; others say that the *ore* (ear) instead refers to a resemblance to peninsular plots of land that border the sound. For the former position, see *Penny Cyclopedia for the Diffusion of Useful Knowledge*, ed. G. Longi (London: Charles Knight, 1837), s.v. "Denmark," 7:398, col. 1.

32. Henry Bradley, "'Cursed Hebenon' (Or 'Hebona')," *Modern Language Review* 15, no. 1 (1920): 85–87.

33. See Chapter 9.

34. See Michel de Montaigne, *Essays*, trans. John Florio (1603), 2.12.276: "The barble fishes . . . will set the line against their backes, and . . . presently saw and fret the same asunder."

35. Humphrey Gilbert cites this proverb in *Gilbert's Voyages and Enterprises* [Edward Hayes's narrative], ed. D. B. Quinn and Neil M. Cheshire (1583; Toronto: University of Toronto Press, 1940), 2:420; see *Dictionary of Newfoundland English*, s.v. "sea."

36. L. Campbell, *Sable Island Shipwrecks, Disaster and Survival at the North Atlantic Graveyard* (Halifax: Nimbus Publishing, 1994).

37. *Orizon* (prayer) was sometimes spelled with the letter *h*. English writers often wrote it out as *horison, horyson*, and *horizon*. In the *Testament of Love*, we read about "Deuoute horisons and prayers to god." (The *Testament* was printed in 1532. At the time, it was thought to have been written by Geoffrey Chaucer. Nowadays, most scholars believe it to be the work of Thomas Usk.) The Second Folio (1632), Third Folio (1664), and Fourth Folio (1685) of *Hamlet* all spell the word as *horizons*. Likewise, *horizon* (bourn) was sometimes spelled without the *h*. The English word *ora*, which means "limit," almost never has the *h* at the beginning. John Trevisa's translation of Ranulph Higden's *Polychronicon* (1387) uses *orisoun* to mean "a boundary, the frontier or dividing line between two regions of being" (*OED*, s.v. "horizon," 2). Trevisa writes, "Mannis soule . . . is i-cleped *orisoun*, as it were þe next marche in kynde bytwene bodily and goostly þinges" (St. John's College, Cambridge, 1869), 2.183.

38. *OED*, s.v. "bourn," 2.

39. Trevisa's *orisoun* (bourn) is akin to the borderline that divides living persons from ghosts in *Hamlet*. Likewise, *orisoun* recalls two kinds of horizons: the "cut" that separates the sea from the sky and the "coast," or horizontal limit (*horos*), that surrounds *terra firma*. In his translation of Bartholomew de Glanville's *De Proprietatibus Rerum* (1398), Trevisa writes, "The circle to þe whiche þe syzte streccheþ and endeþ is calde *Orizon*, as it were þe ende of þe syzte."

40. See Chapter 13.

41. Friedrich Nietzsche, *The Gay Science*, ed. Bernard Williams, trans. Josefine Nauckhoff and Adrian Del Caro (Cambridge: Cambridge University Press, 2001), sec. 343.

CHAPTER 11

1. The English word *halsing* has something like the same meaning as has "necking." Compare John Trevisa's translation of Ranulph Higden's *Polychronicon* (1387) and William Langland's *Piers Plowman*. Ranulph Higden, *Polychronicon* [Rolls], trans. John de Trevisa (1387), 7.139: "Her housbonde halsynges"; and William Langland, *Piers Plowman* (1393), 100.7, 187: "Handlynge and halsynge and al-so þorw cussynge Excitynge oure aiþer oþer til oure olde synne."

2. Terence, *Heauton Timorumenos*, in *Terence in English*, trans. Richard Bernard (Cambridge, 1598), 1:252: "I will say nothing of hausing & kissing."

3. One reads in Helkiah Crooke's Μικροκοσμογραφια: *A Description of the Body of Man* (1615): "The Clitoris is a small body, not continuated at all with the bladder, but placed in the height of the lap," 250.

4. Another etymology has the derivation from *opheleia* (help). See the proper names in Jacopo Sannazaro's poem *Arcadia*.

5. Edward Gibbon points out that certain kinds of rulers committed incest on islands in straits. One example is the twelfth-century Byzantine emperor Manuel I Comnenus. "No sooner did Manuel return to Constantinople, than he resigned himself to the arts and pleasures of a life of luxury: the expense of his dress, his table, and his palace, surpassed the measure of his predecessors, and whole summer days were idly wasted in the delicious isles of the Propontis, in the incestuous love of his niece Theodora." Gibbon, *The Decline and Fall of the Roman Empire* (London: Strahan & Cadell, 1776–1789), chap. 48.

6. Morris P. Tilley, *A Dictionary of the Proverbs in the Sixteenth and Seventeenth Centuries: A Collection of the Proverbs Found in English Literature and the Dictionaries of the Period*, 6th ed. (Ann Arbor: University of Michigan Press, 1950), s.v. "lapwing."

7. William Chillingworth, *The Religion of Protestants* (1638), I.2, sec. 101.91.

8. See Michael Mindlen, *The German Bildungsroman: Incest and Inheritance* (Cambridge: Cambridge University Press, 1997).

9. Just so, one speaks of a gulf of water as a *sinus* and names it as such on many maps of Europe in Shakespeare's day. See Thomas Burnet, *The Theory of the Earth: Containing an Account of the Original of the Earth, and of All the General Changes which it Hath Already Undergone, or is to Undergo, Till the Consummation of All Things* (1684), i:110: "The promontories and capes shoot into the sea, and the sinus's and creeks . . . run as much into the land."

10. A throat, according to the *OED*, is "a narrow passage, esp. in or near the entrance of something." So Sir Walter Scott in his *Diary* entry for August 17, 1814, writes about a small island around the Scapa Flow in the Orkney Islands: "The access through this strait would be easy, were it not for the Island of Graemsay, lying in the very throat of the passage." John Gibson Lockhart, *Memoirs of the Life of Sir Walter Scott* (Edinburgh: Robert Cadell, 1837), 3.6.206.

11. Serpent's Mouth (Boca del Serpiente) and Dragon's Mouths (Bocas del Dragón), both of these being straits between Trinidad and Tobago and Venezuela.

12. Compare Cloaca Maxima, the water passage in ancient Rome that drained the swamps into the Tiber River.

13. John Florio writes in *A Worlde of Wordes* (1598): "Vulva . . . the womb passage." See Thomas Gibson, *The Anatomy of Humane Bodies Epitomized* (1682): "It has passages . . . or the neck of the Bladder, and in Women for the vagina of the Womb," 20.

14. Not only the *reproductive tract*, to which we here devote attention, but also the *digestive tract*, to which we do not (except for the aforementioned reference to cloacas), has a role to play in the bodily geography of *Hamlet*. Consider Hamlet's statement to Claudius that "a man may fish with the worm that hath eat of a / king, and eat of the fish that hath fed of that worm," to which Claudius responds, "What dost thou mean by this?" Hamlet's answer is as cunning as his initial statement: "Nothing but to show you how a king may go a / progress through the guts of a beggar" (act 4, sc. 3). Some members of the audience hear (others not) the reference to "a cat's gut" as one to the strait, the shallow reefs, of the Kattegat (cat's "hole"—the name since medieval times). See Illustration 50. The Øresund opens into the Kattegat on the north, near Jutland. That is where the original story of *Hamlet*, as Saxo Grammaticus tells it, takes place. The English term *gut* is often used to refer to a strait. "The gutt [*sic*] of sea being here but narrow," as the natural philosopher Kenelm Digby writes in *Journal of a Voyage into the Mediterranean* (1665; London: Camden Society, 1868), 9.

15. Pausanias, *Description of Greece* (second century BC), chap. 4, sec. 5.

16. *OED*, s.v. "cleft," 2. See also Geoffrey Chaucer, "The Summoner's Tale," in *The Canterbury Tales* (ca. 1386): "Doun his hond he launcheth to the clifte," 437.

17. Paul Roche translates the line as "This very day will furnish you a birthday and a death." Roche, "Oedipus the King," in *The Oedipus Plays of Sophocles* (1958; New York: Penguin 1991), 229. See also Marc Shell, *The Economy of Literature* (Baltimore: Johns Hopkins University Press, 1978), chap. 3.

18. Sweden took Hven from Denmark by military means in 1658.

19. Johan Ludvig Heiberg, *Hveen, tilforn Danmarks Observatorium* (Copenhagen, 1861).

20. For example, *Eta Carinae and Other Mysterious Stars: The Hidden Opportunities of Emission Line Spectroscopy: Proceedings of an International Conference held at Tycho Brahe's Island, Hven, Sweden, 24–26 August 2000*, ed. Theodore R. Gull, Sveneric Johansson, and Kris Davidson (San Francisco: Astronomical Society of the Pacific, 2001).

21. Christiern Pedersen, *Richt Vay to the Kingdome of Heuine*, trans. John Gau (Malmö: Hochstraten, 1533).

22. Tycho Brahe, *On the New and Never Previously Seen Star* (1573; republished in 1602 and 1610).

23. John Robert Christianson, *Tycho Brahe and His Assistants, 1570–1601* (New York: Cambridge University Press, 1999).

24. Thanks to a dispensation from King Frederick II of Denmark, Tycho established on Hven an observatory and scientific institute that is "estimated to have cost Denmark more than 5% of its Gross National Product, an all-time world record."

25. *OED*, s.v. "orb," IIa.

26. *OED*, s.v. "sphere," 2a.

27. Geoffrey Chaucer, "Hypsipyle and Medea," in *Legend of Good Women* (ca. 1385).

28. The island institute in the Øresund was an influential model for heavenly science and science fiction—for one, the fictional Solomon's House, the state-sponsored institution of higher learning and scientific research on the island of Bensalem, in Francis Bacon's utopian novel *New Atlantis* (1626). For Bacon, his "Atlantic" island institute was to be "the very eye of this kingdom." Palpable throughout is the pun on *eye* meaning "island." The

town of Eye in East Anglia thus derives its name from the Old English word for "island"; the first settlement there was built at the River Dove on marshlands prone to flooding.

29. John Donne, *An Anatomy of the World* ([W. S.] for S. Macham, 1611), lines 205–208. Johannes Kepler (for a time he was Tycho Brahe's assistant) gave Donne a copy of his *De Stella Nova* [On the New Star] (1604) in order for Donne to pass it along to King James.

30. *OED*, s.v. "fret," 1.

31. Frank Kermode, *The Age of Shakespeare* (New York: Random House, 2005), 109. See also J. R. Mulryne and Margaret Shewring, *Shakespeare's Globe Rebuilt* (Cambridge: Cambridge University Press, 1997).

32. Thomas Elyot, *The Image of Gouernance* (1541), xxvii, fol. 60.

33. Thomas Drant's translation of Horace: *Arte of Poetrie, Epistles and Satyrs Englished* (1566).

34. The *OED* defines *shell* as "the ribs of ship."

35. The *OED* defines *shell* in terms of these "spheres."

36. Hamlet then goes on to say (act 3, sc. 1), "Go thy ways to a nunnery," "Get thee to a nunnery," "To a nunnery go," "To a nunnery go, and quickly too," and finally (again) "To a nunnery, go."

37. Danish Vikings famously plundered the abbey on Lindisfarne Island in Northumberland in 793. Likewise, the Donemuthan Convent in Yorkshire, located on an island flat in the Humber Estuary (or perhaps on an island near Wearmouth at the confluence of the Don and Tyne rivers), was plundered by Vikings in 794.

38. Johnny Grandjean Gøgsig Jacobsen, "Monastic 'Islands' in Medieval Denmark: Insular Isolation in Ideal and Practice," in *Isolated Islands in Medieval Nature, Culture and Mind*, ed. Gerhard Jaritz and Torstein Jørgensen (Budapest: Central European University Press, 2011). Relevant here is A. Perron, "Saxo Grammaticus' Heroic Chastity: A Model of Clerical Celibacy and Masculinity in Medieval Scandinavia," in *Negotiating Clerical Identities: Priests, Monks and Masculinity in the Middle Ages*, ed. Jennifer D. Thibodeaux (London: Palgrave Macmillan, 2010).

39. Bertram Colgrave, ed. and trans., *Two Lives of Saint Cuthbert* (Cambridge: Cambridge University Press, 1940), 214–215.

40. Examples include Utstein on Mosterøy Island in Norway and Ionia on Iona Island in Scotland. (On the nearby Eilean nam Ban, or Island of Women, was the celebrated Augustinian nunnery.) See Torstein Jørgensen, "Utstein Monastery: An Island on an Island—or Not?" in Jaritz and Jørgensen, *Isolated Islands*. The dozens of island nunneries elsewhere include Nonnenwerth (near Bad Honnef in the Rhine River), Frauenchiemsee (in the Chiemsee in Bavaria), Alderney (in the Channel Islands), and the Dominican nunnery on Margaret Island (in the Danube River in Budapest). Another example is Insula Convent, established on Öland Island in the Limfjord Sound, which separates the North Jutlandic island of Vendsyssel-Thy from the rest of the Jutland Peninsula. The island is now Oksholm.

41. One such convent was Börringe Priory in Danish Scania, on the island of Byrdingø in Börringe Lake. King Christian III of Denmark gave this convent to the Brahe family in 1536, ten years before Tycho was born.

42. This is one subject of my book *The End of Kinship: "Measure for Measure," Incest, and the Ideal of Universal Siblinghood* (Stanford, CA: Stanford University Press, 1988).

43. That Shakespeare attributed a coast to this land-bound region has puzzled many people since the time of Samuel Johnson. Shakespeare here follows his source: Robert Greene, in *Pandosto: The Triumph of Time* (1588), likewise provides Bohemia with a seacoast, although the integration of the coastline with the plot is not his idea. Compare *Measure for Measure*'s Vienna, which is provided with an imprisoned sea pirate.

44. In *The Winter's Tale*, the oracle at Delphi is located on a small island ("fertile the isle"). Shakespeare was following Robert Greene, who may have been referring to the island of Delos. See Terence Spencer, "Shakespeare's Isle of Delphos," *Modern Language Review* 47, no. 2 (1952): 199–202. Shakespeare sets most of *The Winter's Tale* on the island of Sicily, and when it came to Delphi, he may well have had in mind the way in which Delphi was connected to its mainland by a land canal. For "canal" as the ravine where father and son meet, see Chapter 8.

45. Shakespeare's son's name was Hamnet, sometimes spelled Hamlet. Hamnet died early, at the age of eleven in 1596, like Perdita's little brother Mamillius in *The Winter's Tale*, which mends the problem in the fashion of romance.

46. *Fratricide Punished* [Tragoedia: Der bestrafte Brudermord oder Prinz Hamlet aus Daennemark], 4.1, 130, in *Hamlet: A New Variorum Edition*, ed. Samuel Burdett Hemingway, Hyder Edward Rollins, and Matthias Adam Shaaber, 15th ed. (Philadelphia: Lippincott, 1918). Many literary historians consider the lost writing to have been an acting version used by English players in Germany at the beginning of the seventeenth century.

47. Ibid.

48. Ibid., 5.2, 129.

49. Frederic Williams Hardman, *The Danes in Kent: A Survey of Kentish Place-Names of Scandinavian Origin*, lecture given at Walmer, February 7, 1925 (n.p., 1927).

50. Charles Lyell, *Principles of Geology*, 1:407.

51. Herman Melville writes in *Moby-Dick*: "In what census of living creatures, the dead of mankind are included; why it is that a universal proverb says of them, that they tell no tales, though containing more secrets than the Goodwin Sands."

52. George Byng Gattie, *Memorials of the Goodwin Sands* (London: J. J. Keliher, 1904).

53. John 3:5.

54. *OED*, s.v. "pirate." The twelfth-century *Chronicle of Henry of Livonia* describes the residents of the island of Oesel using a particular kind of ship that he calls *piratica*. The same is also discussed by Saxo.

55. He is like the sailor Odysseus washed naked on the islands of Scheria and Calypso in Homer's *Odyssey*, or like the twin-sibling islanders Sebastian and Viola in Shakespeare's *Twelfth Night*, who are washed ashore in Illyria. (The twins hail from Messaline—a fictional islandological toponymic combination of *Mytilene*, on the island of Lesbos, and *Messina*, on the island of Sicily at the Strait of Messina between Sicily and mainland Italy.) Or he is like Pericles, in Shakespeare's *Pericles, Prince of Tyre*, washed ashore at Pentapolis and awaiting a marine rebirth from his own daughter Marina. (This rebirth, in the island-city of Mytilene, is the great birth, or rebirth, event in Shakespearean drama. "O, come hither," says Pericles to Marina, "Thou that beget'st him that did thee beget" [act 5, sc. 1].) Or he is like the apparently drowned men in *The Tempest*: "You are three men of sin, whom Destiny, / . . . the never-surfeited sea / Hath caused to belch up you; and on this island" (act 3,

sc. 3). *The Tempest* sets the stage, from the very beginning, with these words: "[Scene: A ship at Sea; an island]" (act 1, sc. 1). The play is about a ship, an island, and a ship as an island or an island as a ship of state. To put it otherwise, it is about the way that this insular ship of state is a prison made first by the witch Sycorax and then by the wizard Prospero for such prisoners as the fairy Ariel, the monster Caliban, and the whole motley crew of human beings.

CHAPTER 12

1. Shakespeare's exile play, *As You Like It*, is set in the Forest of Arden. It yokes the Ardennes Forest (also in Thomas Lodge's *Rosalynde*) with the romanticized memory of a youthful Shakespeare, whose mother was an Arden.

2. James Shapiro, *1599: A Year in the Life of William Shakespeare* (London: Faber & Faber, 2006), 275–276.

3. Richard Southern, *The Medieval Theatre in the Round: A Study of the Staging of the Castle of Perseverance and Related Matters*, 2nd ed. (1958; New York: Theatre Arts Books, 1975).

4. There are many other examples. In *Richard II*, Bolingbroke (later King Henry IV) is banished temporarily and Mowbray is banished permanently from "this scepter'd isle" (as John of Gaunt, Bolingbroke's father, calls Britain). In *Romeo and Juliet*, Romeo is banished from Verona. And so on.

5. Plutarch, *Precepts of Statecraft*, in Plutarch, *Moralia*, trans. Harold North Fowler, Loeb Classical Library 321 (Cambridge, MA: Harvard University Press, 1936), vol. 10.

6. English *ostracism* derives from the Greek *ostraka*, referring to the shells or shards by means of which citizens often voted.

7. Plutarch, *On Exile*, 607b, in Plutarch, *Moralia*, trans. Phillip H. de Lacy and Benedict Einarson, Loeb Classical Library 405 (Cambridge, MA: Harvard University Press, 1959), vol. 7.

8. Plutarch, *On Exile*, 607d–e.

9. Ibid., 607c.

10. Ibid., 600e.

11. Ibid., 601d.

12. Lucretius, *On the Nature of Things*, in *The Poem of Empedocles: A Text and Translation*, trans. Brad Inwood, rev. ed. (Toronto: University of Toronto Press, 2001), 168, lines 714–733.

13. The first letter in the word *tingholm* was originally the Old English and Norse thorn, or þ. The thorn remained in English-language alphabetic use into the Middle English period.

14. "The origins of the Viking Ting, a judicial-legislative assembly of freemen, are lost in time but clearly were independent of external influences. The creation of the Althing in 930 and the development of a quasi-democratic constitutional system unique in Europe for its time were offshoots of Norwegian Viking settlers who, it is safe to say, knew nothing of Greek democracy, Roman republicanism, or political theory of philosophy in a formal sense." Robert A. Dahl, *Democracy and Its Critics* (New Haven, CT: Yale University Press, 1989), 32.

15. John Pettingal in *Enquiry into the Use and Practice of Juries among the Greeks and Romans* (London, 1769) writes: "What the Greeks called pragma a Cause, the Saxons and Danes called a Thing or litigated Cause; the Lawyers *Thingmen*, a Judge *Thingrave*" (1.61).

Shakespeare uses the term informally when he writes, in *The Merry Wives of Windsor*, "You shall hear how things go" (act 4, sc. 5).

16. Robert A. Dahl, *On Democracy* (New Haven, CT: Yale University Press, 1998), 18–20. See also "How on Earth Can We Live Together?" Tällberg Foundation Forum, Stockholm, 2010–2011.

17. Siri Ingvaldsen writes: "The medieval parliamentary assembly sites, the Thing Sites, could be regarded as the cradle of democracy in Europe," in *Historical Thing Sites—The Cradle of Democracy: NORA Pre-Project* (Brattahlid, Greenland: Haugland International Research and Development Centre, October 2008). The four NORA regions are Greenland, Iceland, the Faroe Islands, and coastal Norway.

18. So argues Ingvar Anderson, "Early Democratic Traditions in Scandinavia," in *Scandinavian Democracy: Development of Democratic Thought and Institutions in Denmark, Norway and Sweden*, ed. Joseph Albert Lauwerys (Copenhagen: Schultz, 1958), 69–77.

19. William Gershom Collingwood and Jón Stefánsson, *A Pilgrimage to the Sagasteads of Iceland* (Ulverston, UK: W. Holmes, 1899).

20. Israel Gollancz, *Hamlet in Iceland: Being the Icelandic Romantic Ambales Saga* (London: David Nutt, 1898).

21. Hamlet rhymes *king* with *thing* elsewhere, too, as when he says, "Say nothing; no, not for a king . . ." (act 2, sc. 2). Says the Fool to the King in *King Lear*: "Thou art an O without a figure, I am better than thou art now, I am a fool, thou art nothing" (act 1, sc. 4).

22. Gary Johnson, "Mainland, Shetland Archipelago," *Shetlopedia.com*, http://shetlopedia.com/Thing.

23. *Færeyínga saga eller Færøboernes historie i den islandske grundtekst med færøisk og dansk oversættelse*, ed. Carl Christian Rafn, trans. Johan Hen(d)rik Schrøter (Copenhagen: J. H. Schultz, 1832).

24. Section 81: "Thorgney's Speech," in the Icelandic Snorri Sturluson's *Heimskringla* [Chronicles of the Kings of Norway; ca. 1225], trans. Samuel Laing (London, 1844); here taken from *Heimskringla: A History of the Norse Kings* (London: Norroena Society, 1907).

25. Saxo describes the tradition of bog drowning at the beginning of his work. Such drowning is still a common fate in the water-land bogs of Denmark. The Danish director Henrik Ruben Genz's 2008 movie *Terribly Happy* [Frygtelig lykkelig], based on Erling Genz's novel and set in Jutland, makes palpable the old Scandinavian threat.

26. Saxo Grammaticus, *Gesta Danorum*, 4.1.8.

27. This is the inscription on a stone marker erected in 1933 at "Amlads Grave" (Hamlet's Grave) in the town of Ammelhede (near the Kattegat on mainland Jutland in Denmark). A photograph of the stone is in Collection Selechonek. Writes Michael Skovmand (professor of English at the University of Aarhus in Denmark): "A report from a vicar north of Aarhus in Jutland from 1623 describes a locality called Ammelhede where according to local legend Amled lies buried," http://www.pathguy.com/hamlet.htm. Israel Gollancz, *The Sources of Hamlet* (London: Oxford University Press, 1926) mentions this "heath" and the story.

28. In this regard, Hamlet is much like his republican Roman namesake, Brutus. Brutus killed his uncle Tarquin in order that Romans might have liberty. The Latin word *brutus*, like the Danish *amleth*, means "foolish." See Marc Shell, *Children of the Earth: Literature, Politics and Nationhood* (New York: Oxford University Press, 1995), chap. 5.

29. *Gunnlaugs saga ormstungu*, chap. 11; quoted in Frederick Metcalfe, *The Englishman and Scandinavian: A Comparison of Anglo Saxon and Old Norse Literature* (London: Trübner & Co., 1880).

30. "King's Scholars' Pond Sewer" now names the branch of the Tyburn that runs from Buckingham Palace to the Houses of Parliament. The Anglo-Latin-derived verb *sewerare* "appears to mean 'to protect from flood'" (*Rolls of Parliament*, 1314 AD, 1.319/1). The formula enumerating the things placed under the control of the Commissioners of Sewers begins with protective and defining seawalls (*murorum maritimorum*). *Abstracts of Pleas from the Reign of Edward II* (1322 AD). The Elizabethan legal scholar Robert Callis writes that *sewer* is a compound of *sea* and *were* (i.e., *weir*, meaning "defense"). See Callis, *The Reading of the Famous and Learned Robert Callis, esq., Upon the Statute of Sewers, 23 Hen. VIII. c. 5; as it was delivered by him at Gray's Inn, in August, 1622* (London: Printed for William Leak, 1647). The meaning of *sewer* as "an artificial channel or conduit, now usually covered and underground, for carrying off and discharging waste water and the refuse from houses and towns" does not enter the English language until Shakespeare's day (*OED*, s.v. "sewer," n2); see, for example, Shakespeare's *Troilus and Cressida* (act 5, sc. 1).

31. Dorothy L. Sayers, *Thrones, Dominations*, compiled and completed by Jill Paton Walsh (London: Hodder & Stoughton, 1998), 313.

32. Exactly how the institution of the Norse *ting* influenced the Anglo Saxon *witan* and *folkmoot*, the Anglo-Norman *Curia regis*, and the British monarchal parliament is outside the range of the present volume; but see the pages that follow.

33. Quoted in Dorothy Whitelock, "Review of *The Witenagemot in the Reign of Edward the Confessor* by Tryggvi J. Oleson," *English Historical Review* 71 (1956): 642.

34. For the extremes of the arguments, see Hector Munro Chadwick, *Studies on Anglo-Saxon Institutions* (Cambridge: Cambridge University Press, 1905); and Félix Liebermann, *The National Assembly in the Anglo Saxon Period* (Halle: J. Niemeyer, 1913).

35. *OED*, s.v. "thing."

36. Ting Holm Island is also known as Law Ting Holm and as Lawting Holm.

37. The Shetland Archipelago was "sold" to Britain in the fifteenth century, though its ownership was disputed by Britain and Denmark well into Shakespeare's day, especially around the time that James I married Anne of Denmark—in Elsinore. In 1468, King Christian I pledged the Shetlands as security against the payment of the dowry of his daughter Margaret, betrothed to King Christian I; the dowry was not paid.

38. Mainland Island made the news in Shakespeare's day for another reason: ships of the Spanish Armada sought safety there.

39. Norsemen ruled on the Isle of Man from around the year 800 to the middle of the thirteenth century. Eleanor, Duchess of Gloucester, is exiled to the Isle of Man in Shakespeare's *Henry VI, Part 2*.

40. *Dingwall* (*Þingvöllr*) in the Scottish Highlands and *Thingwall* on the Wirral Peninsula. In the Shetland and Orkney archipelagoes, there are also *Gruting, Westing, Lunnasting, Nesting, Aithsting*, and *Sandsting* (and perhaps *Delting*). The *ting* in *Delting* may have had another sense originally, being *taing* or promontory.

41. See David Balfour, *Oppressions of the Sixteenth Century in the Islands of Orkney and Zetland from Original Documents* (Edinburgh, 1859), 7: "Ane letter under the commown seale of Zetland of the electioun of Nichole Ayth . . . to the office of law-

man generale of all Zetland, quhilk is of the dait, In the Ting holm of Ting wale" [July 27, 1532].

42. Richard Moss, "Amateur Archaeologists Find Ancient 'Thyng' in Sherwood Forest," *Culture* 24 (April 25, 2008).

43. Other Scandinavian national *tings*, not insular, include Sweden's *Tingvalla* and Norway's *Tingvoll* or *Storting*. See *The Assembly Project: Debating the Thing in the North*, ed. Alexandra Sanmark, *Journal of the North Atlantic* (forthcoming).

44. *Charter of Cnut* (1032), in *Codex Diplomaticus Aevi Saxonici*, ed. John Mitchell Kemble (London: Sumptibus Societatis, 1839–1848), 4:37.

45. John Cowell, *The Interpreter: or Booke containing the Signification of Words* (1607; London: William Sheares, 1637), sig. Nn2ᵛ/1.

46. Gordan W. J. Gyll, *History of the Parish of Wraysbury, Ankerwycke Priory, and Magna Charta Island; with the History of Horton, and the Town of Colnbrook, Bucks* (London: H. G. Bohn, 1862).

47. John Richard Green, *A Short History of the English People* (London, 1874).

48. Runnymede and Ankerwyke are now on opposite sides of the Thames, but then they were probably one (united) land area. Andrew Brooke, a geomorphologist from the National Rivers Authority, supports this theory: "Ten thousand years ago, the Thames flowed around a series of islands. It had a braided pattern, and only in the last 400 years or so has its main channel been centralized, widened and deepened for the needs of navigation." Brooke, *Channelized Rivers: Perspectives for Environmental Management* (Chichester, UK: Wiley, 1989).

49. "There is a school of thought that says that King Alfred used to have regular council meetings here known as the 'Witan' when he was resident at Old Windsor, just a short distance upriver." Keith Pauling, *Thames Pathway* (Raleigh, NC: Lulu Press, 2009), 147.

50. Discussed by Suzanne Lewis, *The Art of Matthew Paris in the Chronica Majora* (Berkeley: University of California Press, 1987), 204.

51. Matthew the Parisian, *Chronica Majora*, ed. Henry Richards Luard (London: Longman, 1876), 3:30.

CHAPTER 13

1. For the latter, see Chapter 8. For the name, compare the Greek Μεσόγειος (middle of the earth).

2. Many Arab writers called it *baḥr al-Rūm* (the Roman sea). The biblical book of Exodus (23:31) uses the term *yum philistim* (sea of the Philistines).

3. Gaius Sallustius Crispus, *Jugurthine War*, 17.

4. Nicholas Woodsworth, *The Liquid Continent: A Mediterranean Trilogy* (London: Haus Publishing, 2008). The three parts of the trilogy are "Alexandria," "Venice," and "Istanbul."

5. See Chapter 6, as regards Herman Melville and Justinian's *Pandects*. Justinian himself quotes the third-century Roman jurist Aelius Marcianus's *Institutes*, writing: "To begin with, by natural law, the following are common to all: air, flowing water, the sea, and consequently the seashore." William Warwick Buckland, ed., Charles Henry Monro, trans., *The Digest of Justinian* (Cambridge: Cambridge University Press, 1904), 1:40. The shoreline per se remains undefined.

6. Regulations beginning with Alexander VI in 1493, and including the ratification of the Treaty of Tordesillas (1494) and the Treaty of Zaragoza (1529), seem to give one or two nations other than the European countries world dominion.

7. When Robinson Crusoe calls himself "King" of his island—"I was Lord of the whole Manor; or if I pleas'd, I might call my self King, or Emperor over the whole Country which I had Possession of. There were no Rivals"—he seems to recall Grotius's view that islands in the sea belonged to the first inhabitant.

8. Johann Sommerville, "King James VI and I and John Selden: Two Voices on History and the Constitution," in *Royal Subjects: Essays on the Writings of James VI and I*, ed. Daniel Fischlin and Mark Fortier (Detroit: Wayne State University Press, 2002), 290–322. King James died in 1625.

9. John Selden, *Mare Clausum: Of the Dominion, or, Ownership of the Sea* (1618), bk. 1.

10. Marchamont Needham, "Neptune to the Commonwealth of England" (1652), 1127–1128.

11. "Princely Magnificence: Court Jewels of the Renaissance, 1500–1630: Exposition" (London: Debrett's Peerage in association with the Victoria and Albert Museum, 1980), cat. nos. 38 and 117. See also Marc Shell, "Portia's Portrait: Representation as Exchange," *Common Knowledge* 7 (Spring 1998): 94–144.

12. Athelney is bounded partly by the rivers Parret and Tone.

13. James Thomson, *Alfred*, in *The Works of James Thomson* (1763).

14. See, for example, John Nalson, *An Impartial Collection the Great Affairs of State* (London, 1682–1683); see also BM Satires 748.

15. Quoted by Pappus, *Collection*, in *The Oxford Dictionary of Quotations*, 2nd ed. (London: Oxford University Press, 1953), bk. 7, p. 14.

16. Thomson, "An Ode," from *Alfred*, in *Works*, 2:191. The "ode" would become known as "Rule, Britannia!"

17. Thomson, *Works*. These are the closing lines of *Alfred*.

18. HWV 62.

19. WoO 79.

20. Op. 91.

21. Op. 116.

22. Op. 103.

23. WWV 42. Relevant also are Wagner's *Overture for Christopher Columbus*, his stage version of Gluck's *Iphigenia in Aulis*, and his opera *The Flying Dutchman* (1838).

24. Cited in *The Musical Times*, April 1, 1900.

25. So argued the British colony of New South Wales (Australia) in 1834, as reported by Sebastian I. Sobecki, "Introduction: Edgar's Archipelago," in *The Sea and Englishness in the Middle Ages: Maritime Narratives, Identity and Culture*, ed. Sebastian I. Sobecki (Cambridge: D. S. Brewer, 2011), 29. See also "Lewis v. Lambert," in James Dowling, *Select Cases* (Archives of New South Wales), IV 2/3463, 96.

26. Joanne Parker, "Ruling the Waves: Saxons, Vikings, and the Sea in the Formation of the Anglo-British Identity in the Nineteenth Century," in *The Sea and Englishness in the Middle Ages: Maritime Narratives, Identity and Culture*, ed. Sebastian I. Sobecki (Cambridge: D. S. Brewer, 2011).

27. *Johannes Kepler Gesammelte Werke* (Munich: Beck, 1937–), 18:63. See also Charles R. Gibson, *Stories of the Great Scientists* (London: Seeley, Service & Co., 1921).

28. His correspondent hailed from Hallstatt on the freshwater Hallstätter See in Austria.

29. Joshua Gilder and Anne-Lee Gilder, *Heavenly Intrigue: Johannes Kepler, Tycho Brahe, and the Murder behind One of History's Greatest Scientific Discoveries* (New York: Random House, 2005).

30. "I confess that when Tycho died, I quickly took advantage of the absence, or lack of circumspection, of the heirs, by taking the observations under my care, or perhaps usurping them." Arthur Koestler, *The Sleepwalkers* (London: Arkana Books, 1989), 280.

31. The book was published shortly after his own death. *Kepler's Somnium: The Dream, or Posthumous Work on Lunar Astronomy*, trans. Edward Rosen (Madison: University of Wisconsin Press, 1967).

32. Eliza Marian Butler, *The Tyranny of Greece over Germany: A Study of the Influence Exercised by Greek Art and Poetry over the Great German Writers of the Eighteenth, Nineteenth, and Twentieth Centuries* (Cambridge: Cambridge University Press, 1935).

33. Johann Gottfried von Herder, *Another Philosophy of History and Selected Political Writings*, ed. and trans. Ioannis D. Evrigenis and Daniel Pellerin (Indianapolis: Hackett, 2004), 16.

34. Cited in Helmut Peitsch, "Die Rezeption von Reisebeschreibungen über den Pazifik," in "Johann Herders Philosophie der Geschichte der Menschheit," *Herausforderung Herder / Herder as Challenge: Ausgewählte Beiträge zur Konferenz der Internationalen Herder-Gesellschaft Madison*, ed. Sabine Groß (Heidelberg: Synchron, 2006), 98–99.

35. Johann Gottfried von Herder, *Sammtliche Werke*, "Zur Philosophie und Geschichte," Zehnter Teil, Adrastea 2, "Das 18 Jahrhundert," 187–188.

36. Ibid., 189.

37. Kant's late unpublished reflection: *Reflections on Anthropology*, no. 1520, 15:888.

38. The translation appears in August Wilhelm Schlegel, trans., *Shakespeare's dramatische Werke* (Berlin: Johann Friedrich Unger, 1798), vol. 3.

39. Armin von Ungern-Sternberg, *Erzählregionen: Überlegungen zu literarischen Räumen mit Blick auf die deutsche Literatur des Baltikums, das Baltikum und die deutsche Literatur* (Bielefeld, Germany: Aisthesis, 2003).

40. Karl Schlögel, *Im Raume lesen wir die Zeit: Über Zivilisationsgeschichte und Geopolitik* (Munich: Carl Hanser Verlag, 2003), 39.

41. The Greek Phorkys was the son of Pontus (Sea) and Gaia (Earth). *Ho pontos* was the Greek toponym for the Mediterranean. In some ancient mosaics, Phorkys is depicted as a fish-tailed merman with crab-claw forelegs and red-spiked skin, as in the mosaic taken from the late Roman Trajan baths at Acholla, now at the Bardo Museum in Tunis, Tunisia.

42. Johann Gottfried von Herder, *Outlines of the Philosophy of the History of Man* [unfinished], trans. T. Churchill (London, 1880), 21.

43. Johann Gottfried von Herder, *Outlines of the Philosophy of the History of Man*, trans. T. Churchill, 2nd ed., 3 vols. (London, 1803), 309.

44. Ivo Vulkcevich, *Rex Germanorum, Populos Sclavorum: An Inquiry into the Origin and Early History of the Serbs/Slavs of Sarmatia, Germania and Illyria* (Santa Barbara, CA: University Center Press, 2011), 315.

45. Ludwig Kosegarten, as cited in Karl Lappe, *Mitgabe nach Rügen. Den Reisenden zur Begleitung und Erinnerung* (Stralsund, 1818), 92.

46. Ludwig Kosegarten, *Ida of Plessen*, 2 vols. (Dresden, 1800).

47. Cited in Albert Boime, trans., *Art in an Age of Bonapartism, 1800–1815* (Chicago: University of Chicago Press, 1990), 595.

48. Ernst Moritz Arndt, *Erinnerungen 1769–1815* (Berlin: Verlag der Nation, 1985), 70.

49. Roswitha Schieb, "The Island of Rügen as Mythic Site of Germany," *Baltic Sea Library,* http://www.balticsealibrary.info/index.php?option=com_flexicontent&view=items&cid=66:essays&id=153:the-island-of-ruegen-as-mythic-site-of-germany&Itemid=29.

50. Wilhelm Müller, *Muscheln von der Insel Rügen* (1825).

51. Henry Wadsworth Longfellow, ed., *Poems of Places: An Anthology*, 31 vols. (Boston: Houghton Mifflin, 1876–1879). W. W. Story translated the poem "Rügen, the Island Vineta."

52. Hans Henny Jahnn, *Werke und Tagebücher in sieben Bänden*, ed. Thomas Freeman and Thomas Scheuffelen (Hamburg: Hoffmann & Campe, 1974), 7:368–375. See Reinhard Reichstein, "Bornholm in the Work of Hans Henny Jahnn," *Baltic Sea Library*, http://www.balticsealibrary.de/index.php?option=com_flexicontent&view=items&cid=66:essays&id=157:bornholm-in-the-work-of-hans-henny-jahnn&Itemid=29.

53. "However, it is undeniable that that embankment has been made by human hands, and since it could not possibly have served as some kind of fortress, it is highly probable that the lake, embankment and grove served the purpose of some form of worship. This is reinforced by the powerful impression that the wonderful natural environment of the site necessarily leaves on one. The solitary, undisturbed, blackish lake, the dense beeches with their thick foliage, the complete silence, which is only interrupted by the rustle of the thick layer of beech leaves under the feet of the wayfarer, and the mysterious meaning of the space enclosed by the embankment and the lake immerse the soul in a sacred and silent menace. It is hard to imagine another place imbued with such a character of sacredness and reverence." Wilhelm von Humboldt, "Reisetagebücher," in *Gesammelte Schriften* (Berlin: Behr, 1916), 16:284–285.

54. Karl Friedrich von Schlegel, *The Philosophy of History*, trans. James Burton Robertson (1829; London: Saunders & Otley, 1835), 2:41.

55. Rainer Schmitz, ed., *Bis nächstes Jahr auf Rügen. Briefe von Friedrich Schleiermacher und Henriette Herz an Ehrenfried von Willich 1801–1807* (Berlin: Evangelische Verlagsanstalt, 1984), 104–105.

56. The title in German is *Eine Fahrt nach Pommern und der Insel Rügen.*

57. Ibid., 67.

58. Goethe, *Wilhelm Meister's Apprenticeship*, ed. and trans. Eric A. Blackall and Victor Lange (Princeton, NJ: Princeton University Press, 1995), bk. 5, p. 180. A contender for an earlier Bildungsroman would be the twelfth-century Ibn Tufail's *Hayy Ibn Yaqzan* [Philosophus Autodidactus]. See Chapter 7.

59. He also focuses our attention on Usedom, settled by ancient Rügians before the fifth century AD and purchased by King Frederick William I of Prussia in 1720.

60. "Ich brauche nicht von Neuem zu zeigen, was Andre bereits besser gethan als ich es vermöchte, dass wir ein gutes vollgültiges Recht besitzen, Shakespeare für einen *deutschen* Dichter zu erachten. [W]ir wollen den Engländer Shakespeare gleichsam *entenglischiren,*

wir wollen ihn *verdeutschen*, verdeutschen im weitesten und tiefsten Sinne des Worts, d. h. wir wollen nach Kräften dazu beitragen, dass er das, was er bereits ist, ein *deutscher* Dichter, immer mehr im wahrsten und vollsten Sinne des Worts *werde*." Hermann Ulrici, "Jahresbericht [des Präsidenten]," *Shakespeare Jahrbuch* 2 (1867): 2, 3.

61. One of his admiring students at the University of Heidelberg in 1921 was the young Joseph Goebbels, who, although he says he learned much from Gundolf and in fact had wanted to do a doctorate degree with him, saw his works banned in 1933 when Hitler came to power.

62. Published in German as *Land und Meer: Eine weltgeschichtliche Betrachtung* (Leipzig: Reclam Verlag, 1942); translated by Simona Draghici in 1954 as *Land and Sea*, rev. Greg Johnson (Washington, DC: Plutarch Press, 1997), chap. 17.

63. See Alfred Hiatt, "'From Hulle to Cartage': Maps, England, and the Sea," in *The Sea and Englishness in the Middle Ages: Maritime Narratives, Identity and Culture*, ed. Sebastian I. Sobecki (Cambridge: D. S. Brewer, 2011).

CHAPTER 14

1. Richard Wagner, "Die deutsche Oper," *Zeitung für die elegante Welt*, June 10 1834.

2. For the source, see Richard Wagner, "Autobiographic Sketch," in *Richard Wagner's Prose Works*, vol. 1, *The Art-Work of the Future*, trans. William Ashton Ellis (London: Kegan, Paul, Trench, Trübner & Co., 1895).

3. "This voyage [through Norway] I never shall forget as long as I live; it lasted three and a half weeks, and was rich in mishaps. Thrice did we endure the most violent of storms, and once the captain found himself compelled to put into a Norwegian haven. The passage among the crags of Norway made a wonderful impression on my fancy." Ibid.

4. Siegbert Salomon Prawer, *Frankenstein's Island: England and the English in the Writings of Heinrich Heine* (Cambridge: Cambridge University Press, 1986). See also Gerald Opie, "Review of *Frankenstein's Island*—Deutschland: A Not So Sentimental Journey," *Modern Language Review* 83, no. 2 (1988): 524–526.

5. See Willi Goetschel's discussion of Heine's *Reisebilder* [Travel Pictures; 1826] in *The Literary Encyclopedia*, http://german.utoronto.ca/~goetschel/heine_travel_pictures.pdf. See also Heine's fanciful renditions that became sources for operas by Richard Wagner: for example, the story of the Flying Dutchman in Heine, "Aus den Memoiren des Herren Schnabelewopski" [From the Memoirs of Herr Schnabelewopski], in *Der Salon I* (Hamburg: Hoffmann & Campe, 1834).

6. Goetschel, *Literary Encyclopedia*.

7. Anthony Phelan, *Reading Heinrich Heine* (Cambridge: Cambridge University Press, 2007), 171.

8. The third act of Wagner's *Siegfried*, for example, is partly informed by the Old Norse *Poetic Edda*: "Málrúnar skaltu kunna / ef þú vilt, at manngi þér / heiftum gjaldi harm: / þær of vindr, / þær of vefr, / þær of setr allar saman, / á því þingi, / er þjóðir skulu / í fulla dóma fara" [Speech-runes you should know, so that no man / Out of hatred may do you harm: / These you shall wind, these you shall fold, / These you shall gather together, / When the people throng to the *Thing* to hear / Just judgements given.] *Sigrdrífumál*, v.10/11; in the *Codex Regius*.

9. Richard Wagner, "The Wibelungen: World History as Told in Saga," in *Richard Wagner's Prose Works*, vol. 7, *In Paris and Dresden*, trans. William Ashton Ellis (London:

Kegan, Paul, Trench, Trübner & Co., 1898), 259–260. The title of the essay in German is "Die Wiebelungen: Weltgeschichte aus der Saga."

10. Museum Oskar Reinhart am Stadtgarten, Winterthur, Switzerland.

11. See Chapter 6.

12. On rheology, see Chapter 8.

13. *OED*, "Reef," n2.

14. See also "nur der erzielt sich den Zauber, / zum Reif zu zwingen das Gold."

15. Bulstrode Whitelocke, dispatched on a mission to Christina, queen of Sweden, to conclude a treaty of alliance and to ensure the freedom of passage in the Øresund, writes in his diary entry for November 13, 1653, that his "ship . . . made foule water passing over the Riffe near Jutland in Denmark." *The Diary of Bulstrode Whitelocke, 1605–1675*, ed. Ruth Spalding (Oxford: Oxford University Press, 1990), 300.

16. Wagner, "Autobiographic Sketch."

17. On Wagner's Rhine maidens and the Lorelei, see Deryck Cooke, *I Saw the World End: A Study of Wagner's Ring* (Oxford: Oxford University Press, 1979), 139.

18. Heinrich Heine, *Die Loreley*, trans. Robert Clarke (2001).

19. Freiligrath signs the poem "Sankt Goar, 1844." Sankt Goar is by Lorelei way.

20. See Oswald Georg Bauer, *Josef Hoffmann: Der Bühnenbildner der ersten Bayreuther Festspiele* (Munich: Deutscher Kunstverlag, 2008).

21. Specifically, the painter-brothers Gotthold and Marc Brückner and the father-son set designers Karl and Friedrich Brandt. There are no full records of how they reworked the reefs. See Patrick Carnegy, "Designs on the Ring," *Wagner Journal* 4, no. 2 (2010): 41–55.

22. Gottfried Niemann, *Richard Wagner and Arnold Böcklin, oder über das Wesen von Landschaft und Musik* (1904; Whitefish, MT: Kessinger Publishing, 2010).

23. The person who gave this work that title was not the painter; it was the art dealer for the version of 1883.

24. Rainer Karlsch, in *Hitlers Bombe: Die geheime Geschichte der deutschen Kernwaffenversuche* (Munich: Deutsche Verlags-Anstalt, 2006), claims that nuclear weapons were tested at Ohrdruf on Rügen.

25. Hermann Oberth, *Planetenräumen* [By Rocket into Planetary Space] (Munich: R. Oldenbourg, 1923).

26. Peter Schjeldahl, "The Romantic Vision of Caspar David Friedrich: Paintings and Drawings from the USSR," in *Columns and Catalogues* (New York: Figures Press, 1994), 21.

27. Wilfried Daim, who founded the Institute for Political Psychology in Vienna in 1956, wrote *Der Mann, der Hitler die Ideen gab: Von den religiösen Verirrungen eines Sektierers zum Rassenwahn des Diktators* (Munich: Isar Verlag, 1958). The abbreviated title in English is *The Man Who Gave Hitler the Ideas*.

28. For the castle on Rügen, see Dusty Sklar, *Hitler and the Occult* (New York: Dorset Press, 1977), 20. The German-version subtitle of *Ostara* is *Briefbücherei der Blonden und Mannesrechtler*.

29. Gordon Craig, *Theodor Fontane: Literature and History in the Bismarck Reich* (New York: Oxford University Press, 1999).

30. Christian Grawe, *Theodor Fontane: Effi Briest* (Frankfurt: M. Diesterweg, 1988), 212.

31. See Theodor Fontane, *Meine Kinderjahre* (Berlin: F. Fontane, 1894): "Alles [in Swinemünde] war Poesie."

32. See Hans Peter Neureuter, "The Baltic Literary Region: Some Remarks on the Cultural Identity of a European Province," *Baltic Sea Library*, http://www.balticsealibrary .de/index.php?option=com_flexicontent&view=items&cid=66:essays&id=148:the-baltic -literary-region-some-remarks-on-the-cultural-identity-of-a-european-province&Itemid=29.

33. Hermann Oberth, *Primer for Those Who Would Govern* (Clarence, NY: West-Art, 1987).

34. Wagner uses the anachronistic German-language term *Eiland* (island) in *Tristan and Isolde* instead of the more usual *Insel*, but both terms suggest a nesological tradition, Greek and medieval, where such a place as an *Enys* "swims" (nesologically speaking).

35. Richard Wagner, libretto, *Tristan and Isolde*, act 1, sc. 2.

36. Alice Leighton Cleather and Basil Woodward Crump, *Tristan and Isolde: Described and Interpreted in Accordance with Wagner's Own Writings* (London: Methuen, 1905), 27, 92.

37. The full German title is *Tristan und Isolde: Eine Handlung in drei Akten*.

38. Arthur Schopenhauer, *Die Welt als Wille und Vorstellung: Vier Bücher, nebst einem Anhange, der die Kritik der Kantischen Philosophie enthält*, 2nd ed. (Leipzig: Brockhaus, 1844); the first edition was published in 1819.

39. Martin Gregor-Dellin, *Richard Wagner: His Life, His Work, His Century* (London: William Collins, 1983); and Bryan Magee, *The Tristan Chord: Wagner and Philosophy* (New York: Metropolitan Books, 2011).

40. "Married to a Mermaid" from the masque *Alfred*.

41. "The Great Silkie of Sule Skerry (communicated by the late Captain F. W. L. Thomas, R.N., written down by him from the dictation of a venerable lady of Snarra Voe, Shetland)," *Proceedings of the Society of Antiquaries of Scotland* 1 (1852): 86.

42. In his *Teutonic Mythology*, Jacob Grimm remarks on the phenomenon of German midwives, and his translator Stallybrass notes: "It was a popular belief (applied to the Frankish king and gradually distorted) about the union of a wild-woman or mermaid with a Christian hero." Grimm, *Teutonic Mythology*, trans. James Steven Stallybrass, 4th ed. (London: George Bell & Sons, 1992), 1:435n.

43. "*Terra firma*, the Continent, or main Land; so call'd by Geographers." Edward Phillips, *The New World of Words, or, Universal English Dictionary* (London, 1706).

44. John Savage, *A Select Collection of Letters of Antients* (London, 1703), 2.20.

45. Wolfgang Haase and Meyer Reinhold, eds., *The Classical Tradition and the Americas* (Berlin: de Gruyter, 1993), vol. 1, pt. 1; and Polybius, *Histories*, bk. 34.

46. Some people said that ancient "Thule"—already referenced in the ancient period by Strabo in his *Geography* and by Pliny the Elder in his *Natural History*—was the region we now call Iceland (literally "island") in the North Atlantic or the place we now call Saaremaa or Oesel (also literally "island") in the Baltic Sea. Other people say it refers to other islands. A few argue for parts of mainland Europe, including Norway. For the late nineteenth- and early twentieth-century period, the following remark of John Bostock and Henry Thomas Riley, writing in *Natural History of Pliny* (London: George Bell & Sons, 1856–1893), is relevant: "Opinions as to the identity of ancient Thule have been numerous in the extreme" (352). For the meaning of Ultima Thule as "farthest edge, or rim [horizon]," see the Roman poet Virgil, *Georgics*, 1.30.

47. See the literature-oriented essays in Hanjo Berressem, Michael Bucher, and Uwe Schwagmeier, *The Hollow Earth as Concept and Conceit* (Berlin: Lit, 2012); for a study

based partly on German philosophical controversies of the 1920s, especially those involving theories of space planning, see Karl Neupert, *Der Kampf gegen das kopernikanische Weltbild* (Memmingen: Verl. & Dr.-Genossenschaft, 1928).

48. In German: Thule-Gesellschaft (originally Studiengruppe für germanisches Altertum).

49. Others included Hans Frank, Julius Lehmann, Gottfried Feder, Dietrich Eckart, and Karl Harrer. For a more complete list, see Ian Kershaw, *Hitler: 1889–1936 Hubris* (London: Penguin, 2011).

50. See J. A. B. Townsend, "The Viking Society: A Centenary History," in *Saga-Book* 23 (1990). The Viking Society's considerable collection of books and other materials at University College, London, was destroyed during the Blitzkrieg of World War II.

51. The existence of the *Oera Linda* manuscript was made public by Cornelis Over de Linden in the 1860s. By the 1880s, most scholars had dismissed it as a forgery and a fake, but when Herman Wirth's German-language translation appeared in 1933, Nazi ideologues accepted it as genuine. This "endorsement" helped Himmler, Wirth, and Richard Walther Darré—who popularized the phrase *Blut und Boden* (blood and soil)—to a conflict that led to the establishment of the Nazi think tank known as the Ahnenerbe. The Ahnenerbe, claiming that the Aryans came from a North Atlantic or Baltic "Atlantis," formulated plans to send expeditions to such places as Iceland; see Heather Pringle, *The Master Plan: Himmler's Scholars and the Holocaust* (New York: Hyperion, 2006). Although the distinguished scholar Arthur Hübner and other academics again proved that the work was a counterfeit—thus defeating both Wirth and Himmler—it remained influential in popular and political spheres.

52. The report continues: "Visual rays were not suitable because of refraction; but infrared rays had less refraction. Accordingly a party of about ten men under the scientific leadership of Dr. Heinz Fischer, an infrared expert, was sent out from Berlin . . . to photograph the British fleet with infrared equipment at an upward angle of some forty-five degrees." See Gerard P. Kuiper, "German Astronomy during the War," *Popular Astronomy* 54, no. 6 (1946): 262–286.

53. Karl Neupert, *Geokosmos: Weltbild der Zukunft—Forschungen über Weltbau, Natur und Ursprung des Lebens* (Zurich: Gropengiesser, 1942); see also the Postamble.

54. His PhD thesis was titled "Die bruchlose Deformation von Fossilien durch tektonischen Druck und ihr Einfluss auf die Bestimmung der Arten. Beobachtet und bearbeitet an den Pelecypoden der St. Galler Meeresmolasse," University of Zurich, 1929.

55. Insofar as *iceberg* in German (*Eisberg*) means "ice mountain," *S.O.S. Eisberg* links directly with earlier Fanck "mountain movies." These include films like *The Holy Mountain*, with Leni Riefenstahl; *The White Hell of Pitz Palu*, with Leni Riefenstahl and World War I pilot Ernst Udet, who plays himself as a rescuer; and *Storm over Mont Blanc*, likewise starring Riefenstahl and Udet.

56. Quoted by Pappus of Alexandria, *Synagoge* [Collection], bk. 7.

57. The other film was *The Rebel*, directed by Edwin H. Knopf and Luis Trenker, about Napoléon's attempt to seize the Tirol region of the Alps in 1809.

58. Leni Riefenstahl, *Leni Riefenstahl: A Memoir* (New York: St. Martin's Press, 1993), 113, 115. Other parts of the movie were shot in the Alps.

59. See the Postamble.

60. In his *isolario* (island book), written in 1529, Benedetto Bordone's map of the New World displays North America as a large island labeled "Terra del Laboratore." (The name now belongs only to the mainland region of Newfoundland.)

61. See also Chapter 4.

62. Immanuel Kant, *Critique of Pure Reason*, trans. John Miller Dow Meiklejohn (London: Henry G. Bohn, 1855).

63. Immanuel Kant, *Critick [sic] of Pure Reason*, trans. Francis Haywood (London: William Pickering, 1848), 197.

64. Immanuel Kant, *Critique of Pure Reason*, trans. Norman Kemp Smith (New York: St. Martin's Press, 1929).

65. Immanuel Kant, *Critique of Pure Reason*, trans. Paul Guyer and Allen W. Wood (Cambridge: Cambridge University Press, 1998).

66. The love of the right woman—in this case, Riefenstahl—is the main redemptive factor in the German version of *S.O.S. Eisberg*, as in Wagner's later operas.

67. Gibson Gowland, who plays the part of Dragon in *S.O.S. Eisberg*, had just appeared in George B. Seitz's American silent *The Isle of Forgotten Women* (1927) and Lucien Hubbard's semi-talkie (mostly silent) *Mysterious Island* (1929), which was based very loosely on Jules Verne's novel.

68. Ellen Lawrence's lost husband is Dr. Carl Lawrence (played by Rod La Rocque) in the American version; in the German version he is Dr. Karl Lorenz (played by Gustav Diessl).

69. "Wundervoll die Aufnahmen aus dem arktischen Meer, die Niederbrüche der kalbenden Eisberge. Unvergleichlich die grandiose Aufnahme eines Sturmes auf dem Eismeer, die ergreifendsten, packendsten Bilder, seitdem es eine Kinematographie gibt." "S.O.S. Eisberg," *Kinematograph*; cited in Gisela Pichler and Jan-Christopher Horak, *Berge, Licht und Traum: Dr. Arnold Frank und der deutsche Bergfilm* (Munich: GeraNova Bruckmann, 1998), 219.

70. He makes the statement emphatic by adding his handwritten signature.

71. Paul Dessau also did the music for two other of Arnold Fanck's mountain films: *Storm over Mont Blanc* (1930) and *The White Flame: New Miracles of the Snowshoe* (1931).

72. See Lion Feuchtwanger, *Two Anglo-Saxon Plays: "The Oil Islands," "Warren Hastings,"* trans. Willa Muir and Edwin Muir (London: M. Secker, 1929).

73. Paul Dessau, "Musik der Gründerjahre: Ein Interview," *Theater der Zeit* 13, no. 12 (1958): 19–20; cited in Elaine Kelley, "Imagining Richard Wagner: The Janus Head of a Divided Nation," in *Imagining the West in Eastern Europe and the Soviet Union*, ed. György Péteri (Pittsburgh: University of Pittsburgh Press, 2010), 144.

74. Consider, however, that Daniel Defoe actually did not set his story on Juan Fernández. The full subtitle of his book is *The Life and Strange Surprizing Adventures of Robinson Crusoe, of York, Mariner: Who lived Eight and Twenty Years, all alone in an un-inhabited Island on the Coast of America, near the Mouth of the Great River of Oroonoque; Having been cast on Shore by Shipwreck, wherein all the Men perished but himself. With An Account how he was at last as strangely deliver'd by Pirates.*

75. See Janine Hansen, *Arnold Fanck's "Die Tochter des Samurai": Nationalsozialistische Propaganda und japanische Filmpolitik* (Wiesbaden: Harrassowitz, 1997). The Japanese-German movie *Atarashiki tsuchi* [Die Tochter des Samurai] was made in 1937.

76. Udet, who did not approve of Hitler's military policies, killed himself in the offices of the Luftwaffe. For the personal circumstances, see Armand van Ishoven, *The Fall of an Eagle: The Life of Fighter Ace Ernst Udet* (London: W. Kimber, 1979).

77. Quoted in John Clare, "Eagle for Hire," *Maclean's*, May 15, 1948.

78. Hanns Eisler, "Bericht über die Entstehung eines Arbeiterliedes," in *Musik und Politik, Schriften 1924–1948*, ed. Günter Meyer (Munich: Rogner & Bernhard, 1973), 274–280.

79. Quoted earlier in the epigraph to the current section. Bertolt Brecht, "Bei Durchsicht meiner ersten Stücke," in *Gesammelte Werke*, 20 vols. (1954; Frankfurt: Suhrkamp, 1967), 17:952. "Bei Durchsicht meiner ersten Erst ist da noch Land, aber schon mit Lachen, die zu Tümpeln und Sunden werden; dann ist nur noch das schwarze Wasser weithin, mit Inseln die schnell."

80. There would be not only blimps and balloons but also the sort of tethered floating island that Russian architect Alexander Asadov puts forward—for example, the Aerohotel. Writes Asadov: "The project is an alternative for the man-made islands." See Mike Chino, "The Floating Aerohotel: A Modern Aquatecture Marvel," http://inhabitat.com /aerohotel-by-alexander-asadov.

81. Francesco Lana de Terzi, *Prodromo, overo, Saggio di alcune invenzioni nuove premesso all'arte maestra* [To Test Some Premised New Inventions of the Master Artist] (Brescia, 1670).

82. *OED*, s.v. "aeronautical," etymology.

83. R. Hooke, ed., *Philosophical Collections, containing an Account of Such Physical, Anatomical, or other Mathematical and Philosophical Experiments and Observations, as have Lately come to the Publishers Hands* (1679–1682), 1.18 heading.

84. "Aerostatics: Anecdotes of Mr. Blanchard," *London Magazine: Enlarged and Improved* 3 (1784): 391: "Mr. Blanchard . . . during the late war, formed a *flying boat* [my emphasis], which he intended for carrying the dispatches from Brest to Paris, but as this did not answer his expectations, he was obliged to relinquish the idea of elevating himself above the clouds."

85. H. G. Wells, *The War in the Air; and Particularly How Mr. Bert Smallways Fared While It Lasted*, illus. Eric Pape (London: George Bell & Sons, 1908), 151. At a main turning point in the story, Smallways is stranded on Goat Island in the midst of Niagara Falls.

86. For a translation of the fifth and sixth chapters of *Prodromo*, see *The Aerial Ship by Francesco Lana-Terzi*, ed. Thomas O'Brien Hubbard and John Henry Ledeboer (London: King, Sell & Olding, 1910).

87. Charles Frederick Snowden Gamble, *Story of a North Sea Air Station* (London: Oxford University Press, 1928), xxii, 400.

88. Scott Anthony, "Aviation and Enchantment: Snowden Gamble and the Public Education Programmes of Imperial Airways in 1930s Britain" (lecture), *Sacred Modernities* 18 (September 2009); mp3 file downloaded from beemp3.com.

89. The Jesuit father wrote that "God will never allow that such a machine be built . . . because everybody realises that no city would be safe from raids. . . . Iron weights, fireballs and bombs could be hurled from a great height." See Richard Hallion, *Taking Flight: Inventing the Aerial Age from Antiquity through the First World War* (New York: Oxford University Press, 2003).

90. Arthur K. Kuhn, "Aërial Navigation and Its Relation to International Law," *Proceedings of the American Political Science Association* 5 (1908): 84.

91. "Book Notes," *Political Science Quarterly* 27 (1912): 176, in a discussion of J. F. Lycklama à Nijeholt, *Air Sovereignty* (The Hague: Martinus Nijhoff, 1910).

92. The crossing was first scheduled for 1913, but was postponed when war broke out until 1919.

93. Élie Halévy, *History of the English People in the Nineteenth Century*, trans. E. I. Watkin (London: Ernest Benn, 1952), 6:582.

94. See Andrew Wawn, *The Vikings and the Victorians: Inventing the Old North in Nineteenth-Century Britain* (Cambridge: D. S. Brewer, 2000).

95. December 8, 1908; cited in Alfred M. Gollin, *No Longer an Island: Britain and the Wright Brothers, 1902–1909* (Stanford, CA: Stanford University Press, 1984), 410.

96. Cited in Gollin, *No Longer an Island*, 335.

97. Walter Wellman, *The Aerial Age: A Thousand Miles by Airship over the Atlantic Ocean; Airship Voyages over the Polar Sea; the Past, the Present and the Future of Aerial Navigation* (New York: A. R. Keller, 1911).

98. Walter Wellman, *The German Republic* (New York: E. P. Dutton, 1916).

99. Walter Wellman, "A Tragedy of the Far North," in *The White World: Life and Adventures within the Arctic Circle Portrayed by Famous Living Explorers*, ed. Rudolf Kersting for the Arctic Club (New York: Lewis, Scribner, 1902).

100. Zeppelin postage stamps are studied by "aero-philatelists" as a special area of interest. See Charles Kiddle, *German Aviation: Zeppelins* (Alton, UK: World Poster Stamps, 2008), based on earlier work by Kuno Sollors, who was president of the International Federation of Aero-Philatelic Societies. See especially Sollors, *Zeppelin, Parseval und andere Luftschiffe* (self-published, 1973).

101. These postal issues from the Third Reich were overstamped "Chicago fahrt Weltausstellung 1933," or "Chicago World Exhibition Flight 1933."

102. Others included Hans Frank, Julius Lehmann, Gottfried Feder, Dietrich Eckart, and Karl Harrer. For a more complete list, see Ian Kershaw, *Hitler: 1889–1936 Hubris* (London: Penguin, 2011).

103. Ruth Heller, *A Sea within a Sea: Secrets of the Sargasso* [children's book] (New York: Grosset & Dunlap, 2000).

104. Based on Crittenden Marriott's novel *The Isle of Dead Ships: A Tale of the Sargasso Sea* (London: Readers Library Publishing, 1909).

105. For details, see Emily Jane Brontë, *Gondal Poems*, ed. Helen Brown and Joan Mott (Oxford: Shakespeare Head Press, 1938), 5–8.

106. "A Different Perspective: Vaniman the Acrobatic Photographer," curated by Alan Davies with help from Alan Tierney, State Library of New South Wales, Sydney, 2001.

107. Francis Bacon, *Sylva Sylvarum* (London: William Lee, 1627), sec. 790 (margin).

108. First published in the *Atlantic Monthly* (December 1869–February 1870), the story was later collected in Edward Everett Hale, *The Brick Moon and Other Stories* (Boston: Little, Brown, 1899).

109. Sam Moskowitz, "The Real Earth Satellite Story," in *Explorers of the Infinite: Shapers of Science Fiction* (Cleveland: World Publishing, 1963).

110. Relevant here are other works by Hale such as "The Man without a Country," *Atlantic Monthly* (December 1863), and *Sybaris and Other Homes* (Boston: Fields, Osgood, 1869), which is about a colony of ancient Sybarians discovered on an Italian island.

111. William Sloane Kennedy, "Edward Everett Hale," *Century Illustrated Monthly Magazine* 29 (n.s., 7) (November 1884–April 1845).

112. The song collector and performer Frank Warner found this song in Edward Boatner's small hymnal *Spirituals Triumphant, Old and New* (Nashville: Sunday School Publishing Board, National Baptist Convention, 1927).

113. Colin Larkin, ed., *The Encyclopedia of Popular Music*, 4th ed., 10 vols. (New York: Muze, 2006), s.v. "London, Laurie."

114. Bolt's script includes these famous lines, spoken by Thomas More to his daughter Meg: "When a man takes an oath, Meg, he's holding his own self in his own hands. Like water. [cups his hands] And if he opens his fingers then he needn't hope to find himself again."

CHAPTER 15

1. The one near Oberröblingen in Saxony-Anhalt is 7,000 years old. According to Robert Ganslmeier of the State Museum of Prehistory in Halle, Germany, it is 100 meters long.

2. Martin Schmidt, "Reconstruction as Ideology: The Open Air Museum at Oerlinghausen, Germany," in *The Constructed Past: Experimental Archaeology, Education, and the Public*, ed. Peter G. Stone and Philippe G. Planel (London: Routledge, 1999), 146.

3. Martin Griepentrog, "'Frischer Wind' in der musealen 'Leichenhammer': Zur Modernisierung kulturhistorischer Museen von Jahrhundertwende bis zum Nationalsozialismus," *Geschichte in Wissenschaft und Unterricht* 42 (1991): 153–173.

4. *Die deutsche Vorgeschichte: Eine hervorragend nationale Wissenschaf* (Würzburg: C. Kabitzsch, 1914); cited in Bettina Arnold, "The Past as Propaganda: How Hitler's Archaeologists Distorted European Prehistory to Justify Racist and Territorial Goals," *Archaeology* (July–August 1992): 30–37.

5. Bettina Arnold, "The Past as Propaganda: Totalitarian Archaeology in Nazi Germany," *Antiquity* 64 (1990): 464–478.

6. By 1990, most people recognized that these reconstructions were "politically biased," but just how and why remains unexplored. See Gunter Schöbel, *Hans Reinerth Forscher, nationalsozialistischer Funktionär, Museumsleiter, Leben und Werdegang, was blieb?* (2001); cited in Urs Leuzinger, "Experimental Archeology," in *Living on the Lake in Prehistoric Europe: 150 Years of Lake-Dwelling Research*, ed. Francesco Menotti (London: Routledge, 2004), 239.

7. See Adolf Max Vogt, "Le Corbusier and Swiss Lake Dwellings of the Neolithic Age," in *Das Bauwerk und die Stadt: Aufsätze für Eduard F. Sekler* [The Building and the Town: Essays for Eduard F. Sekler], ed. Wolfgang Böhm (Vienna: Böhlau Verlag, 1994), 319.

8. Quoted in Andrew Sherratt, "The Importance of Lake Dwellings," in *Living on the Lake*, 272.

9. Genesis 1:19, King James version.

10. For example, the "Burga Water Island" dun in the Shetlands. For the "duns" generally, see Ian Armit, *The Archaeology of Skye and the Western Isles* (Edinburgh: Edinburgh University Press, 1996). See also D. W. Harding, "Crannogs and Island Duns: An Aerial

Perspective," in *Archaeology from the Wetlands: Proceedings of the 11th WARP Conference, Edinburgh 2007*, ed. J. Barber, C. Clarke, A. Crone, A. Hale, J. Henderson, R. Housley, and A. Sheridan (Edinburgh: Society of Antiquaries of Scotland), 267–273.

11. Theodore William Moody, Dáibhí Ó Cróinín, Francis X. Martin, and Francis John Byrne Dineen, *A New History of Ireland: Prehistoric and Early Ireland* (Oxford: Oxford University Press, 2008).

12. For example, the Principality of Sealand off the coast of Suffolk in the North Sea and the Republic of Rose Island off the coast of Italy in the Adriatic Sea.

13. Paul Rainbird, *The Archeology of Micronesia* (Cambridge: Cambridge University Press, 2004), 92–98. Eugene O'Curry, *On the Manners and Customs of the Ancient Irish*, ed. William Sullivan Kirby (London: Williams & Norgate, 1873), 3:3: "The Dun was of the same form as the *Rath*, but consisting of at least two concentric circular mounds or walls, with a deep trench full of water between them."

14. M. Pinedo-Vasquez, M. L. Ruffino, C. Padoch, and E. S. Brondízio, eds., *The Amazon Várzea: The Decade Past and the Decade Ahead* (New York: Springer, 2011).

15. Herodotus, *Histories*, 5.16–17.

16. See Marc Shell, *Polio and Its Aftermath: The Paralysis of Culture* (Cambridge, MA: Harvard University Press), 99–109.

17. In Newfoundland, a "waterman" is "a supernatural figure inhabiting the sea or 'salt water.'" *Dictionary of Newfoundland English*, s.v. "water man." Memorial University Folklore and Language Archive, Manuscript Collection: D. James; Corner Brook; 1971 (M 71-103).

18. Christina Fredengren, *Crannogs: A Study of People's Interaction with Lakes, with Special Reference to Loch Gara in the North West of Ireland* (Bray, Ireland: Wordwell, 2009).

19. "In grassland regions he plants trees around his homes, towns, and farms. . . . When man settles in the forest he replaces most of it with grasslands and croplands, but leaves patches of the original forest on farms and around residential areas. . . . Man depends on grasslands for food, but likes to live and play in the shelter of the forest." Eugene P. Odum, *Ecology* (New York: Holt, Rinehart & Winston, 1963).

20. *Terp* means "village" in Old Frisian. It is cognate with the English-language *thorp* (cf. German *dorf*), meaning "hamlet" or "ham," and connotes water-surrounded land: islands.

21. "'The Dry Salvages,'" comments Eliot—presumably *les trois sauvages*—"is a small group of rocks, with a beacon, off the N.E. coast of Cape Ann, Massachusetts. . . ." The poem was published in the *British Weekly*. Eliot sailed not only on the Mississippi River but also in the waters off Massachusetts.

22. Lyndall Gordon, *T. S. Eliot: An Imperfect Life* (New York: Norton, 2000), 336–337.

23. Lois A. Cuddy, *T. S. Eliot and the Poetics of Evolution: Sub/Versions of Classicism, Culture, and Progress* (Lewisburg, PA: Bucknell University Press, 2000).

24. It appeared in England as *Look, Stranger* (London: Faber & Faber, 1936) and in the United States as *On This Island* (New York: Random House, 1937).

25. See Nicholas Jenkins, *The Island: W. H. Auden and the Regeneration of England* (Cambridge, MA: Harvard University Press, 2009).

26. Compare M. F. Howley, "Newfoundland Name-Lore," *Newfoundland Quarterly* 9, no. 1 (1909): 9: "The fishermen are accustomed, in foggy weather, to find their bearings

by carefully listening to the rout of the sea on the shore, which they (very correctly) call rote, or rut."

27. *Callimachus, Hymns and Epigrams. Lycophron. Aratus*, trans. A. W. Mair and G. R. Mair, Loeb Classical Library 129 (Cambridge, MA: Harvard University Press, 1921), 909.

28. Howley, "Newfoundland Name-Lore," 9.

29. Ibid.

30. Thomas James, *Strange and Dangerovs Voyage of Captaine Thomas Iames: In his Intended Discouery of the Northwest Passage into the South Sea* (London: I. Partridge, 1633), 8.

31. Jay W. Baird, "Hitler's Muse: The Political Aesthetics of the Poet and Playwright Eberhard Wolfgang Möller," *German Studies Review* 17 (1994): 269–286.

32. For example, the German term *Taething*.

33. Emanuel Gebauer, *Fritz Schaller: Der Architekt und sein Beitrag zum Sakralbau im 20. Jahrhundert* (Cologne: Bachem, 2000). See also Gebauer, "Das Thing und der Kirchenbau: Fritz Schaller und die Moderne (1933–1974)," PhD diss., Johannes Gutenberg University of Mainz, 1995.

34. "The Thing and the Church: Fritz Schaller and Modernity from 1933 to 1974" contains chapters on the construction sites at the beginning of the *Thing* of National Socialism.

35. See Roger Manvell and Heinrich Fraenkel, *Dr. Goebbels: His Life and Death* (London: Heinemann, 2010).

36. Wolfgang von Moers-Messmer, *Der Heiligenberg bei Heidelberg: Ein Führer durch seine Geschichte und seine Ruinen* [The Holy Mountain near Heidelberg: A Guide to Its History and Its Ruins] (Community Protection eV Heiligenberg, 1987).

37. Joseph Goebbels, *Heidelberger Volksblatt* 144 (June 24, 1935). "In diesem monumentalen Bau haben wir unserem Stil und unserer Lebensauffassung einen lebendigen plastischen und monumentalen Ausdruck gegeben. . . . Diese Stätten sind in Wirklichkeit die Landtage unserer Zeit. . . . Es wird einmal der Tag kommen, wo das deutsche Volk zu diesen steinernen Stätten wandelt, um sich auf ihnen in kultischen Spielen zu seinem unvergänglichen neuen Leben zu bekennen." Ibid.

38. William Niven, "The Birth of Nazi Drama? Thing Plays," in *Theatre under the Nazis*, ed. John London (Manchester: Manchester University Press, 2000), 54–95.

39. Rainer Stommer, *The Staged National Community. The "Thing" Movement in the Third Reich* (Marburg: Jonas, 1985). See also Henning Eichberg, "The Nazi *Thingspiel*: Theater for the Masses in Fascism and Popular Culture," *New German Critique* 11 (Spring 1977), 133–150.

40. Published by Langen-Müller, Berlin, 1936. See Karl-Heinz Schoeps, *Literature and Film in the Nazi Reich* (Rochester, NY: Camden House, 2004).

41. Glen W. Gadberry, "Eberhard Wolfgang Möller's Thingspiel *Das Frankenburger Würfelspiel*," in *Massenspiele: NS-Thingspiel, Arbeiterweihespiel und olympisches Zeremoniell*, ed. Henning Eichberg, Michael Dultz, Glen Gadberry, and Günther Rühle, Problemata 58 (Stuttgart-Bad Cannstatt: Frommann-Holzboog, 1977), 238. See also Gadberry, "E. W. Möller and the National Drama of Nazi Germany: A Study of the Thingspiel and of Möller's *Das Frankenburger Würfelspiel*," master's thesis, University of Wisconsin, 1972.

42. Published by W. J. Schnell, Warendorf, 1924.

43. Published by W. J. Schnell, Warendorf, 1923.

44. Published by G. Reinhold, Dresden, 1930.

45. See Martin Heidegger, *Vorträge und Aufsätze* (Pfullingen: Günter Neske, 1954), 173. Concerning the seminar, see Heidegger, *Die Frage nach dem Ding: Zu Kants Lehre von den transzendentalen Grundsätzen*, ed. Petra Jaeger (Frankfurt: Klostermann, 1984).

46. Martin Heidegger, *Poetry, Language, Thought*, trans. Albert Hofstadter (New York: Harper & Row, 1971), 174; quoted in Fred Moten, "The Case of Blackness," *Criticism* 50, no. 2 (2008): 182.

47. According to Bruno Latour, "The liberal model of politics needs to be complemented by a new *Dingpolitik* in which objects rejoin people in processes of representation." Latour is careful to note that in practice these two modes of representation have always been mixed. Still, it is difficult not to take his manifesto for a *Dingpolitik*.

48. This was followed by another essay in 1922, "Politische Theologie," in which Schmitt gives further substance to his authoritarian theories, analyzing the concept of "free will," which was influenced by Christian/Catholic thinkers. The book begins with Schmitt's famous, or notorious, definition: "Sovereign is he who decides on the exception." By "exception," Schmitt means the appropriate moment for stepping outside the rule of law in the public interest. The book's title derives from Schmitt's assertion that "all significant concepts of the modern theory of the state are secularized theological concepts": political theory addresses state sovereignty in much the same manner that theology treats God.

49. Thomas B. Macaulay, *History of England* (1849), 1.1.90.

50. David Lindsay Keir, "The Case of Ship-Money," *Law Quarterly Review* 52 (1936): 546–574.

51. Ioannis D. Evrigenis, *Fear of Enemies and Collective Action* (Cambridge: Cambridge University Press, 2007), 60, 65.

52. Friedrich Nietzsche, "Was den Deutschen abgeht," in *Götzendämmerung* (Stuttgart: A. Kröner, 1990), 122.

53. See Brecht's *Svendborger Gedichte* (1937).

54. Michael E. Geisler, "In the Shadow of Exceptionalism: Germany's National Symbols and Public Memory after 1989," in *National Symbols, Fractured Identities: Contesting the National Narrative*, ed. Michael E. Geisler (Lebanon, NH: University Press of New England, 2005), 74–75.

55. The pertinent voyage and island are described in *The First Voyage Made to the Coasts of America* (1584), and the mission settlement is mapped in the fourth edition of the astronomer Thomas Hariot's *Briefe and True Account of the New Found Land of Virginia* (1590).

56. See William Strachey, *A Voyage to Virginia in 1609*, ed. Louis B. Wright (Charlottesville: University of Virginia Press, 1964).

57. Published by W. Barret, London, 1610.

58. See Hobson Woodward, *Brave Vessel: The True Tale of the Castaways Who Rescued Jamestown and Inspired Shakespeare's "The Tempest"* (New York: Viking, 2009).

59. Or perhaps they suffered some other setback. See W. Stahle, M. K. Cleaveland, D. B. Blanton, M. D. Therrell, and D. A. Gay, "The Lost Colony and Jamestown Droughts," *Science* 280 (1998): 564–567.

60. See Margaret Laurie, "Did Purcell Set *The Tempest*?" *Proceedings of the Royal Musical Association* 90 (1963–1964): 43–57.

61. *Our Island Home* takes its name from English poet laureate Alfred Lord Tennyson's "The Lotus Eaters" (1832), where mariners are isolated from the rest of the world by

partaking of the lotus, thus recalling the Greek island-dwelling *lotophages* whom Odysseus encounters in Homer's *Odyssey*, bk. 9.

62. Azorín [José Martínez Ruiz], in *Dicho y hecho* (Barcelona: Destino, 1957), pretends to search for Sancho's island.

63. Chorus of Sailors and Relatives, *H.M.S. Pinafore*, lines 543–545.

64. George Keith Chesterton, "False Theory and the Theatre," in *On Lying in Bed and Other Essays*, ed. Alberto Manguel (1923; Calgary: Bayeux Arts, 2000), 147–148.

65. "Flaunt out O sea your separate flags of nations! / Flaunt out visible as ever the various ship-signals! / But do you reserve especially for yourself and for the soul of man / One flag above all the rest." Walt Whitman, *Leaves of Grass* (1855).

66. Op. 10.

67. Consider Churchill's works—for example, his radio speech, "This Century of Tragedy and Storm," given at the Dorchester Hotel in London on July 4, 1950; and his books *Thoughts and Adventures: amid These Storms* (London: Thornton Butterworth, 1932) and *The Gathering Storm* (Boston: Houghton Mifflin, 1948).

68. Op. 33.

69. In particular: "Letter XXII: Peter Grimes."

70. Op. 50.

71. "To the British Empire the Nore Mutiny was what a strike in the fire-brigade would be to London threatened by general arson." Herman Melville, *Billy Budd*, chap. 3.

72. George Ernest Manwaring and Bonamy Dobrée, *The Floating Republic: An Account of the Mutinies at Spithead and the Nore in 1797* (New York: Harcourt, Brace, 1935).

73. Prologue, *Billy Budd*, 33. (The libretto is included with the CD of a 1968 production with Britten conducting the London Symphony Orchestra, the Ambrosian Opera Chorus led by John McCarthy, and Peter Pears playing the part of Edward Fairfax Vere.)

74. Ibid., 45.

75. Ibid., 67.

76. Op. 11.

77. The original title of Auden's typescript was *Thirty-One Poems*. Auden's British publisher Faber & Faber asked him to supply another title, but he was traveling in Iceland and inaccessible, so the publisher titled the book *Look, Stranger!* Auden asked his American publisher, Random House, to use the title *On This Island*.

78. Quoted in Donald Mitchell, *Britten and Auden in the Thirties: The Year 1936*, 2nd ed., Aldeburgh Studies in Music 5 (Woodbridge, UK: Boydell Press, 2000), 91.

79. Larchfield Academy is now the Lomond School in Helensburgh, Scotland.

80. At the entrance to the saltwater Gare Loch—a sea inlet.

81. "Sit down awhile, / And let us once again assail your ears, / That are so fortified against our story, / What we two nights have seen" (act 1, sc. 1).

82. Ilan Kelman, *Assessment of UK Deaths, 1953*, Cambridge University Centre for Risk in the Built Environment (CURBE), Cambridge, August 17, 2009.

83. See Royal Flaxman, with Dean Parkin, *Wall of Water: Lowestoft and Oulton Broad during the 1953 Flood* (Lowestoft, UK: Rushmere, 1993); the Dutch author Jan de Hartog's *Little Ark* (1953) depicted the flood and was made into James B. Clark's film of the same name, concerning a houseboat, in 1972.

84. March 24, 1938, House of Commons, London.

85. July 21, 1951, Woodford, in the London suburbs.

86. Cyril Ionides and J. B. Atkins, *A Floating Home*, illus. Arnold Bennett (London: Chatto & Windus, 1918), 101.

87. Abu al-Hasan Ali ibn al-Husayn ibn Ali al-Mas'udi, *The Meadows of Gold and Mines of Gems: The Abbasids*, trans. Paul Lunde and Caroline Stone (London: Kegan Paul International, 1989). Compare Koranic surah 11:44.

88. Cizre is surrounded by the Tigris to the north, east, and south.

89. Israel Zangwill, *Ghetto Tragedies* (New York: Macmillan, 1899).

90. Ian Kershaw, *Hitler: 1936–1945 Nemesis* (New York: Norton, 2000), 320–322.

91. Under Queen Elizabeth I the plays were seen as "popery" and banned. Despite this, a play cycle was performed in 1568 and the cathedral paid for the stage and beer, as in 1562. The plays were performed again, over four days, in 1575. Shakespeare was born in 1564, and there are records of Chester plays being performed in England in the 1560s and 1570s. Despite an active campaign to suppress them by Protestant reformers who saw them as papist, the last cycle in York, in North Yorkshire, was performed in 1576; the last in Chester, in Cheshire on the Welsh border, in 1600.

92. Igor Stravinsky, *Dialogues and a Diary* (Garden City, NY: Doubleday, 1963), quoted in Eric Walter White, *Stravinsky: The Composer and His Works*, 2nd ed. (Berkeley: University of California Press, 1979), 526. Robert Craft wrote the libretto for *The Flood*.

93. Acts of the Apostles 27:36, 27:31.

94. Psalm 104.

95. Genesis 1:9–1:10.

96. Psalm 107. Certain verses have been changed in modern hymnals, partly to accommodate religious and national differences as well as various branches of the military.

97. The Hebrew word for "ark" in the story of Noah is *tebah*. The only other place this word is used has to do with the reed basket that contains the baby Moses. There is another word for ship, *oniyah*, as in the Jonah story. "Dove" is the meaning of Jonah's name.

98. Paul White, *The Cornish Smuggling Industry* (Redruth, UK: Tor Mark, 1997).

99. John Bridcut, *Britten's Children* (London: Faber & Faber, 2006).

100. On Orff's ambiguous relationship with the Nazi regime, see Michael H. Kater, *Composers of the Nazi Era: Eight Portraits* (New York: Oxford University Press, 2000).

101. Carl Orff, with Gunild Keetman, *Musik für Kinder* (1930–1935, reworked 1950–1954).

102. Op. 34.

103. Op. 45.

104. Op. 74, no. 4.

105. Eric Crozier, "Buxton Festival Program Notes" (1981), quoted in Claire Seymour, *The Operas of Benjamin Britten: Expression and Evasion* (Woodbridge, UK: Boydell Press, 2007), 119.

POSTAMBLE

1. The herald's opening words at the ancient Greek constitutional assembly (Pnyx).

2. On Demosthenes' rhotacism, see Marc Shell, *Stutter* (Cambridge, MA: Harvard University Press, 2005).

3. The Greek term *ekklēsia* derives from a word meaning "to call out," as when the Herald calls out for someone to speak.

4. John Dryden translates the relevant passage from Plutarch's *Demosthenes*: "His inarticulate and stammering pronunciation, [Demosthenes] overcame and rendered more distinct by speaking with pebbles in his mouth." *Plutarch's Lives of Illustrious Men*, trans. John Dryden, rev. A. H. Clough (Boston: Little, Brown, 1876), 609.

5. On the term *rut* (or *rote*), meaning the "rock-voice of the sea," see Chapter 15.

6. "He used likewise at some certain times to go down to the shore at Phalerum [near Athens], to the end that, being accustomed to the surges and noise of the waves, he might not be daunted by the clamours of the people, when he should at any time declaim in public." Pseudo-Plutarch, *Lives of the Ten Orators*, trans. Charles Barcroft, in *Plutarch's Morals*, with various translators (London: R. Bentley, 1690).

7. See Chapter 8.

8. Demosthenes, *On the Navy* [Περὶ τῶν Συμμοριῶν].

9. Compare the prophecy concerning Tyre in Isaiah 23.

10. Mogens Herman Hansen, *Athenian Democracy in the Age of Demosthenes* (Somerset, NJ: Wiley-Blackwell, 1991).

11. See, for example, Demosthenes' "Reply to the Thebans," 18.70.

12. The island was thus known at one time as *Eirene* (Peace).

13. Livy, *Ab urbe condita*, ed. R. S. Conway and C. F. Walters (Oxford, 1914), 2.1.9.

14. This was Agrippina's third marriage.

15. Claudius defended Cicero against charges made in Asinus Gallus's *De Comparatione patris et Ciceronis*.

16. See Shell, *Stutter*.

17. See Tacitus, *Annals*, 12.66; Cassius Dio, *Roman History*, 61.34; and Suetonius, "Life of Claudius," in *The Lives of the Twelve Caesars*, 44.

18. Plutarch, *Demosthenes*, 5.5.

19. Shell, *Stutter*, 169–170.

20. This title, which means "language of poetry," indicates the first part of the author's *Prose Edda*.

21. That is, between Aegir and Bragi.

22. For the view that he was a sailor, see Israel Gollancz, *The Sources of Hamlet* (London: Oxford University Press, 1926), 9.

23. On maelstroms, see Chapters 5 and 9.

24. Giorgio de Santillana and Hertha von Dechend, *Hamlet's Mill: An Essay on Myth and the Frame of Time* (Boston: Gambit, 1969), chap. 6.

25. See Matthias Jochumsson, *1616–1916: On the Tercentenary Commemoration of Shakespeare—Ultima Thule Sendeth Greeting—An Icelandic Poem*, trans. Israel Gollancz (London: Oxford University Press, 1916).

26. Israel Gollancz, *Hamlet in Iceland: Being the Icelandic Romantic Ambales Saga* (London: David Nutt, 1898).

27. Frank Efraim Martinus [Frank Martinus Arion], *De laatste vrijheid* (Amsterdam: De Bezige Bij, 1995).

28. Martinus, *De laatste vrijheid*, 98; see also Doris Hambuch, "Walcott versus Naipaul: Intertextuality in Frank Martinus Arion's *De laatste vrijheid* (The Ultimate Freedom)," *Journal of Caribbean Literature* 3, no. 2 (2001): 91.

29. Charles Darwin, "On Certain Areas of Elevation and Subsidence in the Pacific and Indian Oceans, as deduced from the Study of Coral Formations," *Proceedings of the Geological Society of London* 2 (1837): 552–554.

30. Quoted by Pappus of Alexandria, *Synagoge* [Collection], bk. 7.

31. Philip Kearey and Frederick J. Vine, *Global Tectonics*, 2nd ed. (Boston: Blackwell Science, 1996).

32. One early image of the Earth as an "ocean of magma" "insulated" by a shell of solid rock can be found in the work of Pierre Louis Antoine Cordier (1777–1861). See Davis A. Young, *Mind over Magma: The Story of Igneous Petrology* (Princeton, NJ: Princeton University Press, 2003), 59.

33. Earlier was Alfred Wegener, "Die Entstehung der Kontinente" [The Origin of the Continents], *Geologische Rundschau* 3, no. 4 (1912): 276–292; see *Dr. A. Petermanns Mitteilungen aus Justus Perthes' Geographischer Anstalt* 63, 185–195, 253–256, 305–309.

34. See Chapter 11.

35. Old High German *scolla*, meaning "earthy crust," is cognate with English *shell*. See *OED*, s.v. "skull," etymology.

36. This sort of *chassé-croisé* occurs "on the ground" where island chain meets the seismic belt. One such place is Thwart-the-Way Island (also known as Sangiang) in the Sunda Strait of Indonesia. The toponym *Thwart*, or *dwars*, comes from the Old Norse term meaning "across" or "athwart." The English sometimes call the place Thwartway Island. The Sunda Strait, where Thwart-the-Way lies, separates the island of Java from the island of Sumatra and joins the Java Sea (on the east) with the Indian Ocean (on the west). Engineers are now building the world's longest suspension bridge between Sumatra and Java by way of Thwart-the-Way. It is a droll location for a bridge. Thwart-the-Way is located at the intersection of two immense areas of tectonic plate movement. First is the Pacific Ring of Fire, the horseshoe-shaped, circum-Pacific seismic belt that includes three-quarters of the world's volcanoes. Second is the Alpide Belt, a mountainous range that stretches from Indonesia out to Europe. The number and size of the volcanic eruptions around this *chassé-croisé* are staggering. The loudest two sounds heard in modern history were those of the eruptions of Krakatoa (1883) and Tambora (1815). The eruption at Toba created Samosir Island; Samosir is the world's largest island (actually an extreme peninsula) in a lake (here, Toba) on an island (Sumatra) in an ocean (Indian).

37. Anja Wendt, Reinhard Dietrich, Jens Wendt, Mathias Fritsche, Valery Lukin, Alexander Yuskevich, Andrey Kokhanov, Anton Senatorov, Kazuo Shibuya, and Koichiro Doi, "The Response of the Subglacial Lake Vostok, Antarctica, to Tidal and Atmospheric Pressure Forcing," *Geophysical Journal International* 161, no. 1 (2005): 41–49.

38. Some geomatics experts now hypothesize that Lake Vostok contains one "bedrock island" protruding into *terra firma*; see David Whitehouse, "Russia to Resume Vostok Drilling," *BBC News*, May 25, 2005. Many subterranean lakes and rivers, located in caves, likewise have islands, among them Melissani Lake (on the island of Cephalonia in the Ionian Archipelago in Greece), where one finds a small islet near the cave entrance.

39. See Robert H. Tyler, "Strong Ocean Tidal Flow and Heating on Moons of the Outer Planets" (letter), *Nature* 456 (December 11, 2008): 770–772.

40. The quotation defining *orogen* is from the *Quarterly Journal of the Geological Society of London* 108 (1953): 2; it was introduced as early as Leopold Kober, *Der Bau der Erde* (Berlin: Gebrüder Borntraeger, 1921), 21.

41. See the following: (1) David S. Ferris's review of Henry Sussman, *Aesthetic Contract: Statutes of Art and Intellectual Work in Modernity* (Stanford, CA: Stanford University Press, 1997), in *Studies in Romanticism* 41, no. 1 (2002): 123–128; (2) Jonathan Bordo, "The Homer of Potsdamerplatz—Walter Benjamin in Wim Wenders' *Sky over Berlin / Wings of Desire*, a Critical Topography," *Images: A Journal of Jewish Art and Visual Culture* 2 (2008): 85–108; (3) Ryan James Melsom, "West Coast Apocalyptic: A Site-Specific Approach to Genre," PhD diss., Queen's University, Kingston, Ontario, 2011, 216; (4) publicity for Robert Bond and Jenny Bavidge, eds., *City Visions: The Work of Iain Sinclair* (Newcastle, UK: Cambridge Scholars Publishing, 2007): "[It] offer[s] an unnerving and inventive critical topography"; and (5) Claudio Fogu and Lucia Re, "Italy in the Mediterranean Today: A New Critical Topography," *California Italian Studies* 1, no. 1 (2010): 1: "The ancient Romans were the first to give geopolitical unity to—and to claim as *theirs*, i.e. *Mare Nostrum*—a body of water that previously had been only given local names by each of its coastal communities"; "Italy in the Mediterranean" is the title of the first thematic issue of *California Italian Studies*.

42. See J. E. Malpas, *Place and Experience: A Philosophical Topography* (Cambridge: Cambridge University Press, 1999), 1–19; and Malpas, *Heidegger's Topology: Being, Place, World* (Cambridge, MA: MIT Press, 2006).

43. For the circumscribed definition, see Cindi Katz, "On the Grounds of Globalization: A Topography for Feminist Political Engagement," *Signs* 26, no. 4 (2001): 1213–1234.

44. Some claim that Isaiah Bowman's reasons for helping Conant close the department at Harvard boiled down to his homophobic view that it was harboring a nest of "vice, nepotism, and pederasty." Neil Smith, *American Empire: Roosevelt's Geographer and the Prelude to Globalization* (Berkeley: University of California Press, 2003), 438.

45. Neil Smith, *American Empire*, 443. Some "biographical" details of the demise of geography at Harvard are reported by Smith in "Academic War over the Field of Geography: The Elimination of Geography at Harvard, 1947–1951," *Annals of the Association of American Geographers* 77, no. 2 (1987): 155–172. The notion of "area studies" began at Columbia University and, within a decade or two, most other major universities, but geography did not last long even there. There is still, in 2012, as I write these words, no geography department at Harvard.

46. Robert E. Dickinson, *The Makers of Modern Geography* (New York: Praeger, 1969), 64. Some of Ratzel's work was cited earlier by way of his American follower Ellen Churchill Semple.

47. Friedrich Ratzel, "Inselvölker und Inselstaaten: Eine politisch-geographische Studie," *Beilage zur Allgemeinen Zeitung* 251–252 (1895). For further such references, see Harriet Grace Wanklyn, *Friedrich Ratzel: A Biographical Memoir and Bibliography* (Cambridge: Cambridge University Press, 1961).

48. See Chapter 6.

49. This idea Ernst Haeckel brought to Germany from England in his chorological *Anthropogenie, or Evolution of Man* (1874), which concerns the relationship between human ontogeny and phylogeny. The chorology of organisms is the doctrine of the geographical and topographical distribution of animal and vegetable species. Haeckel, *The Evolution of Man: A Popular Exposition of the Principal Points of Human Ontogeny and Phylogeny* (1879), 1.4.74.

50. Friedrich Ratzel, *Sein und Werden der organischen Welt* [Being and Becoming in the Organic World] (Leipzig: Gebhardt & Reisland, 1869).

51. Friedrich Ratzel, *Städte- und Kulturbilder aus Nordamerika* (Leipzig: F. A. Brockhaus, 1876).

52. Friedrich Ratzel, *Die Vereinigten Staaten von Nordamerika* (Munich: R. Oldenbourg, 1878, 1880).

53. Friedrich Ratzel, *Schneedecke, besonders in deutschen Gebirge* (Stuttgart: J. Engelhorn, 1889).

54. Friedrich Ratzel, *Anthropo-Geographie; oder Grundzüge der Anwendung der Erdkunde auf die Geschichte*, 2 vols. (Stuttgart: J. Engelhorn, 1882, 1891).

55. Allen Chadwick, "Review of *Routes and Roots: Navigating Caribbean and Pacific Island Literatures* by Elizabeth M. DeLoughrey (Honolulu: University of Hawaii Press, 2007)," *Journal of New Zealand Literature* (June 1, 2007): 183.

56. Guy Debord, "Introduction à une critique de la géographie urbaine," *Les Lèvres Nues* 6 (September 1955). For the English translation, see Debord, "Introduction to a Critique of Urban Geography," in *Situationist International Anthology*, trans. Ken Knabb (Berkeley, CA: Bureau of Public Secrets, 1981).

57. August Schleicher, Ernst Haeckel, and Wilhelm Bleek, *Linguistics and Evolutionary Theory: Three Essays*, ed. Konrad Koerner, introduction by J. Peter Maher (1863–1867; Amsterdam: J. Benjamins, 1983).

58. Conant was a chemist who became president of Harvard in 1933 and then served as chairman of the National Defense Research Council and helped to ramp up the Manhattan Project.

59. Geography disappeared almost in the same way that numismatics did, which likewise pays attention to cultural differences in a material context. See Marc Shell, *Wampum; or, The Origins of American Money* (Urbana: University of Illinois Press, forthcoming).

60. "*Lebensraum* is probably the best known of all twentieth century German political terms." Woodruff D. Smith, "Friedrich Ratzel and the Origins of Lebensraum," *German Studies Review* 3, no. 1 (1980): 51.

61. As in David Morrison's review of the third edition of *Geistige Strömungen der Gegenwart* (1904) by the German philosopher and Nobel Prize winner Rudolf Eucken, published in *Mind* 24 (1905): 266.

62. Karl Haushofer, *Grenzen in ihrer geographischen und politischen Bedeutung* [Borders in Their Geographic and Political Significance] (Heidelberg: K. Vowinckel, 1939).

63. Ola Tunander, "Swedish-German Geopolitics for a New Century—Rudolf Kjellén's *The State as a Living Organism*," *Review of International Studies* 27, no. 3 (2001): 451–463.

64. Friedrich Ratzel, *Politische Geographie* (Munich: R. Oldenbourg, 1897). See Gerry Kearns, *Geopolitics and Empire* (Oxford: Oxford University Press, 2009).

65. Rolf Hobson, *Imperialism at Sea: Naval Strategic Thought, the Ideology of Sea Power, and the Tirpitz Plan, 1875–1914* (Boston: Brill, 2002), 292–294.

66. See Halford Mackinder, *Britain and the British Seas* (New York: D. Appleton & Co., 1902). See also Sadao Asada, *From Mahan to Pearl Harbor: The Imperial Japanese Navy and the United States* (Annapolis, MD: Naval Institute Press, 2006). For Mahan's original

work, see Alfred Thayer Mahan, *Influence of Sea Power upon History, 1600–1783* (Boston: Little, Brown, 1890).

67. Friedrich Ratzel, "The Inter-Ocean Canal through Central America" [Der inter-ozeanische Kanal durch Mittelamerika], *Beiträge zur Allgemeine Zeitung* [Five Supplementary Articles] (1880).

68. Friedrich Ratzel, "Die Seemacht: Eine polischer-geographische studie," *Wissenschaftliche Beilage zur Leiziger Zeitung* 123, no. 4 (1896): 489–495.

69. Friedrich Ratzel, *Das Meere als Quelle der Völkergrösse: Eine politisch-geographische Studie* (Munich: R. Oldenbourg, 1900).

70. Halford Mackinder, *Our Own Islands: An Elementary Study in Geography* (London: G. Philips, 1907).

71. Halford Mackinder, *The Rhine: Its Valley and History* (New York: Dodd, Mead, 1908).

72. Halford Mackinder, "The Geographical Pivot of History," *Geographical Journal* 23, no. 4 (1904): 421–437. See also Pascal Venier, "The Geographical Pivot of History and Early 20th Century Geopolitical Culture," *Geographical Journal* 170, no. 4 (2004): 330–336.

73. *OED*, s.v. "pivot," A3.

74. *OED*, s.v. "pivot," A1.

75. *OED*, s.v. "pivot," compounds, "pivot ship."

76. Halford Mackinder, *Democratic Ideals and Reality* (New York: Henry Holt, 1942), 64–65.

77. The "Heartland" was then presumably the area of the (old) Russian Empire; see Illustration 79.

78. Derwent Whittlesey, "Haushofer: The Geopoliticians," in *Makers of Modern Strategy*, ed. Edward Mead Earle (Princeton, NJ: Princeton University Press, 1941), 388–411. See also Whittlesey, *German Strategy of World Conquest* (New York: Farrar & Rinehart, 1942).

79. Karl Haushofer, "Dai Nihon: Betrachtungen über Groß-Japans Wehrkraft, Weltstellung und Zukunft" [Reflections on Greater Japan's Military Strength, World Position, and Future], PhD diss., Munich University, 1913.

80. See Louis Pauwels and Jacques Bergier, *The Morning of the Magicians*, trans. Rollo Myers (New York: Stein & Day, 1964); see also Richard L. McGaha, "Setting the Demon Free: Karl Haushofer, Rudolf Hess, the Thule Society and Hitler in Munich, 1918–1920," master's thesis, University of Calgary, 2002.

81. Haushofer, for example, was close friends with Hitler's deputy führer Rudolf Hess. See Henning Heske, "Karl Haushofer: His Role in German Politics and in Nazi Politics," *Political Geography* 6 (1987): 135–144.

82. Karl Haushofer, *Weltmeere und Weltmächte* [World Seas and World Powers] (Berlin: Zeitgeschichte Verlag, 1937).

83. On those claims, see Mika Luoma-Aho, "Geopolitics and *Gross* politics: From Carl Schmitt to E. H. Carr and James Burnham," in *The International Political Thought of Carl Schmitt: Terror, Liberal War, and the Crisis of Global Order*, ed. Louiza Odysseos and Fabio Petito (Abington, UK: Routledge, 2007), 39.

84. Nicholas J. Spykman, "Geography and Foreign Policy II," *American Political Science Review* 32, no. 2 (1938): 236.

85. Bas Umali (of the Philippines) "calls for the dismantling of the Philippine nation-state and the implementation of an 'archipelagic confederation' in its place." Umali, "Archi-

pelagic Confederation: Advancing Genuine Citizens' Politics through Free Assemblies and Independent Structures from the Barangay & Communities," http://www.anarkismo.net /article/2923?userlanguage=ht&save_prefs=true.

86. See Chapter 1.

87. Nicholas Woodsworth, *The Liquid Continent: A Mediterranean Trilogy* (London: Haus Publishing, 2008). The three parts of the trilogy are "Alexandria," "Venice," and "Istanbul."

88. David Harvey, "David Harvey on the Geography of Capitalism, Understanding Cities as Polities and Shifting Imperialisms," *Theory Talks*, October 9, 2008, http://www .theory-talks.org/2008/10/theory-talk-20-david-harvey.html.

89. David Harvey, *Spaces of Capital: Towards a Critical Geography* (New York: Routledge, 2001), 143.

90. Ibid., 334.

91. David Harvey, *Megacities: Lecture 4* (Amersfoort, The Netherlands: Twynstra Gudde Management Consultants, 2001), 56.

92. The phrase is from Max Caspar, *Kepler*, trans C. Doris Hellman (New York: Dover, 1993), 181–185. The full title is *Tertius Interveniens, das ist Warnung an etliche Theologos, Medicos vnd Philosophos, sonderlich D. Philippum Feselium, dass sie bey billicher Verwerffung der Sterngguckerischen Aberglauben nict das Kindt mit dem Badt ausschütten vnd hiermit jhrer Profession vnwissendt zuwider handeln* [translated by Hellman as "*Tertius Interveniens*, that is warning to some theologians, medics and philosophers, especially D. Philip Feselius, that they in cheap condemnation of the star-gazer's superstition do not throw out the child with the bath and hereby unknowingly act contrary to their profession"] (Frankfurt: Godtfriedt Tampachs, 1610).

93. The island is about 450 meters long east to west, about 275 meters wide at the west end, and 175 meters wide upriver at the east end, where the university was located. The lagoon is now usually known as "Frisches Haff."

94. *Ape* means "river" in Old Prussian; the term is *upė* in modern Lithuanian. See Christian Gottlieb Mielcke, *Littauisch-deutsches und deutsch-littauisches Wörter-Buch* (Königsberg: Im Druck und Verlag der Hartungschen Hofbuch druckerey, 1800), 307.

95. Mielcke's *Wörter-Buch* is a reworking of Philipp Ruhig's earlier dictionary (1747).

96. Other extinct Baltic languages are Curonian, Galindian, and Sudovian.

97. *OED*, "Over," n1.

98. William Wallace, *Kant* (Edinburgh: W. Blackwood, 1887), 6.

99. For details, see Klaus Garber, *Das alte Königsberg: Erinnerungsbuch einer untergegangenen Stadt* (Cologne: Böhlau Verlag, 2008).

100. Plutarch, *Parallel Lives: The Life of Marcellus*, trans. John Dryden, 14.

101. Ibid.

102. Ibid.

103. Carl Schmitt, *Land und Meer: Eine weltgeschichtliche Betrachtung* (Leipzig: Reclam Verlag, 1942); translated by Simona Draghici in 1954 as *Land and Sea*, rev. Greg Johnson (Washington, DC: Plutarch Press, 1997), chap. 17.

104. Ellsworth Huntington, "The Geographer and History," *Geographical Journal* 43 (1914): 19.

Name Index

Italic page numbers indicate material in illustrations.

Abbas the Great, 71

Adeimantus, 112

Adler, Selig, 285n39

Adramelek of Byblos, 114

Aelfric of Eynsham, 177

Aelius Marcianus, 309n5

Aeolus, 107

Aeschines, 231

Aeschylus, 104, 143

Agamemnon, 105–106

Agatharchides, 256n62

Agrippina, 160, 233, 326n14

Ahab, 214

Alberich, 200, *201*, 203

Alcinous, 107

Alexander the Great, 28, 60, 232

Alexander VI (pope), 46

Alexandra of Denmark, *155*

Alfred the Great, 28, 125, 179, 185–187, 201, 216, 223

Allen, Thomas, 165

al-Muqaddasī the Jerusalemite, 66

Aloadae, 139

Amleth, 151, 175–176, 233

Amundsen, Roald, 47

Anderson, Wes, 229

Angul, 127

Anne of Denmark, 131–132, 153, 292n19, 295n47, 295n49, 308n37

Antiochus, 160

Antonio (*The Merchant of Venice*), 63, 69, 170–171

Antonio (*The Tempest*), 172

Antonioni, Michelangelo, 96, *97*, 284n23

Apollonius of Tyre, 170

Aratus, 219

Archimedes, 1, 2–3, 17, 29, 83, 94, 98, 123, *124*, *166*, 206, 214, 235, 245–248, *246*

Arden, Mary, 172

Aretades of Knidos, 110

Ariel, 306n55

Arion, Frank Martinus (pseud. Frank Efraim Martinus), 233, 235

Aristophanes, 110

Aristotle, 14, 106, 114–116, 290n71

Arkstee, Johann Caspar, *67*

Arnaud, Ramón de, 96

Arndt, Ernst Moritz, 191

Arne, Thomas Augustine, 185–187, 222–225

Arnot, W. G., 288n34

Asadov, Alexander, 207, 318n80

Ashurbanipal, 145

Asiak, 266n11

Asinius Gallus, 233

Athelstan, 128

Atkins, John Black, 129, 227

Atlas, 121, *123*

Auden, W. H., 218–220, 226, 324n77

August, Bille, 49

Ayth, Nichole, 178, 308–309n41

Bacon, Francis, 40, 98, 103, 211, 303–304n28

Baertz, Caspar, 71, 274n71

Balanchine, George, 228

Baldacchino, Godfrey, 253n4

Ballantyne, R. M., 97

Barbosa, Duarte, 71

Barbour, John, 300n22

Bartholomeus Anglicus, 38

Bates, Paddy Roy, 279n30

Beer, Gillian, 255n42

Beethoven, Ludwig van, 187

Behemoth, 36, *37*

Bell, Gertrude, *40*

Belleforest, François de, 126, 127, 134, 151

Bellenden, John, 299n19, 300–301n28

Bennett, Arnold, 103

Bennett, Raine Edward, 7

Beowulf, 125, 152

Bergman, Ingmar, 50–53, 268n41, 269n66

Berlioz, Hector, 156, 300n27

Bernard, Richard, 160

Bernegger, Matthias, 188

Berry, John, 99–100

Beurling, George Frederick "Buzz," 207

Billings, Marland Pratt, 237–238

Billington, Michael, 53

Bishop, Malcolm, 285n44

Blake, William, 148, *150*, 214, 230

Blanchard, Jean-Pierre, 208

Blechen, Carl, 196

Bloch, Ernst, 100

Boas, Franz, 44, 265n9

Böcklin, Arnold, 51, 197, *198*

Boece, Hector, 299n19

Boethius, 28

Bolingbroke, 306n4

Bolt, Robert, 214

Boorman, John, 97

Borca, Federico, 111

Bordone, Benedetto, 22–23, 58, 61, 271n28, 317n60

Börne, Ludwig, 195

Boroughs, John, 156

Botero, Giovanni, 266n12

Bougainville, Louis-Antoine de, 73

Bowman, Isaiah, 237–238, 328n44

Brahe, Tycho, 23, 116, 126, 153, 165, 166, 188, 253n5, 292n21, 303n24, 304n29, 304n41

Branagh, Kenneth, *143*

Brandt, Friedrich, 314n21

Brandt, Karl, *198*, 314n21

Brangäne, 200

Brathwaite, Kamau, 19–20, 35

Braudel, Fernand, 4, 110

Braun, Georg, *164*, 165

Breca, 125

Brecht, Bertolt, 202, 206–207, 223

Brendan, Saint, 38, *39*

Bridge, Frank, 225

Bright, James Wilson, 124

Britten, Benjamin, 223, 225–226, 228–230, 324n73

Brontë, Charlotte, 210

Brooke, Andrew, 309n48

Browne, Thomas, 115, 290n78

Bruckner, Gotthold, *198*

Bruckner, Max, *198*

Bruni d'Entrecasteaux, Antoine Raymond Joseph de, 73

Buck, Pearl, 64

Buitrago, Fanny, 95

Buondelmonti, Cristoforo, 21, 22

Burke, Edmund, 8, 36

Burnet, Thomas, 302n9

Burroughs, Edgar Rice, 286n49

Burstein, Stanley Mayer, 256n62

Burton, Richard Francis, 106

Butler, Eliza Marian, 189

Butler, Samuel, 103, 285n49

Bynkershoek, Cornelius, 185

Calchas, 105

Caliban, 20, 96

Calypso, 107

Canute IV, 294n38

Canute the Great, 79, 128–129, 131, 176, 178, 294n38

Carlyle, Thomas, 231

Carpaccio, Vittore, 62, *62*

Carson, Rachel, 1, 217

Carteret, Philip, 73

Cartier, Jacques, 267n29, 270n4

Carus, Carl Gustav, 196

Catherine of Valois, 295n46

Cavelier, René-Robert, Sieur de La Salle, 267n29

Cavendish, Margaret, 160

Caxton, William, 38, 291n7

Ceccarelli, Paola, 111

Cecil, William, 125

Cécille, Jean-Baptiste, 73

Cervantes, Miguel de, 225

Chabon, Michael, 285n39

Chaloner, James, 137

Chambers, Ephraim, 26

Charlemagne, 141

Charles I of England, 221–222

Charrière, Henri, 279n34

Charron, Andrea, 45, 47

Chaucer, Geoffrey, 165, 301n37, 303n16

Chesterton, G. K., 225

Chillingworth, William, 161

Choi Young-hwan, 77

Chomsky, Noam, 15

Christian I of Denmark, *130*, 308n37

Christian III of Denmark, 304n41

Christian IV of Denmark, 131, 268n38

Christina of Sweden, 314n15

Churchill, Winston, 227, 229, 324n67

Cicero (Marcus Tullius Cicero), 154, 156, 233, 300n21, 326n15

Cimon, 113

Circe, 107

Clare, Angel, 7

Clare, John, 318n77

Clark, George Thomas, 297n85

Clarke, Catherine, 145

Claudian, 123

Claudius (*Hamlet*), 131, 135–136, 158, 160, 166–167, 170–171, 174, 176, 232–233, 300n22, 303n14

Claudius (Roman emperor), 232–233, 326n15

Clay, Diskin, 98

Coates, Richard, 277n8

Cole, Ann, 277n8

Collingwood, William Gershom, 174, *174*, 202

Columbus, Christopher, 46, 81, 183–184

Conant, James B., 237–238, 328n44, 329n58

Connolly, Thomas E., 285n39

Constantakopoulou, Christy, 286n2

Cook, James, 46, 73, 267n25

Cordier, Pierre Louis Antoine, 327n32

Cortés, Hernán, 58, 103

Coryate, Thomas, 63

Cousineau, Marie-Hélène, 266n11

Crabbe, George, 226

Craig, Edward Gordon, 292–293n22

Cresques, Abraham, 273n61

Cresques, Jehuda, 273n61

Crozier, Eric, 230

Crusoe, Robinson, 17–18, 23, 76, 310n7

Cuthbert, Saint, 185

Daim, Wilfried, 314n27

d'Albedyhll, Eleonora Charlotta, 291n6

Dan, 127

Dangun, 74

Darré, Richard Walther, 316n51

Darwin, Charles, 31–34, 94, 97, *234*, 235, 283n21

Darwin, Erasmus, 32, *32*

Davenant, William, 224

da Vinci, Leonardo, *162*

Davis, John, 267n29

Davis, Thomas R. A. H., 280n44

Debord, Guy, 239

Debussy, Claude, 225

Dedalus, Stephen, 298n3

Defoe, Daniel, 5, 7, 23, 76, 96, 103, 262n42, 317n74

Deleuze, Gilles, 22, 257n71

Della Dora, Veronica, 270n18

DeLoughrey, Elizabeth, 20, 239

Demetrius Poliorcetes, 288n44

Demosthenes, 231–233, *232*

Dening, Greg, 23

Dent-Young, Alex, 64

Dent-Young, John, 64
Depraetere, Christian, 21
Derrida, Jacques, 18
Descartes, René, 2–3
Dessau, Paul, 206–207, 317n71
de Vere, Edward, 125
Dickinson, Robert, 238
Diessl, Gustav, 317n68
Digby, Kenelm, 303n14
Digges, Thomas, 165
Dinocrates of Rhodes, 60
Diodorus of Sicily (Diodorus Siculus),
    3n1, 104, 110, 259n105, 286nn2–3
Donne, John, 35, 165, 304n29
Douglas, Gavin, 299n19
Douglas, William O., 264n37
Drake, Francis, 267n28
Drant, Thomas, 167
Drummond, William, 121
Dryden, John, 3, 224
Duff, David, *155*
Duncan, 52
Duracotus, 188
d'Urfey, Thomas, 224
Dutta, Madhusree, 59
Dykes, John B., 229

Earhart, Amelia, 205
Eckart, Dietrich, 210
Eco, Umberto, 23–24
Eden, Richard, 13
Edith (daughter of King Athelstan), 128
Edward VII, *155*
Egerton, Francis, 267n26
Egil, 132–133
Eisenstein, Sergei, 144, *145*
Ekwall, Eilert, 277n8
Eleanor, Duchess of Gloucester, 308n39
Elgar, Edward, 225
Eliot, T. S., 218–220
Eliott, Andrew, 218
Elizabeth I, 142, 221, 228, 325n91
Elyot, Thomas, 166
Empedocles, 173
Engels, Friedrich, 29

Eratosthenes, 116
Eric Bloodaxe, 132–133
Eric of Denmark. *See* Eric of Pomerania
Eric of Pomerania, 131, 156, 223, 294n45
Erik the Red, 49
Essarg (Tuirbe Tramár), 294n35
Eucken, Rudolf, 240
Euhemerus of Messene, 97
Euler, Leonhard, 15
Euripides, 104
Eutropius, 297n93
Evans, T. M., 138

Faik, Sait (Adalı), 270n5
Fanck, Arnold, 202–203, 206–207, 238
Feng, 134, 175
Fermi, Enrico, 284n35
Feuchtwanger, Lion, 206
Field, Todd, 52
Flaherty, Robert, 93–94, 266n11
Fletcher, John, 224
Fletcher, Robert, 111
Florio, John, 161, 302n13
Fontane, Theodor, 200
Forster, E. M., 226
Fortinbras, 52, 135–136, 143, 167, 176
Foster, George Eulas, 254n27
Fowler, William, 282n2
Frank, Hans, 210
Franks, Jill, 4
Fraser, Christine Marion, 97
Frederick, Duke, 172
Frederick I of Denmark, 223
Frederick II of Denmark, 131, 296n70,
    303n24
Frederick William I of Prussia, 312n59
Frege, Friedrich Ludwig Gottlob, 14
Freiligrath, Ferdinand, 192, 197
Friedrich, Caspar David, 43, *43*, 196, 199,
    221
Friedrich, Marquess of Ely, 187
Frissell, Varick, 43, *44*, 203, 266n11
Frobisher, Martin, 267n29
Fuca, Juan de, 267n28
Fuseli, Henry, 148, *149*

Gaia, 190, 311n41

Galaup, Jean-François de, 73

Garfunkel, Art, 262n4

Garnett, Tay, 206

Garrick, David, 224

Gautier, Henri, 202, *203*

Geddes, Robert, 217

Gefjun, 123, 291n6, 299n17

Gelling, Margaret, 277n8

Genz, Henrik Ruben, 307n25

George I (Georg Ludwig), 187

Gérard, Jean Ignace Isidore (pseud. J. J. Grandville), *102*, 208, 210, *211*

Gertrude, 134, 137–138, 160, 168

Geruthe, 131

Gibbon, Edward, 302n5

Gibson, Thomas, 161, 302n13

Gielgud, John, 53, 269n67

Gilbert, Humphrey, 158, 267n29

Gilbert, W. S., 224–225, 229

Gillis, John R., 4–5, 252n20

Gilpin, William, 78

Glanville, Bartholomew de, 301n39

Glissant, Édouard, 35

Glover, Judith, 288n34

Godwin, Edward, 292n22

Goebbels, Joseph, 220, 313n51

Goethe, Johann Wolfgang von, 61, 86, 119, 161, 189–190, 192

Golding, William, 96

Goldsworthy, Andy, 261n31

Gollancz, Israel, 147, 174, 233

Gonzalo, 170

Goschen, George (Lord Goschen), 15

Gowland, Gibson, 317n67

Grandville, J. J. (Jean Ignace Isidore Gérard), *102*, 208, 210, *211*

Green, John Richard, 178

Greenaway, Peter, 53

Greene, Robert, 146–147, 305n43

Gregoras, Nicephorus, 100

Greville, Fulke, 69

Griffith, Paul A., 20

Grotius, Hugo, 184–185, 222, 310n7

Grove, Richard, 4, 252n17

Gudfred of Denmark, 141

Guérin, Nicolas-François, 73

Guildenstern, 157–158, 167, 175, 299n11

Gulliver, *102*, 103

Gundolf, Friedrich, 193, 313n41

Gunnlaugr, 176

Guthrie, Woody, 47

Guyer, Paul, 205

Gylfi, 291n6

Haakon VI of Norway, 130

Hackert, Philipp, 196

Hadrian, 114

Haeckel, Ernst Heinrich, 283n21, 328n49

Hale, Edward Everett, 211–214, *212*

Hale, John, 114

Hale, Matthew, 266n12

Halévy, Élie, 209

Hamblet, 151

Hamel, Hendrick, 73, 76, 275n106

Hamilton-Paterson, James, 90

Hamlet (*Hamlet*), 13, 23, 50–52, 101–103, 121, 123, 126–129, 151–154, 159, 160–163, 232, 299n11, 303n14

Hamlet (*Hamlet in Iceland*), 233

Hamnet, 305n45

Handel, George Frideric, 187

Harald I, 175

Hardy, Oliver, 99

Hardy, Thomas, 7

Harper, Stephen, 46–47

Harrison, William, 110

Hartog, Francis, 114

Harvey, David, 8, 242–243

Haushofer, Karl, 241, 242, 330n81

Haywood, Francis, 204–205

Hedlund, Torsten, 50

Hegel, Georg Wilhelm Friedrich, 18–20, 255n46, 283n7

Heidegger, Martin, 221

Heine, Heinrich, 195–196

Heisenberg, Werner, 19

Helen of Troy, 104–105, 107, 190

Helga, 176

Helgesen, Geir, 173

Helgi, 151

Helot, Louis, 73

Henricus Martellus, 101

Henry III of England, 179

Henry IV of England, 131, 306n4

Henry V of England, 295n46

Henry VIII of England, 100–101

Henry of Huntingdon, 128

Heraclitus, 18–19, 28–30, 107, 114–115, 236

Hercules, 121, 214, 237

Herder, Johann Gottfried von, 14, 26,
    189–190

Hermione, 168

Herodotus, 25, 79, 107–109, 110–115,
    263n29

Herschel, John, 211

Herschel, William, 16, 28, *29*, 260n14

Herz, Henriette, 191–192

Herzl, Theodor, 100

Herzog, Werner, 280n40

Hesiod, 104, 106–107

Hess, Rudolf, 202, 210, 330n81

Hieron, 246

Higden, Ranulph, 301n37, 302n1

Hillyer, V. M., *30*, *154*

Himmler, Heinrich, 202–203, 316n51

Hitchcock, Alfred, 76

Hitler, Adolf, 51, 197, 199–200, 203,
    206–207, 210, 214, 222–223, 313n51

Hobbes, Thomas, 20, 87, 222

Høeg, Peter, 48, 267n36

Hoffmann, Josef, 197

Hoffmann von Fallersleben, August Hein-
    rich, 222–223

Hogenberg, Frans, *164*, 165

Holbein, Ambrosius, 101, *101*

Holland, Philemon, 266n12

Homer, 98, 104, 106–107

Hong Jong-chan, 73

Hooke, Robert, 207–208

Horatio, 134, 156, 161, 169

Hornby, William, 59

Horwendil, 134

Houseman, John, 99

Howells, William Dean, 93

Hrafn, 176

Hubbard, Lucien, 317n67

Hübner, Arthur, 316n51

Hudson, Henry, 267n29

Humboldt, Alexander von, 28, 260n14

Humboldt, Wilhelm von, 191, 192, 312n53

Hume, David, 237

Hunt, Courtney, 282n56

Huntington, Ellsworth, 247

Hussey, Greta, 266n16

Hutton, James, 290n71

Huxley, Aldous, 93

Ibn al-Nafis, 96, 284n25

Ibn Hawqal, Muhammad Abu'l-Qasim,
    24–25

Ibn Tufail, 96, 284n25

Idrisi, Muhammad al-, *22*, 28

Illtud, Saint, 129

Inagaki, Hiroshi, *133*, 295n57

Ingvaldsen, Siri, 307n17

Ionides, Cyril, 129, 227

Iphigenia, 104–105, *106*, 287n8

Isabella, 167

Ivalu, Madeline, 266n11

Jacob of Ancona, 69

Jahnn, Hans Henny, 191

James, Thomas, 220

James I of England. *See* James VI of
    Scotland

James III of Scotland, 132, 308n37

James VI of Scotland, 52, 71, 131–132, 153,
    184–185, 188, 222, 292n19, 295n47,
    304n29, 308n27, 310n8

Jang, Chul-soo, 276n108

Jang Bogo, 77

Jaques, 172

Jasconius, 38

Jeong Seon, 77

Jesus, 229

Jocasta, 163

Johanan, Rabbi, 263n17

John (king of England), 178–179, 233

John of Gaunt, 186, 306n4

Johnson, Martin, 93
Johnson, Osa, 93
Johnson, Robert, 224, 266n12, 305n43
Johnson, Samuel, 305n43
Johnston, James B., 151
Jones, Mary, 29
Jones, Raymond F., 29
Jonson, Ben, *122*, 295n49
Josephson, Erland, 53
Jourdain, Sylvester, 224
Joyce, James, 298n3
Julius Caesar, 151
Justinian, 85–86, 273n19, 309n5

Kaempfer, Engelbert, 73
Kahanu, Noelle, 280n45
Kant, Immanuel, 1, 16–17, 28–29, 86,
    147–148, 189–190, 194, 201, 204, 243,
    247, 254nn31–32, 255n39
Katchor, Ben, 285n39
Kennedy, William Sloane, 214
Kepler, Johannes, 116, 188, 253n5, 304n29
Kermode, Frank, 166
Kim, Uichol, 173
Kim Dae-seung, 77
Kingsley, Charles, 230
Kjellén, Rudolf, 240
Knopf, Edwin H., 316n57
Ko, Mohammed Dirhem, 68
Kojiro, Sasaki, *133*, 295n57
Koller, 129, 134–136
Kolodny, Émile, 110–111
Kosegarten, Ludwig Gotthard, 190–191
Kossinna, Gustaf, 216
Kraken, 35–38, 262n9
Kurosawa, Akira, 269n67
Kurwenal, 200

Laemmle, Carl, Jr., 206–207
Laemmle, Carl, Sr., 206–207, 214
Laertes, 52, 138, 150, 176, 269n58
Lagarde, Paul de, 228
Laius, 161, 163
Lamothe, Arthur, 47
Lana de Terzi, Francesco, 207–208

Lang, Fritz, 197
Langland, William, 302n1
La Rocque, Rod, 317n68
Latour, Bruno, 323n47
Laube, Heinrich, 192, 194
Laurel, Stan, 99
Lawrence, Carl, 317n68
Lawrence, D. H., 26
Lawrence, Ellen, 205, 317n68
Lear, King, 295n53
Le Corbusier, 216
Lee Hae-jun, 76
Leonidas, 104
Leontes, 168
Lestringant, Frank, 256n60
Leviathan, 20, 35–37, *37*, 193, 222, 262n10,
    263n17
Liblin, Marc, 284n26
Liebenfels, Jörg Lanz von, 200
Livy, 60, 156
Locke, Matthew, 224
Longfellow, Henry Wadsworth, 191
Lorenz, Karl, 317n68
Lowell, James Russell, 262n9
Lucian of Samosata, 98
Lucius Junius Brutus, 151, 232, 298n2
Lucius Tarquinius Superbus, 151, 232
Lucretius, 41, 173
Lux, Thomas, 216
Luynes, Duc de, 2
Lyell, Charles, 169
Lyotard, Jean-François, 17

Macaulay, Thomas B., 222
MacDonald, James, 74
MacFarlane, Walter, 140
Machiavelli, Niccolò, 60
Mackey, Nathaniel, 20
Mackinder, Halford, 239, *241*, 241–243
MacLean, George Edwin, 166
Macrobius, 28
Maelgwn Gwynedd, 293–294n35
Magellan, Ferdinand, 13, 71
Magnus (brother of Frederick II), 296n70
Magnus, Olaus, 141–142

Mahan, Alfred Thayer, 240–241

Maktoum, Mohammed bin Rashid Al
    (Sheikh Mohammed), 79–80

Mallet, David, 185

Mallet, Paul Henri, 132

Mamillius, 305n45

Mandelbrot, Benoit, 15

Manucci, Niccolao, 57, 71

Manuel I Comnenus, 302n5

Marchand, Étienne, 73

Marcus Junius Brutus, 151

Margaret (daughter of Christian I), 132,
    308n37

Margaret I of Denmark, 129–131, 168,
    294n42

Marina, 305n55

Massinger, Philip, 224

Matthew of Paris, 179

Maude, H. E., 280n45

Mawer, Allen, 288n34

McCarthy, John, 324n73

McLuhan, Marshall, 2, 3n2

Meiklejohn, John Miller Dow, 204

Meister, Wilhelm, 192

Melford, George, 266n11

Melville, Herman, 21–23, 36, 38–39,
    40, 85–86, 92, 93–94, 226, 263n28,
    278n25, 305n51, 324n71

Menemachus, 173

Merivale, Charles, 44

Merkus, Henricus, *67*

Meursius, Johannes, 292n21

Michael Palaeologus, 257n73

Mielcke, Christian Gottlieb, 243–245

Mieszko, 128

Mill, John Stuart, 13

Millais, John Everett, 156, 300n27

Milton, John, 70

Minsheu, John, 151

Miriam, 229

Mitford, Mary Russell, 95

Mohammed VI, 84

Moll, Herman, 23, *27*, 103, 277n2, 286n56

Möller, Eberhard Wolfgang, 220–221

Moncrif, François-Augustin Paradis de, 141

Montalvo, Garci Rodríguez de, 103

Montigny, Louis Charles Nicholas Maxi-
    milian de, 73

Moore, Thomas, 39

More, Thomas, 59–61, 94, 98, 100–103,
    101, 112, 214

Morold, 200

Morris, Harry, 229

Morrison, David, 240

Moses, 229

Mowbray, 306n4

Muir, Edwin, 282n2

Müller, Wilhelm, 191

Munk, Kaj, 51

Münster, Sebastian, 13, 26

Murray, Gilbert, 113

Musashi, Miyamoto, *133*, 295n57

Nahum, 145

Nahyan, Khalifa bin Zayed Al (Sheikh
    Khalifa), 79

Napoléon, 27, 316n57

Nares, George Armstrong, 282n59

Nashe, Thomas, 29, *122*, 167

Necho II, 25, 259n106

Needham, Marchamont, 184

Nero, 160, 232–233

Nettleford, Rex, 20

Neupert, Karl, 203

Nevsky, Alexander, 144

Nicodemus, 170

Nietzsche, Friedrich, 17, 159

Nikitin, Afanasy, 70

Noah, 229

Noah, Mordecai Manuel, 227, 285n39

Northcliffe, Lord, 207, 209

Nowell, Laurence, 125

Oberth, Hermann, 199–200

Odum, Eugene, 217

Odysseus, 98, 111, 305n55

Oechelhäuser, Wilhelm, 192

Oedipus the Tyrant, 161, 163

Offa of Mercia, 141

Olaf (son of Margaret I), 130–131, 294nn42–43

Olaf of Sweden (Olof Skötkonung), 130–131, 175, 294n42

Old Denmark, 129, 134–136

Old Norway, 129, 134–136

Old Oligarch (Pseudo-Xenophon), 113

Oliver, Michael, 279n31

Ophelia, 138, 140, 154–156, 159, 160–161, 167–168

Orff, Carl, 230, 325n100

Ortelius, Abraham, 121

Osric, 161, 167

Ostrogorsky, Moisey, 183

Paine, Thomas, 2, 3n2

Paltock, Robert, 262n42, 286n49

Pappus of Alexandria, 1, 2

Paramount, King, 225

Parker, Richard, 226

Parry, William Edward, 43

Pastreich, Emanuel, 275–276n107

Paul, Saint, 229

Pausanias, 163

Payami, Babak, 71

Pears, Peter, 324n73

Pedersen, Christiern, 126, 163

Perdita, 168, 305n45

Periander, 112, 288n44

Pericles, 114, 161, 305n55

Pettingal, John, 306n15

Phaestis, 115

Philip II of Macedon, 232

Philip II of Spain (and Aragon), 299n16

Philippa (daughter of Henry IV), 131

Philoctetes, 89

Philo of Alexandria, 60

Phorkys, 190, 311n41

Pindar, 105, 111

Piri Reis, 63–66, *64*, *65*

Plato, 14, 18–19, 97–98, 100–101, 115, 247

Plato Comicus, 110

Plautus, Caspar, *39*

Pliny the Elder, 215, 239, 266n12, 315n46

Plumb, Hay, 148, *149*

Plutarch, 2–3, 28, 172–173, 231, 233, 245–247

Poe, Edgar Allan, 147

Polo, Marco, 68, *68*, 267n28, 272n55

Polonius, 158

Polybius, 202

Pontanus, John Isaac, 292n21

Pontoppidan, Erik, 36

Pontus, 190, 311n41

Popeye, 36

Porter, Cole, 111

Posilge, Johann von, 294n40, 294n42

Powtee, 265–266n11

Préault, Antoine-Augustin, 155–156, *156*

Preller, Friedrich the Elder, 196

Prince of Arragon, 299n16

Proctor, Richard Anthony, 29, 260n14

Prospero, 53, 147, 172, 247, 306n55

Pseudo-Plutarch, 326n6

Pseudo-Xenophon (Old Oligarch), 113

Ptolemy (Claudius Ptolemy), 28, 31

Ptolemy II, 25, 259nn107–108

Pullum, Geoffrey K., 265n9

Purcell, Henry, 224

Purchas, Samuel, 42, 70

Purvis, Andrea, 98

Putsch (Johannes Bucius Aenicola), *88*

Pyrard, François, 67–68, *69*

Pytheas, 202

Queequeg, 93

Quine, Willard V., 14

Raedwald of East Anglia, 171

Raleigh, Sir Walter, 158, 224

Ransen, Mort, 282n56

Rasmussen, Knud, 47

Rasoulof, Mohammad, 87

Rastell, John, 151–152

Ratzel, Friedrich, 87, 238–240, 283n21

Ray, Nicholas, 265–266n11

Reed, Thomas German, 225

Reinerth, Hans, 216
Resnais, Alain, 269n67
Rhys, Jean, 210
Richards, Thomas, 84
Riedelsheimer, Thomas, 261n31
Riefenstahl, Leni, 205, 206–207, 316n55,
    317n66
Riel, Jørn, 266n11
Ries, Ferdinand, 187
Risdon, Tristram, 110
Rodney, Walter, 20
Rolls, Charles Stewart, 209
Romeo, 306n4
Roosevelt, Franklin Delano, 100, 205–206,
    214, 229
Rorick, 134–135
Rorty, Richard, 17, 282n3
Rosa, Giorgio, 279n32
Rosenberg, Alfred, 202, 210, 216
Rosencrantz, 157–158, 167, 299n11
Rosenkrantz, Jørgen, 126
Rosenkranz, Karl, 255n46
Ross, John Robert, 15
Rousseau, Jean-Jacques, 17
Roze, Pierre-Gustave, 74, 78, 275n97
Ruesch, Hans, 265–266n11
Ruhig, Philipp, 331n95
Rukn al-Din Mahmud, 68
Ruskin, Effie, 300n27
Ruskin, John, 300n27
Rutherfurd, Edward, 297n100

Safire, William, 251n2
Sahlins, Marshall, 4
Salbancke, Joseph, 69
Sallust (Gaius Sallustius Crispus), 22, 183
Samarqandi, Abd al-Razzaq, 68–70
Sancho Panza, 225
Sannazaro, Jacopo, 302n4
Santa Cruz, Alonso de, 22
Santillana, Giorgio de, 147
Sarris, Andrew, 50, 268n41
Saxo Grammaticus, 125–127, 129–131,
    133–134, 143, 151, 153, 160, 175, 187, 189,
    292n17, 296n76

Sayers, Dorothy L., 176
Schaffner, Franklin J., 279n34
Schaller, Fritz, 220
Schinkel, Karl Friedrich, 196
Schjeldahl, Peter, 199
Schlegel, August Wilhelm von, 148, 190
Schlegel, Karl Friedrich von, 191
Schleiermacher, Friedrich, 191–192
Schmeider, Felicitas, 293n28
Schmidt, Arno, 284n24
Schmitt, Carl, 15, 36, 38, 121, 193, 218,
    220–222, 242, 247, 323n48
Schnabel, Johann Gottfried, 284n24
Schoenberg, Arnold, 206
Schopenhauer, Arthur, 201
Schröder, Friedrich Ludwig, 150
Schütz, Wilhelm von, 220
Scott, Sir Walter, 27, 152, 302n10
Sebastian, 305n55
Seitz, George B., 317n67
Selden, John, 132, 184–185, 222
Semple, Ellen Churchill, 59, 87, 239,
    278n9
Senior, Duke, 172
Sesostris, 25, 259n105
Seward, Thomas, 32
Shadwell, Thomas, 224
Shakespeare, William, 52, 62, 68, 96, 97,
    101–103, 116, 121–123, 125, 126, 131–132,
    136, 142, 147, 153, 163–166, 170, 172,
    177–178, 184, 197, 223–224, 305nn43–
    45, 308n37, 325n91
Shapiro, Sidney, 64
Shute, John, 292n16
Shylock, 63, 69
Sihtric, 128
Simon, Paul, 262n4
Simonides, 106, 287n10
Simplicius of Cilicia, 290n74
Sinbad, 36, 66
Skovmand, Michael, 307n27
Smith, John, 25
Smith, Norman Kemp, 204–205
Smithson, Harriet, 300n27
Snæbjörn, 233

Snowden Gamble, Charles Frederick, 208

Socrates, 115, 173

Sophie of Denmark, 292n19

Sophocles, 104, 163

Sophy, Lady, 225

Stanford, Charles Villiers, 225

Stanislavski, Constantin, 292n22

Stephano, 96

Stewart, George Rippey, 287n14

Stewart, Jimmy, 76

Stiffe, A. W., 67

Stoker, Bram, 2

Strabo, 112, 116, *213*, 214, *239*, 315n46

Strachan, John, 84–85

Strachey, William, 224, 272n52

Strauss, Johann, 187

Strauss, Joseph Baermann, 78

Stravinsky, Igor, 228

Strindberg, August, 50–51

Sturluson, Snorri, 52, 233, 291n6

Suess, Eduard, 7

Sullivan, Arthur, 224–225, 229

Surrey, Lord, 103

Sweyn Forkbeard, 128

Swift, Jonathan, 23, *102*, 103, 208, 225, 262n42

Sycorax, 306n55

Sylvester, Joshua, 255n44

Tadao, Shizuki, 73

Tarkovsky, Andrei, 268n47

Tennyson, Alfred, 47

Terry, Ellen, 292n22

Themistocles, 112–113

Theopompus, 97–98

Thévet, André, 22, *213*, 214

Thompson, Daniel Pierce, 264n37

Thomson, James, 185, 187

Thoreau, Henry David, vii, 41, 94, 218

Thorfinn, Earl of Orkney (Thorfinn Sigurdsson), 52

Thorgny the Lagman, 175

Thorkelin, Grímur Jónsson, 125

Thucydides, 109, 110, 112–114

Tian Hu, 64

Tieck, Ludwig, 284n24

Timon, 152

Tiresias, 163

Tolias, Georgios, 256n60

Torres-Saillant, Silvio, 20

Tourneur, Maurice, 210

Trenker, Luis, 316n57

Trevisa, John, 301n37, 301n39, 302n1

Tuan, Yi-fu, 23

Tubal, 63

Tuirbe Tramár (Essarg), 294n35

Twain, Mark, 154, 245, 300n20

Twyne, Laurence, 170

Twyne, Thomas, 170

Tyssot de Patot, Simon, 285–286n49

Tzetzes, John, 2

Udet, Ernst, 205–207, 218, 316n55, 318n76

Ullmann, Liv, 53

Ulloa, Francisco de, 103, 267n28

Ulrici, Hermann, 192

Umali, Bas, 242, 330n85

Usk, Thomas, 301n37

Utopus, 59–60, 100, 112, 216, 283n13

Valdemar IV of Denmark, 130

Valdemar the Conqueror (Valdemar II), 144

Vaniman, Chester Melvin, 209–210

Varen, Bernhard, 31

Värnlund, Rudolf, 52

Varthema, Ludovico di, 69

Vashishtiputra Shri Pulumavi, 114

Vaus, John, 300n28

Vedel, Anders Sørensen, 126, 292n19

Venerable Bede, 128, 168

Venn, John, 1, 13–15, 27

Vere, Edward Fairfax, 226

Vernadsky, Vladimir, 7

Verne, Jules, 14, 15, 39, 264n33

Vespucci, Amerigo, 63

Vigneault, Gilles, 47

Viola, 305n55

Voltaire, 66, 98, 141

von Braun, Wernher, 199
von Dechend, Hertha, 147

Wagner, Richard, 136, 187, 192, 194–201, *198*
Wagner de Bosquet, Eloise, 96
Walcott, Derek, 238
Wallace, William, 243
Wallis, Samuel, 73
Walton, Francis R., 2
Warner, Frank, 214
Wartislaw VII (father of Eric of Pomera-
    nia), 294n45
Wartislaw Jagiello, 294n45
Watling, John, 81
Weaver-Hightower, Rebecca, 4
Wegener, Alfred Lothar, 204, 236
Weibull, Lauritz, 296n74
Weill, Kurt, 206
Weinbaum, Batya, 261n34
Welles, Orson, 99, 269n67
Wellman, Walter, 209–210
Wells, H. G., 208, 209
Werner, Abraham, 290n71
Wertmüller, Lina, 96
Whewell, William, 13
Whitelocke, Bulstrode, 314n15
Whiting, William, 98, 228–229
Whitman, Walt, 225
Whittlesey, Derwent, 238

Wieland, Christoph Martin, 148, 192
Wilde, Oscar, 93
Wilkes, Charles, 265n6
Wilkins, Peter, 286n49
William of Normandy, 38
William the Conqueror, 294n38
Williams, Ralph Vaughan, 225
Willich, Ehrenfried von, 192
Wilson, Woodrow, 242
Wirth, Herman, 316n51
Wittgenstein, Ludwig, 14
Wood, Allen, 205
Wright, Thomas, 16, 28
Wyss, Johann David, 96

Xavier, Saint Francis, 71, 274n71
Xerxes I of Persia, 60, 79, 107–109, *109*,
    111–112, 115, 216, 232

Yoon Sang-ho, 73
Yorick, 101, 158
Yoshikawa, Eiji, 295n57

Zangwill, Israel, 228
Zerkaulen, Heinrich, 221
Zeus, 109, 111
Zheng He, 70
Zuber, Jean Henri, 74
Zupitza, Julius, 166

Place Index

Italic page numbers indicate material in illustrations.

Aarhus, 36
Abadan Island, 66
Abraxas, 59
Abu Dhabi, 269n1
Abu Musa, 70
Acholla, 311n41
Actium, 114
Acushnet River, 86
Aden, Gulf of, 270n3
Adriatic Sea, 87, 279n32, 321n12
Aeaea, 107
Aegean Archipelago, 22
Aegean Sea, 107, 111, 114
Aegina, 270n15, 288n44
Aeolia, 107
Aeolian Archipelago, 96
Aerohotel, 318n80
Ahnenerbe, 216, 316n50
Aithsting, 308n40
Akwesasne, 91
Åland, 19, 143, 178
Alappuzha, 63
Alaska, 100, 285n39
Albertina, 243–245, *244*
Alcatraz, 279n33
Aldeburgh, 226, 230
Alde River, 226
Alderney, 258n84, 304n40
Alexander Archipelago, 100
Alexander Island, 282n60
Alexandria, 58
Alluttoq Island, 48, 49

Alpide Belt, 327n36
Alps, 7, 316nn57–58
Altino, 60
Altruria, 93
Alvastra, 216
Amager, 49
Ammelhede, 307n27
Amorgos, 287n10
Amsterdam, 59, 61, 63
Amur River, 281n47
Androscoggin River, 264n37
Angel's Resort Island (Seonyudo), 77
Anglesey, 258n84
Anian, Strait of, 267n28
Ankerwyke, 178, 309n48
Ankoko, 281n48
Antarctica, 43, 46, 48, 78, 92, 236, 282n60,
    283n15
Antigua and Barbuda, 278n26
Aqaba, Gulf of, 72
Aquileia, 60
Arabian Gulf, 79
Arabian Peninsula, 70, 72, 283n20
Arabian Sea, 58
Aragon, 299n16
Aral Sea, 90
Ararat, 100, 227
Ararat, Mount, 227, *228*
Archipelago of San Andrés, Providencia
    and Santa Catalina, 281n49
Archipelago Sea, 266n19
Arctic, 45, 46, 47, 78, 275n105

Arctic Ocean, 99, 264n44

Arctic Straits, 270n3

Ardennes Forest, 306n1

Argentina, 89, 270n6, 281n51, 282n60

Argun River, 281n47

Arkona, 190–191

Armorica, 31

Arroio Invernada, 281n48

Asopus River, 114

Astrakhan Kremlin Island, 57

Atammik, 49

Athelney, 185, 310n12

Athens, 79–80, 109–114, 232, 233

Athos, Mount, 21, 60, 79, 100–101,
    107–111, 115, 129, 216, 232, 270n18

Atlantic Ocean, 85, 92, 205, 210, 236,
    287n15

Atlantis, 31, 97, 101, 137, 202, 210, 216,
    316n51

Attica, 106, 113, 290n78

Aulis, 105–106, *106*

Australia, 7, 26, 99, 242, 249, 279n26,
    279n33, 281n50, 310n25

Austral Islands, 284n26

Austria, 269n1, 282n55, 311n28

Austronesia, 256n66

Avalon, 201

Avalon Peninsula, 219

Avon, 278n12

Axe, 278n12

Ayeyarwady Delta, 58

Ayles Ice Island, 43

Ayutthaya, 63

Bab-el-Mandeb Strait, 70, 72, 270n6

Bad Honnef, 304n40

Badmadow, 70

Baengnyeong, 75

Baffin Island, 49

Bahamas, 2, 278n26

Bahrain, 90, 278n26

Baja California, 267n28

Bajo Nuevo, 281n49

Baker, 280n45

Baltic Sea, 19, 143, 148, 153, 169, 200,
    287n15, 297–298n105, 315n46

Baltic Straits, 270n3

Bam Island (Bamseom), 75–77

Bamseom (Bam Island), 75–77

Bandar Seri Begawan, 63

Bangkok, 63

Bangladesh, 58, 63, 90, 281n50

Baranof Island (Sitka Island), 100, 285n39

Barataria, 225

Barbados, 278n26

Bardsey, 258n84

Barisal, 63

Bar Island, Maine, 261n27

Bartlow Burial Mounds, 139

Basra, 63, 64–66

Bavaria, 304n40

Bay of Fundy, 26, 31, 97, 219

Bayreuth, 197

Beagle Channel, 270n6

Beaufort Sea, 91

Bedloe's Island, 89

Belmont, 63

Benátky, 188

Benin, 280n46

Bensalem, 40, 211, 303n28

Bentley Subglacial Trench, 48, 267n35

Berghof, 197

Bergonund's Island, 133

Bering Strait, 78

Bermuda, 224, 272n52, 278n9

Bidasoa River, 89

Big Diomede Island, 78

Bigeumdo, 73

Bikini Atoll, 99

Bird Island, 281n48

Bjäre Peninsula, 50

Black Sea (Euxine), 24, 107, 258n94,
    287n8, 287n14

Black Water (Kālā Pānī), 279n33

Boa Vista, 269n1

Boca del Serpiente (Serpent's Mouth),
    302n11

Bocas del Dragón (Dragon's Mouths),
    302n11

Boeotia, 104, 106

Bogomerom Archipelago, 282n55

Bohemia, 168, 305n43

Bolivia, 264n37

Bolshoi Ussuriysky, 89

Bombay (Mumbai), 58–59, 269n1

Börgermoor, 207

Borneo, 89, 281n47

Bornholm, 50, 136, *137*, 191, 295n53,
    296n69

Börringe Lake, 304n41

Börringe Priory, 304n41

Bosham, 293n33

Bosporus, 24, 107, 270n5

Bothnia, Gulf of, 19, 143

Botswana, 281n46

Boundary Lake, 89

Brännö Island, 125

Brattahlid, 49

Brazil, 59, 90, 281n48

Bressay, 258n84

Bright Island, 103

Bristol Channel, 129, 141

Britain (as an island), 6, 15, *15*, 52, 53, *122*,
    124, 125, 141, 153, 166, 169, 187, 193, 194,
    219, 221, 224, 226, 227, 233, 294n38,
    295n51, 297n93, 299n10, 306n4

British Columbia, 236

British Isles, 14, *15*, 124–129, 177–178,
    185–186, 224

Brittany, 31, 200, 201, 294n35

Brobdingnag, 286n51

Bruges, 63

Brühl, 187

Brunei, 89

Bubiyan, 66, 70, 272n43

Budapest, 304n40

Bulgaria, 269n1, 281n52

Burga Water Island, 320n10

Busan, 76, 78

Byrdingø, 304n41

Calauria (Poros), 110, 232–233

Calcutta, 58, 264n40

Caldey Island, 129

California, 103, 258n93

California, Gulf of, 267n28, 286n56

Calypso, 305n55

Cambodia, 281n50

Cameroon, 281n46

Canada, *15*, 26, 42, 45–47, 48, 49, 78, 89,
    90–92, 97, 100, 132, 227, 236, 264n37,
    264n44, 266n18, 277n2, 278n9, 282n53

Canadian Arctic Archipelago, 46–47,
    264n44

Cape Krio (Triopion), 111

Cape Verde, 278n26

Capri, 107, 233

Caribbean, 19–20, 35, 233, 235

Carillon Island, 85

Caroline Islands, 217

Carthage, 111

Caspian Sea, 24–25, 57

Cat Island, 277n1

Cephalonia, 327n38

Cephissus River, 113

Cernofca, 281n52

Chad, 282n55

Chad, Lake, 282n55

Chafarinas Islands, 281n52

Chalcidice Peninsula, 115

Chalcis, 104, 106, 115, 290n80

Champlain, Lake, 57

Channel Islands, *15*, 270n15, 277n8,
    295n53, 304n40

Cheonghae (Wando), 77

Cheshire, 325n91

Chester, 325n91

Chiemsee, 304n40

Chile, 89, 270n6, 282n60

China, 58, 59, 63, 64, 74–77, 89, 267n29,
    270n6, 275n105, 281n47

Chios, 106

Chittagong, 58

Christianshavn, 49

Cimbrian Peninsula, 141

Cizre, 227, 325n88

Clipperton Atoll, 95, 96

Cloaca Maxima, 302n12

Cocos (Keeling) Islands, 235, 281n50

Coiba, 279n33

Cologne, Electorate of, 187

Colombia, 269n1, 279n33, 281nn48–49

Colonsay, 258n84

Colossus of Prora, 199

Comino Island, 96

Comoros, 278n26

Conakry, 269n1

Concordia (Portogruaro), 60

Conejo Island, 281n49

Congo, Democratic Republic of, 280n46

Cook Islands, 95, 280n44

Copenhagen, 49, 63, 178

Cordova Island, 282n53

Corfu, 270n15

Corinth, 112, 289n46

Corinth, Gulf of, 110, 112, 288n44

Cork, 66

Cork, County, 272n41

Cornwall Island (Kawehnoke), 90–92

Corsica, 270n15

Corsico Bay, 280n46

Crapeau Point, 220

Crete, 89, 270n15

Croatia, 269n1, 281n52

Cuba, 89, 278n26

Cumae, 106

Curaçao, 235

Cyclades, 287n10

Cyprus, 89, 106, 270n15, 278n26, 299n16

Daecheong, 75

Dagelet, 73

Dagö Island, 296n72

Dalkey, 258n84

Dannevirke, 141

Danube, 58, 141, 281n52, 304n40

Dardanelles (Hellespont), 24, 79, 107–109, *108*, 111, 129, 232, 287n14

Davis Strait, 49

Deben River, 111

Decelea, 113

Deer Isle, 97

Dejima, 73, 87

Delos, 305n44

Delphi Island, 168, 305n44

Delting, 308n40

Denmark, 46, 48–49, 79, 125–132, *155, 158,* 269n58, 281n51, 291n82, 291n6 (chap. 9), 292n21, 294n39, 295n53, 299n9, 299nn17–18, 303n18, 307n25, 307n27

Dessau, 191

Detroit, 57

Devil's Island, 279nn33–34

Devon, 110

Dingwall, 308n40

Diolkos, 112

Disko Bay, 49

Diua Damasciaca, 70

Djibouti, 70, 72

Doi. *See* Jeju

Dokdo, 75

Dominica, 278n26, 281n48

Dominican Republic, 89, 277n7

Donemuthan Convent, 304n37

Donghwa Island, 77

Dongping Lake, 271n37

Don River, 278n12, 304n37

Douarnenez, 201

Double Road Point (Double Rote Point), 219–220

Doumeira, 70

Dove, River, 258n84, 304n28

Dover, 169

Dover, Strait of, 170

Dragon's Mouths (Bocas del Dragón), 302n11

Drepane (Scheria), 107, 305n55

Dry Salvages, 218, 321n21

Dubai, 79–80

Dublin, 258n84, 268n42

Duchy of Madgeburg, 187

Dursey, 258n84

Dutch Antilles, 235

Duwamish, 70

Dwyryd, River, 279n34

Dyfi, River, 293n35

Earth, 13–14, 23, 28–30, *29, 30,* 42, 194, 202–204, *203, 204,* 209, 210–214, *211, 213, 234,* 235–237, *237,* 239, 245–246, 248

East Anglia, 111, 171, 226, 258n84, 304n28

Eday, 258n84

Egypt, 25, 58, 60, 72, 144–145

Eider, 293n31

Eight Hundreds Li Liangshan Lake, 271n37

Eilean nam Ban (Island of Women), 31, 304n40

Eirene, 326n12

Eismitte, 204

Ellesmere Island (Umingmak Nuna), 46, 92, 267n26

Ellis Island, 87

El Salvador, 281n49

Elsinore, 36, 51, 52, 57, 123, 131, 136, 150, 153–154, 156–158, *158*, 163, *164*, 165, 166, 169, 176, 183, 253n5, 269n58, 287n15, 292n19, 299n17, 308n37

"Empress Island" (Britain), 15, 254n27

Emsland, 207, 215

Enceladus (moon of Saturn), 237

Enewetak Atoll, 99, 280n43

England (as an island), 15, *15*, 140–141, 185–186, 207, 209, 227–228, 247

English Channel, 141, 146, 170, 208, 219

Enys Samson, 201

Ephesus, 107

Equatorial Guinea, 280–281n46

Eretria, 107

Erewhon, 103

Eritrea, 70

Esk, 278n12

Estonia, 296n72

Euboea, 104, 107–109, 113, 115, 286nn2–3

Euphrates, 66, 261n30

Euripus Strait, 14, 104–106, *105*, 109, 115–116, 154, 286n3, 290n76, 290n78, 300n21

Europa (moon of Jupiter), 236–237

Europa Island, 281n52

Euxine (Black Sea), 24, 107, 258n94, 287n8, 287n14

Exe, 278n12

Eye, 258n84, 304n28

Eynsham, 177

Eysysla. *See* Oesel

Failaka, 66

Fairhaven, Massachusetts, 278n24

Falkland Islands, 281n51

Fangataufa Atoll, 96

Far North (Canada), 42

Fårö, 50, 53, 269n66

Faroe Islands, 52, 131, 175, 178, 281n51, 294n39, 295n53, 307n17

Federsee, 216

Fehmarn Belt, 223

Fehmarn Island, 223

Fenit, 38

Fiji, 278n26, 281n50

Finland, 46, 90, 266n19, 294n39

Firth of Clyde, 226

Floreana Island, 96

Flores, 269n1

Forest of Arden, 172, 306n1

Fox Point Beach, 31, 261n31

Foyle River, 277n7

France, 89, 95, 143, 150, 261n28, 278n9, 279n33, 281n52, 284n26

Frauenchiemsee, 304n40

French Guiana, 279nn33–34

French Polynesia, 96, 284n26

Frioul Archipelago, 202

Frisches Haff (Vistula Lagoon), 243, 331n93

Frisian Islands, 152

Frjóey (Fair Isle), 299n16

Funen Island, 223, 299n18

Fyn, 207

Gabon, 280n46

Galápagos, 96

Galilee, Sea of, 229

Ganges Delta, 58

Ganghwa, 74–75, 78, 275n102

Ganryu Island, 132, *133*, 295n57

Gardar, 49, 268n38

Gare Loch, 324n80

Gavnø Island, 168

Geatland, 125

Geomundo, 73

German Democratic Republic (East Germany), 206

Germany, 51, 89, 137, 150, 183, 187, 188–193, 194–195, 199, 207, 220, 221, 233, 235, 242, 269n1, 296n73, 328n49

Gerrish Island, 262n8

Ghoramara, 282n54

Gibraltar, 67, 190, 270n6, 281n51

Gibraltar, Strait of, 184, 190, 259n6, 270n6

Gilbert Islands, 280n43

Globe Theatre, 101, 121, 142, 146–147, 166, 172, 195, 197, 214, 240

Glorioso Islands, 281n52

Goat Island, 318n85

Goldajärvi, Lake, 90

Golden Gate Bridge, 78

Golden Triangle (Jiangnan), 270n10

Gondal, 210

Goodwin Sands (the Goodwins), 169–170, 305n51

Gorgona Island, 279n33

Gothenburg, 52, 125

Gotland Island, 50–51, 268nn46–47

Graemsay, Island of, 302n10

Grand Canal (Venice), 62

Grand Island, 227, 285n39

Grand Manan Archipelago, 26, 91, 282n53

Grand Manan Island, 26, 219

Grand Seogang Bridge, 76

Great Belt, 223, 299n18

Great Britain, *15*, 79, 99, 178, 214

Great Lakes, 85, 267n29

Greece, 6, *15*, 22, 60, 89, 90, 107, 109, 110, 189, 277n3, 327n38

Green Island, 72

Greenland, 26, 46, 48–49, 92, 131, 178, 203–205, 267n35, 294n39, 307n17

Greifswald, 221

Grenada, 278n26

Gruting, 308n40

Guam, 280n43

Guanabara Bay, 59

Guanahani, 2, 81

Guantánamo Bay, 89

Guatemala, 269n1

Guernsey, 258n84

Guiana, *27*

Haarlem, 292n21

Habomai Archipelago, 281n47

Haggier Mountains, 283n20

Hague, The, 218n47

Haiti, 89, 277n7, 281n49, 282n53

Haldia, 58

Halicarnassus, 107, 110–111

Hallands Väderö, 50

Halle, 192

Hallstatt, 311n28

Hallstätter See, 311n28

Ham (London suburb), 142–143

Hamburg, 63

Hamlet's Mill, 233

Hanish Islands, 70

Hanover, 190

Han River, 74–77

Hans Island (Tartupaluk), 46, 90, *91*, 91–92

Harper's Meadow, 264n37

Hawaii, 90

Hayling Island, 128

Heartland, 242–243, 330n77

Hebrides, 97

Hedeby, 293n31

Heidelberg, 220–221

Heligoland, 195, 222

Hellespont (Dardanelles), 24, 79, 107–109, *108*, 111, 129, 232, 287n14

Helsingborg, 51, 52

Helsingør, 51. *See also* Elsinore

Heptanesia, 58–59

Hirado, 73

Hiroshima, 280n43

Hispaniola, 89, 277n7

Hokkaido, 275n105

Holderness, 138, 296n78

Holland, 184

Honduras, 281n49

Hongdo, 73

Hong Kong, 269n1

Honshu, 132

Hoo Peninsula, 288n32

Hormozgán, 272n46

Hormuz, 15, 57, 59, 60, 66–72, *67*, *68*, 169, 257n80, 272n51, 273n61, 273n66, 274n71, 274n77

Hormuz, Strait of, 78

Housay, 258n84

Houyhnhnms, 286n51

Hovs Hallar, 50

Howland, 280n45

Huangyan Island (Scarborough Shoal), 281n47

Hudson Bay, 57, 93, 131–132

Hudson River, 86

Hull Island, 261n28

Humber Estuary, 15, 304n37

Hungary, 282n55

Hunter Island, 281n52

Hven, 23, 116, 153, 163–166, *164*, 188, 253n5, 291n82, 292n21, 299n17, 303n18, 303n24

Iberia, 71

Iceberg Alley, 49

Iceland, 45, 52, 131, 174, *174*, 176, 178, 202, 210, 278n26, 281n51, 294n39, 307n17, 315n46, 316n51

Igaliku, 49

Iki Island, 78

Île d'If, 279n33

Île Petite, 81–83

Île Tristan, 201

Ilissus River, 113

Illyria, 23, 258n88, 305n55

Ilulissat Icefjord, 48

İmralı Island, 279n33

Incheon, 76

India, 58, 59, 63, 71, 90, 132, 279n33, 281n50

Indian Ocean, 66, 225, 327n36

Indonesia, 33, 63, 89, 224, 256n66, 264nn39–40, 281n47, 327n36

Indus Delta, 58

Inland Sea, 132, 295n58

Inner Niger Delta, 270n9

Insula Convent, 304n40

Intermontane Islands, 236

Iona Island, 304n40

Ionia, 304n40

Ionian Archipelago, 327n38

Ios, 106

Iran, 66, 70, 78, 272n46, 274n79

Iraq, 31, *40*, 63, 66, 70, 261n30, 272n43

Irbe Strait, 136

Ireland, *15*, 52, 66, 89, 132, 201, 268n42, 272n42, 277n7, 281n51

Ireland's Eye, 258n84, 268n42

Isla de Perejil. *See* Parsley Island

Island of Women (Eilean nam Ban), 31, 304n40

Isle of Dogs, *122*

Isle of Ely, 187

Isle of Man, *15*, 153, 177, 308n39

Isle of Wight, 86, 128

Isles of Scilly, 201

Isola dei Pescatori, 269n1

Israel, 72, 144–145

Isthmus of Perekop, 287n8

Itaipu, 90

Italy, 106, 269n1, 279n32, 305n55, 321n12

Ithaca, 106

Iwo Jima, 280n43

Jakarta, 63

Jakobshavn Isbrae, 48

Jamaica, 278n26

James River, 224

Jamestown Island, 25, 224

Japan, 72–73, *75*, 75–76, 78, 132, 225, 242, 275n105, 278n26, 281n47

Japan, Sea of, 72, 75, 132, 275n105

Jarvis, 280n45

Java, 327n36

Java Sea, 327n26

Jazirat al-Ma'danus. *See* Parsley Island

Jazirat al Maqlab (Telegraph Island), 67

Jeju (Doi, Quelpart), 73, 74–75, 76, 275n90

Jersey, 258n84

Jiangnan (Golden Triangle), 270n10

Jizera River, 188

Johnson's Island, 279n33

Jomsborg, 137, 296n74

Juan Fernández Archipelago, 207, 317n14

Judi, Mount, 227

Jutland, 5, 52, 125, 134, 141, 144, 153,
175–176, 223, 299n18, 303n14, 304n40,
307n25, 307n27, 314n15

Kachatheevu Island, 281n50
Kālā Pānī (Black Water), 279n33
Kaliningrad, 245. *See also* Königsberg
Kalmar Strait, 129, 143
Kanmon Straits, 132
Karachi, 58, 67
Karafuto Island (Sakhalin Island),
275n105, 281n47
Kareol, 201
Karlsgraben, 141
Kattegat, 125, *154*, 303n14, 307n27
Kawehnoke (Cornwall Island), 90–92
Kazakhstan, 90, 281n47
Kenya, 70
Key Biscayne, 217
Khasab, 274n77
Khor Ash-Sham Fjord, 67
Killiniq Island, 89
King Fahd Causeway Embankment
"40:66," 90
King George Island, 78
King Sejong Station, 78
King's Scholars' Pond Sewer (Tyburn),
176, 308n30
Kiribati, 278n26
Kish, 71, 272n46
Kneiphof, 16, 243–245, *244*
Knidos, 111, 113, 256n62
Koh Ses, 281n50
Koh Ta Kiev, 281n50
Koh Thmey, 281n50
Koh Tonsay, 281n50
Ko-Koun-to Archipelago, 74
Königsberg, 16, 190, *243*, 243–245
Korea, 72–79, 275n102, 275n106
Korea Strait, 72, 75, 78, 269n1, 270n3
Kosa Tuzla Island, 281n47
Krakatoa, 33, 154, 300n20, 327n36
Kronborg Castle, 165
Kuril Archipelago, 281n47
Kuwait, 66, 272n43

Kwajalein, 280n43
Kyushu, 132

Labrador Peninsula, 89
Lac Brûlé, 85
Lachine Rapids, 57
Lajes, 59
Lake District (Great Britain), 295n51
Lambay, 258n84
Lands End, 42–43, 277n2
Langeland, 299n18
La Pérouse Strait, 275n105
Laptev Sea, 46
Laputa, *102*, 208, 286n51
Leila. *See* Parsley Island
Leipzig, 187
Lemnos, 89
Lena Delta, 58
Lennox, 270n6
Lesbos, 305n55
Lesser Sunda Islands, 264n39
Lete Island, 280n46
Leyte, 280n43
Liancourt Rocks, 281n47
Liang, Mount, 64
Liangshan Marsh, 64
Liaodong, *75*
Ligitan, 281n47
Lijiang, 63
Lilliput Island, 286n51
Limfjord Sound, 304n40
Lincoln Sea, 92, 282n59
Lindau, 269n1
Lindisfarne Island, 168, 185, 261n28,
304n37
Line Islands, 280n45
Lisca Bianca, 96, *97*, 284n23
Lithuania, 279n34
Little Belt, 223, 299n18
Little Diomede Island, 78
Little Island, 66, 272n41
Liverpool, 21
Livonia, 297–298n105
Loch of Tingwall, 177
Lofoote, 298n112

Lofoten Archipelago, 147
Lohachara Island, 282n54
Lolland, 223
Lolwe, 70
Lomea, 170
Londonderry, 205, 277n7
Lorelei, 196–197
Los Monjes Archipelago, 281n48
Lower Saxony, 207, 215
Lowestoft, 226, 227
Lübeck, 136, 296n69
Luggnagg, 286n51
Lulworth Cove, 148
Lundy, 258n84
Lunnasting, 308n40
Luzon, 280n43, 281n47

Maalhosmadulu Island, 67
Macclesfield Bank, 281n47
Mackenzie River, 78
Madagascar, 228, 278–279n26, 281n52
Magellan, Strait of, 270n7
Maggiore, Lake, 269n1
Magna Carta Island, 178, 233
Mahee Island, 66, 272n42
Maine, 262n8, 264n37
Mainland Island, 26–27, 177, 308n38
Maktoum's World, 79–80, *80*
Malacca, 63
Malacca, Strait of, 270n5
Malaku Archipelago, 224
Malawi, 280n46
Malaysia, 89
Maldives, 67, 269n1, 279n26, 281n50
Malé, 269n1
Malmö, 292n18
Malozhemchuzny, 281n47
Malta, 270n15, 279n26
Maltese Archipelago, 96
Manchester, 63
Manhattan, 57, 86
Manicouagan, 47, 258n85
Mani Peninsula, 110
Manitoba, 89
Manra Island, 280n45

Maracaibo, 2, 63
Marajó, 217
March of Brandenburg, 200
Marcotis, Lake, 60
Mardi Archipelago, 21
Mare Internum, 22. *See also* Mediterranean
    Sea
Margaret Island, 304n40
María Madre, 279n33
Marianas, 280n43
Marie Byrd Land, 48
Maritsa River, 89
Marmara, Sea of (Propontis), 107, 270n5
Marquesas, 23
Marseilles, 202
Marshall Islands, 99, 279n26, 280n43,
    282n53
Martinique, 35
Massachusetts, 218, 321n21
Massachusetts Bay Colony, 218
Massacre River, 277n7
Matthew Island, 281n52
Mauritius, 279n26
Mawson Sea, 43
Mbamba Bay, 280n46
Mbanie, 280n46
Mediterranean Sea, 22, 24, 25, 60, 104, 111,
    114, 183–184, 189–190, 272n52, 311n41
Melanesia, 256n66
Melissani Lake, 327n38
Melita, 229
Memphremagog, Lake, 89
Menai, 258n84
Meropis, 98
Messaline, 305n55
Messina, 305n55
Messina, Strait of, 305n55
Meuse, 58, 223
Mexcaltitán, 269n1
Mexico, 33, 269n1, 279n33, 282n53
Mexico, Gulf of, 85
Mexico City, 57–58
Micronesia, 23, 33, *33*, 63, 217, 256n66,
    279n26
Middle Rocks, 281n47

Migingo, 70

Milky Way, 28, 29, *29*, 163

Millingen aan de Rijn, 58

Mindanao, 280n43

Minerva Reefs, 87, 279n31, 281n50

Minicoy Island, 281n50

Ministers Island, 261n28

Miquelon, 278n9

Mississippi, 58, 85, 321n21

Mitala Island, 264n38

Miyoun Island (Perim Island), 72, 270n6

Mongla, 58

Montebello Islands, 99

Montreal, 57, 85, 270n4

Mont Saint-Michel, 261n28

moon, 98, 211, *212*, 214

Morocco, Kingdom of, 83–84

Moruroa Atoll, 96

Moskenstraumen, 147

Mosterøy Island, 304n40

Mumbai (Bombay), 58–59, 269n1

Musandam Peninsula, 71, 274n77

Myanmar, 58, 281n50

Mytilene, 305n55

Naf River, 281n50

Nagasaki, 73

Namibia, 281n46

Nan Madol, 23, 63, 257n80

Nares Strait, 282n59

Narsholmen Peninsula, 268n47

Nasiriyah, *40*

Nauru, 279n26, 279n33

Navassa Island, 281n49, 282n53

Neay, 258n84

Nendrum Monastery, 272n41

Nesebar, 269n1

Nesting, 308n40

Netherlands, 58, 78, 89, 191, 227, 235, 281n50

Neuruppin, 200

Neusiedler See, 282n55

New Atlantis, 98, 103

New Bedford, Massachusetts, 86

Newfoundland, 43–45, 86, 89, 158, 204–205, 217, 219, 235, 265n8, 266n15, 268n42, 317n60, 321n17, 321nn26–27

Newfoundland and Labrador, 89

New Guinea, 89

New Hampshire, 35, 264n37

New Hormuz, 69

New Moore Island, 282n54

New Orleans, 58

New Penzance, 229

New South Wales, 310n25

New World, 15, 46, 204–205, 209, 214, 224, 317n60

New York, 264n37

New York City, 57, 89

New Zealand, 279n26, 282n53

Niagara Falls, 318n85

Niagara River, 227

Nibelheim, 203

Nicaragua, 281n49

Niger, 280n46

Niger Delta, 58

Nigeria, 58, 282n55

Nikumaroro, 280n45

Nile, 25, 58, 60, 114

Nonnenwerth, 304n40

Norderney, 195

Nore, 169, 226, 278n25

North Aegean Sea, 107

North Atlantic, 315n46, 316n51

North Dakota, 89

Northeast Passage, 78

Northern Hemisphere, 63, 195

Northern Ireland, *15*, 66

Northern Pirates, 281n50

North Jeolla Province, 276n111

North Korea (Democratic People's Republic of Korea), 72, 75, 76, 275n105, 281n47

North Rock, 282n53

North Sea, 15, 46, 53, 153, *154*, 169, 170, 171, 176, 193, 195, 197, 223, 227, 287n15, 293n31, 321n12

North Sea Empire, 124–135, 153

North Sea Region, 132, *137*, 168, 174, 189
Northumberland, 304n37
Northwest Passage, 43, 45–47, 78, 220,
    267n28, 267n32, 270n4, 282n59
Norway, 50, 90, 129–134, *154*, 175, 195,
    283n15, 291n6, 294n39, 294n42,
    304n40, 307n17, 309n43, 313n3, 315n46
Nottinghamshire, 178
Nova Scotia, 31
Novaya Zemlya Archipelago, 99
Ntem River, 281n46
Nueva, 270n6
Nukuoro Atoll, 33, *33*
Nuljarfik, 203
Nunalivut, 47
Nunavik, 266n11
Nunavut, 89, 92
Nuremberg, 207
Nyasa, Lake, 280n46

Oahu, 90
Oare, 300n25
Oberröblingen, 320n1
Oceania, 217
Oesel (Eysysla, Saaremaa), 19, 136, 256n50,
    296nn70–71, 305n54, 315n46
Offa's Dyke, 141
Ogijima, 295n58
Ogygia, 107
Ohrdruf, 314n24
Okavanga Delta, 258n93
Okhotsk, Sea of, 275n105
Okinawa, 280n43
Oksholm, 304n40
Öland Island, 129, 143, 304n40
Olympus, Mount, 139
Oman, 78, 274n76
Øresund, 67, 123, 127, 141, 153, 156–158,
    158, 161–166, *164*, 196, 240, 287n15,
    294n39, 299n17, 301n31, 303n14,
    303n28, 314n15
Orkney Islands, 26, 31, 52, 93, 131–132, 153,
    201, 258n84, 269n63, 294n39, 295n53,
    299n15, 302n10, 308n40, 320n10

Ornö, 50
Orona Atoll, 280n45
Oronsay, 258n84
Osaka, 63
Ouse, River, 15, 84
Oxfordshire, 177
Oxney (Isle of Oxen), 49, 258n84
Oyasi, 70

Pacific Ring of Fire, 235, 327n36
Padua, 60
Paeonia, 217
Pakistan, 58
Palau, 279n26, 280n43
Palembang, 63
Panama, 279n33
Panama Canal, 240–241, 270n3
Panarea, 96
Panchaea, 97
Pangaea, 236
Papua New Guinea, 89
Paracel Islands, 281n47
Parret, 310n12
Parsley Island (Isla de Perejil, Jazirat al-
    Ma'danus, Leila, Tura), 83–84, 259n6,
    277n2, 281n52
Patrae, 289n68
Peace Dam, 275n104
Pearl Harbor, 90
Pedra Branca, 281n47
Peenemünde, 199
Peipus, 144
Peleliu, 280n43
Peloponnesus, 21, 110
Peñón de Alhucemas, 281n52
Peñón de Vélez de la Gomera, 281n52
Penrhyndeudraeth, 279n34
Pentapolis, 305n55
Penzance, 229
Perim Island (Miyoun Island), 72, 270n6
Pershore, 300n25
Persia, 60, 66–68, 71, 90, 104, 107–114,
    129, 143–144, 184, 216, 232, 286n3
Persian Gulf, 59, 66–68, *68*, 79, 169

Peru, 264n37

Pescatori, Isola dei, 269n1

Peter I Island, 283n15

Petermann Ice Island, 43

Petite Île, 82–83

Petty Place, 82–83

Pfahlbaumuseum, 216

Phalerum, 326n6

Pharos, 60

Pheasant Island, 89

Philippines, 38–39, 263n28, 279n26,
    280n43

Phoenicia, 60, 114

Phoenix Islands, 280n45

Piave River, 269n1

Picton, 270n6

Pillars of Heracles, 25

Piraeus, 112–113

Pitcairn Island, 95

Plataea, 114

Pnyx, 233

Pobeda Ice Island, 43, 265n6

Poland, 89, 130, 279n34, 294nn41–42,
    294n45, 296nn72–73

Polynesia, 256n66

Pomègues, 202

Pomerania, 137, *137*, 143, 190, 296n76,
    297–298n105

Pomeranian Islands, 137

Ponte della Moneta (Bridge of Money), 62

Ponte di Rialto (Rialto Bridge), 62, *62*

Pontus, 287n14

Po River, 269n1

Poros (Calauria), 110, 232–233

Portmeirion, 279n34

Portogruaro (Concordia), 60

Port Said, 58

Portsea Island, 86, 225, 226

Portsmouth, New Hampshire, 35

Portsmouth, Portsea Island, 225–226

Portugal, 13, 46, 63, 70, 279n34

Pouldavid Estuary, 201

Poulo Condor, 279n33

Prasias, Lake, 217

Pregel River, 16, 243

Prince Patrick Island, 43, 277n2

Princes' Islands, 270n5

Principality of Sealand, 87, 279n30, 321n12

Prison Island, 279n33

Propeller Island, 39

Propontis (Sea of Marmara), 107, 270n5

Prora, 199

Province Island, 89

Prussia, 190, 195, 245

Puerto Rico, 96

Pukapuka, 95

Puvirnituq, 266n11

Q (island), 89

Qaanaaq, 268n39

Qalhât, 70

Qaruh, 70

Qeshm, 272n46

Qubban Island, 66

Quebec, 89

Québec's Eye, 258n85

Quelpart. *See* Jeju

Quita Sueño Bank, 281n49

Rajin-Sonbong, 75, 275n105

Ralik Chain, 99

Randers, *140*

Rapa Iti, 284n26

Raquette, 91

Rarotonga Island, 100

Ras Doumeira, 70

Ras Kamboni, 70

Recife, 269n1

Red Sea, 25, 72, 144–146, *146*, 269n1,
    270n3, 270n6

Reggane, 96

Remba, 70

Rendlesham, 111

René-Levasseur Island, 258n85

Republic of Minerva, 87, 279n30

Republic of Rose Island, 87, 279n31

Republic of Venice, 23

Respubliko de la Insulo de la Rozoj,
    279n32

Revillagigedo Archipelago, 33

Rhine-Meuse Delta, 58
Rhine River, 183, 304n40
Rhodes, 270n15
Rialto, 61–63
Rialto Bridge (Ponte di Rialto), 62, *62*
Riau Islands, 270n5
Richelieu River, 57
Riga, Gulf of, 19, 136
Rikers Island, 279n33
Rimini, 279n32
Ringiti, 70
Rink Glacier, 204
Rio–Antirrio Bridge, 110
Rio de Janeiro, 59
Rio Grande, 282n53
Rion, Strait of, 110
Rio Quarai, 281n48
Riviera del Brenta, 269n1
Rivière du Nord, 81–83, 85
Rivo Alto, 62
Roanoke Island, 224
Robben Island, 89, 279n33
Robinson Crusoe Island, 207
Robinson Crusoeland, 99
Rockall Island, 52, 269n61, 281n51
Rock of Gibraltar, 67
Rocky Tale, 281n52
Rokovoko, 93
Romania, 58, 281n52
Rome, 121, 183, 302n12
Romsdal Peninsula, 50
Roosevelt Island, 279n33
Rosenön, 50
Rost Islands, 298n112
Rotterdam, 78
Royal Prussia, 294nn41–42
Rub' al Khali, 277n2
Rügen, 18–19, 190–192, 194, 196, 199–200,
    220, 255n46, 314n24, 314n28
Rukwanzi, 280n46
Runnymede, 178, 233, 309n48
Ruppiner See, 200
Russia, 46, 58, 75, 78, 89, 90–91, 279n33,
    281n47
Ryck River, 221

Saaremaa. *See* Oesel
Sable Island, 158, 301n36
Sacramento River, 258n93
Saint Helena, 279n33
Saint Illtud, 129
Saint Kilda, 95
Saint Kitts and Nevis, 279n26
Saint Lawrence, Gulf of, 85, 267n28
Saint Lawrence River, 57, 85, 91, 267n29
Saint Lucia, 279n26
Saint Martin Island, 89, 235
Saint Petersburg, 63
Saint Regis, 91
Saint Vincent and the Grenadines, 279n26
Saipan, 280n43
Sakhalin Island (Karafuto Island),
    275n105, 281n47
Salamis, 106, 112, 144
Salvington, 222
Samar, 280n43
Samoa, 279n26
Samosir Island, 327n36
Samsø, 299n18
Sanafir Island, 72
San Benedicto Island, 33
Sanday, 258n84
Sandsting, 308n40
Sandwich Islands, 282n60
Sangiang (Thwart-the-Way), 236, 327n36
San Joaquin River, 258n93
Sankt Goar, 314n19
San Salvador, 81, 277n1
Santa Cruz del Islote, 269n1
Santa Maria di Nazareth Island, 89
Santo Antônio, 269n1
São Tomé and Príncipe, 279n26
Sardinia, 270n15
Sardis, 173
Šarengrad, 281n52
Sargasso Sea, 14–15, 210
Saronic Gulf, 112, 288n44
Saudi Arabia, 70, 72, 90
Saxony, 192, 215
Saxony-Anhalt, 320n1
Scania, 153, 292n18, 299n17, 304n41

Scapa Flow, 302n10

Scarborough Shoal (Huangyan Island),
    281n47

Scheria (Drepane), 107, 305n55

Schlei Estuary, 293n31

Schleswig, 223

Scotland, *15*, 131, 132, 153, 177, 195, 216,
    227, 295n53, 299n19, 304n40

Scottish Highlands, 308n40

Scythia, 114

Sealand, 19. *See also* Zealand

Seal Island, 282n53

Sebatik, 89

Sedudu River, 280–281n46

Senkaku Islands, 281n47

Seonyubong, 77

Seonyudo (Angel's Resort Island), 77

Seoul, 72–79

Seoul Floating Island, 77

Serbia, 281n52

Serpent's Mouth (Boca del Serpiente),
    302n11

Serranilla, 281n49

Severn Channel, 295n53

Seychelles, 279n26

Shadwan Island, 72

Shandong Peninsula, 77

Shandong Province, 271n37

Shanghai, 275n105

Sharjah, 70

Shatt al-Arab, 64, 66

Sheppey Island, 169, 178, 226, 258n84

Sherwood Forest, 178

Shetland Islands, 27, 131, 132, 153, 177,
    177, 191, 201, 294n39, 295n53, 308n37,
    308n40, 320n10

Shoreditch, 142

Sicily, 168, 173, 258n88, 270n15, 290n80,
    305n44, 305n55

Sidon, 114

Sigulu, 70

Sile River, 269n1

Singapore, 269n1, 279n26, 281n47

Sinsido, 73

Siorapaluk, 268n39

Sipadan, 281n47

Sitka Island (Baranof Island), 100, 285n39

Sjælland, 19. *See also* Zealand

Skagerrak, 52, *154*

Slaughden, 226

Smailholm Tower, *152*

Smith Sound, 46, 268n42

Snape, 171

Socotra Archipelago, 283n20

Socotra Rock, 75, 281n47

Sohar, 274n76

Solent, 86

Solomon Islands, 95, 279n26

Solovki, 279n33

Somalia, 70

South Africa, 89, 279n33

Southampton, 128

Southern Hemisphere, 43, 46

South Georgia Island, 281n51, 282n60

South Korea (Republic of Korea), 72, 75,
    77, 78, 79, 281n47

South Orkney, 282n60

South Pacific, 96, 99, 225

South Sandwich Islands, 281n51

South Seas, 225

South Shetland Islands, 282n60

Soviet Union, 99

Spain, 36, 46, 81, 83–84, 89, 221, 270n6,
    279n34, 281nn51–52, 299n16

Spiekeroog, 199

Spinalonga, 89, 280n40

Spithead, 86, 226, 278n25

Spitsbergen, 78

Spratly Islands, 281n47

Sri Lanka, 279n26, 281n50

Srinagar, 63

Stagira, 115

Steep Holm, 295n53

Stepney Marsh, *122*

St. Helena, 27

Stiltsville, 217

St. Mary's Harbour, 219–220

Stockholm, 51–52, 63, 295n53

Stockholm Archipelago, 50–51

Stokesay, 258n84

Storting, 309n43
St. Pierre, 278n9
Strangford Lough, 66
Strymon, River, 144
Suez, Gulf of, 72
Suez Canal, 240
Suffolk, 87, 110, 171, 279n30, 321n12
Sulina, 58
Sulu Archipelago, 38, 263n28
Sumatra, 235, 327n36
Sumbawa Island, 264n39
sun, 98
Sundarbans, 90
Sunda Strait, 154, 327n36
Suriname, 235
Sutton Hoo, 110, 171, *171*
Suzhou, 63–64
Svendborg Sound, 223
Swains Island, 87, 282n53
Sweden, 50–52, 90, 131, *154*, *158*, 163, 190,
    196, 268n41, 291n82, 291n6 (chap.
    9), 294n39, 295n53, 296n70, 296n72,
    299n17, 303n18
Sweyn Holm, 295n53
Swinemünde, 200
Switzerland, 269n1
Szczecin Lagoon, 200

Taihu, Lake, 64
Taiwan, 270n6, 270n15, 279n26, 281n47
Taiwan Strait, 270n6
Tambora, 264n39, 327n36
Tanzania, 280n46
Tarawa Atoll, 280n43
Tarsus, 229
Tartupaluk (Hans Island), 46, 90, *91*,
    91–92
Tåsinge, 223
Tauris, 105, 287n8
Tel Aviv, 100
Telegraph Island (Jazirat al Maqlab),
    66–67
Temwen Island, 23
Tenochtitlán, 58, *58*
Termination Land, 265n6

Ternate, 224
Terra, 77
Terra del Laboratore, 317n60
Texcoco, Lake, 58, *58*
Thames, 121–123, 129, 142, 169, 177–178,
    309n48
Thames Estuary, 226
Thanet, 169
Thebes, 145
Thermopylae, 104, 112, 286n3
Thingwall, 308n40
Thorney Island, 121, 128–129, 176, 258n84,
    293n33, 294n36
Thrinacia, 107
Thule, 188, 202–203, 210, 315n46
Thwaites Iceberg Tongue, 43
Thwart-the-Way (Sangiang), 236, 327n36
Thynghowe, 178
Tiber River, 302n12
Tideway, 142
Tidore, 224
Tierra del Fuego, 89, 270nn6–7
Tigris, 66, 261n30, 325n88
Tikopia, 95
Tiksi, 58
Timor, 89
Timor Leste, 89
Tinganes, 175
Tinghøjen, 139, *140*
Ting Holm Island, *177*, 177–178, 308n36
Tingvalla, 309n43
Tingvoll, 309n43
Tinian, 280n43
Tiran, Straits of, 72
Tiran Island, 72
Tirol, 316n57
Titicaca, Lake, 264n37
Toba, 327n36
Tokelau, 282n53
Tokelau Archipelago, 87
Tombo Island, 269n1
Tondi Kwara Barou, 280n46
Tone, 310n12
Tonga, 279n26, 281n50
Tongli, 63

Tonlé Sap, 216
Torres Strait, 270n3
Tórshavn, 175
Toura, 84
Tranquebar, 132
Treene River, 293n31
Trent, 15
Treriksröset, 90
Treviso, 60
Trimouille, 99
Trinidad and Tobago, 279n26, 302n11
Trinity Bay, 268n42
Triopion (Cape Krio), 111
Trogir, 269n1
Tsushima Island, 75, 78
Tuamotu Archipelago, 96
Tumen River, 75
Tunbs, 70
Tunulliarfik Fjord, 49
Tura. *See* Parsley Island
Turkey, 89, 279n33
Tuvalu, 279n26
Tyburn, 129, 176, 308n30
Tyne, 304n37
Tyre, 60, 232, 271n19, 326n9
Tyrrhenian Sea, 233

Udaipur, 63
Uganda, 70, 280n46
Ukatny, 281n47
Ukraine, 281n47
Ulhas River, 58
Ulsan, 78
Ultima Thule, 202, 210, 315n46
Umbagog, 264n37
Umingmak Nuna (Ellesmere Island), 46, 92, 267n26
Umm al Maradim, 70
United Arab Emirates, 70, 78–80, *80*
United Kingdom, *15*, 89–90, 95, 270n6, 279n33, 281n51, 282n60
United States, 46, 89, 90, 91, 96, 99, 100, 264n37, 264n44, 279n33, 281n49, 282n53
universe, 260n12, 260n14
Unteruhldingen, 216

Upemba, Lake, 264n38
Upernavik, 48
Uppsala, 51
Uraniborg, 23, 165
Uros, 264n37
Uruguay, 281n48
Usedom, 89, 137, 190, 199–200, 312n59
Usk, 278n12
Ussuri, 281n47
Utopia (*Utopia*), 59–60, 61, 94, 100–101, *101*, 216
Utopia (*Utopia Limited*), 225
Utstein, 304n40
Uzbekistan, 90, 281n47

Val-David, 81–83, 85
Vanuatu, 93, 279n26, 281n52
Vatnajökull, 45
Vendsyssel-Thy, 304n40
Venetian Lagoon, 89
Venezuela, 2, 63, 257n81
Venice, 15, 23, 59, 60–66, *62*, *64*, *65*, 69, 71, 79, 217, 225, 257n80, 269n1
Venice, Florida, 257n81
Venise-en-Québec, 63, 257n81
Vermont, 89, 264n37
Verona, 306n4
Viborg, *140*
Vieques, 96
Vietnam, 279n33, 281n47, 281n50
Villegagnon, 59
Vineta, 137, 296n74
Virginia Colony, 25
Visby, 268n46
Vista, 77
Vistula Lagoon (Frisches Haff), 243, 331n93
Viva, 77
Volga River, 57
Vosges, 284n26
Vostok, Lake, 236, 327n38
Vozrozhdeniya, 90, 281n47
Vukovar, 281n52

Wake Island, 282n53
Walden Pond, 24

Wales, *15*, 85, 201, 279n34

Walrus Island, 93

Wando (Cheonghae), 77

Warbah, 66, 70

Watling Island, 81, 277n1

Wear, 278n12

Wearmouth, 304n37

Weddell Sea, 46

Westing, 308n40

Westminster, 80

Whaleback Island, 35, 262n8

White Cliffs of Dover, 141

White Head Island, 26

Wido, 74

Wiek, 136, 296n72

Windsor, 300n25

Wirral Peninsula, 308n40

Wiske, 278n12

Wittow, 191

Wolin, 137, 200, 296n73

World Island, *241*, 241–242

Wuzhen, 63

Yangon, 58

Yangtze, 58, 64, 270n10

Yellow River, 271n37

Yellow Sea, 72, 75

Yemen, 70, 72, 270n6, 283n20

Yeongjong, 76

Yeonpyeong Island, 75, 281n47

Yeouido, 76

Yijiangshan Islands, 270n6

Yongyu, 76

York, 325n91

Yorkshire, 84, 138, 304n37, 325n91

Yukon, 78

Zanzibar, 279n33

Zealand, 19, 36, 49, 51, 123, 125–127,
129–130, 136, 143, 153, 158, 160, 166,
169, 170, 223, 291n6, 292n21, 299n17

Zetland, 178

Zhenbao Island, 281n47

Zhestky, 281n47

Zhouzhuang, 63

The authorized representative in the EU for product safety and compliance is:
Mare Nostrum Group
B.V Doelen 72
4831 GR Breda
The Netherlands

www.ingramcontent.com/pod-product-compliance
Lightning Source LLC
Chambersburg PA
CBHW061753260326
41914CB00006B/1091